Digital Signal Processing 101

Digital Signal Processing 101

Everything you Need to Know to Get Started

Second Edition

Michael Parker
Intel Corporation

Newnes is an imprint of Elsevier
The Boulevard, Langford Lane, Kidlington, Oxford OX5 1GB, United Kingdom
50 Hampshire Street, 5th Floor, Cambridge, MA 02139, United States

Library of Congress Cataloging-in-Publication Data
A catalog record for this book is available from the Library of Congress

British Library Cataloguing-in-Publication Data
A catalogue record for this book is available from the British Library

ISBN: 978-0-12-811453-7

For information on all Newnes publications visit our website at
https://www.elsevier.com/books-and-journals

Publisher: Jonathan Simpson
Acquisition Editor: Tim Pitts
Editorial Project Manager: Charlotte Kent
Production Project Manager: Sue Jakeman
Designer: Mark Rogers

Typeset by TNQ Books and Journals

Contents

Acknowledgments .... *xiii*

Introduction .... *xv*

Chapter 1: Numerical Representation *1*

1.1 Integer Fixed Point Representation 2
1.2 Fractional Fixed Point Representation 4
1.3 Floating Point Representation 7

Chapter 2: Complex Numbers and Exponentials *9*

2.1 Complex Addition and Subtraction 10
2.2 Complex Multiplication 10
2.3 Polar Representation 12
2.4 Complex Multiplication Using Polar Representation 12
2.5 Complex Conjugate 15
2.6 The Complex Exponential 16
2.7 Measuring Angles in Radians 18

Chapter 3: Sampling, Aliasing, and Quantization *21*

3.1 Sampling Effects 22
3.2 Nyquist Sampling Rule 25
3.3 Quantization 27
3.4 Signal to Noise Ratio 29

Chapter 4: Frequency Response *31*

4.1 Frequency Response and the Complex Exponential 31
4.2 Normalizing Frequency Response 32
4.3 Sweeping Across the Frequency Response 34
4.4 Example Frequency Responses 35
4.5 Linear Phase Response 37
4.6 Normalized Frequency Response Plots 38

Chapter 5: Finite Impulse Response (FIR) Filters *41*

5.1 Finite Impulse Response Filter Construction 41
5.2 Computing Frequency Response 45

5.3 Computing Filter Coefficients 49
5.4 Effect of Number of Taps on Filter Response 52

Chapter 6: Windowing **59**
6.1 Truncation of Coefficients 59
6.2 Tapering of Coefficients 60
6.3 Sample Coefficient Windows 61

Chapter 7: Decimation and Interpolation **65**
7.1 Decimation 65
7.2 Interpolation 69
7.3 Resampling by Noninteger Value 71

Chapter 8: Infinite Impulse Response (IIR) Filters **75**
8.1 Infinite Impulse Response and Finite Impulse Response Filter Characteristic Comparison 76
8.2 Bilinear Transform 78
8.3 Frequency Prewarping 80

Chapter 9: Complex Modulation and Demodulation **83**
9.1 Modulation Constellations 83
9.2 Modulated Signal Bandwidth 86
9.3 Pulse-Shaping Filter 87
9.4 Raised Cosine Filter 90

Chapter 10: Discrete and Fast Fourier Transforms (DFT, FFT) **99**
10.1 Discrete Fourier Transform and Inverse Discrete Fourier Transform Equations 100
10.2 First Discrete Fourier Transform Example 102
10.3 Second Discrete Fourier Transform Example 103
10.4 Third Discrete Fourier Transform Example 104
10.5 Fourth Discrete Fourier Transform Example 105
10.6 Fast Fourier Transform 108
10.7 Filtering Using the Fast Fourier Transform and Inverse Fast Fourier Transform 113
10.8 Bit Growth in Fast Fourier Transforms 114
10.9 Bit Reversal Addressing 115

Chapter 11: Digital Upconversion and Downconversion **117**
11.1 Digital Upconversion 118
11.2 Digital Downconversion 121
11.3 Intermediate Frequency Subsampling 122

Chapter 12: Error-Correction Coding **129**
12.1 Linear Block Encoding 130
12.2 Linear Block Decoding 132

12.3 Minimum Coding Distance 134
12.4 Convolutional Encoding 135
12.5 Viterbi Decoding 137
12.6 Soft Decision Decoding 144
12.7 Cyclic Redundancy Check 146
12.8 Shannon Capacity and Limit Theorems 146

Chapter 13: Matrix Inversion 149

13.1 Matrix Basics 149
13.2 Cholesky Decomposition 151
13.3 4 × 4 Cholesky Example 155
13.4 QR Decomposition 157
13.5 Gram–Schmidt Method 158
13.6 QR Decomposition Restructuring for Parallel Implementation 160

Chapter 14: Field-Oriented Motor Control 163

14.1 Magnetism Basics 163
14.2 AC Motor Basics 165
14.3 DC Motor Basics 166
14.4 Electronic Commutation 167
14.5 AC Induction Motor 171
14.6 Motor Control 173
14.7 Park and Clark Transforms 177

Chapter 15: Analog and Time Division Multiple Access Wireless Communications 183

15.1 Early Digital Innovations 184
15.2 Frequency Modulation 185
15.3 Digital Signal Processor 186
15.4 Digital Voice Phone Systems 187
15.5 Time Division Multiple Access Modulation and Demodulation 187

Chapter 16: CDMA Wireless Communications 191

16.1 Spread Spectrum Technology 191
16.2 Direct Sequence Spread Spectrum 192
16.3 Walsh Codes 193
16.4 Concept of Code Division Multiple Access 195
16.5 Walsh Code Demodulation 195
16.6 Network Synchronization 199
16.7 RAKE Receiver 200
16.8 Pilot Pseudorandom Number Codes 200
16.9 Code Division Multiple Access Transmit Architecture 201
16.10 Variable Rate Vocoder 202
16.11 Soft Handoff 203
16.12 Uplink Modulation 204

16.13 Power Control 205
16.14 Higher Data Rates 206
16.15 Spectral Efficiency Considerations 207
16.16 Other Code Division Multiple Access Technologies 207

Chapter 17: Orthogonal Frequency Division Multiple Access Wireless Communications 209

17.1 WiMax and Long-Term Evolution 209
17.2 Orthogonal Frequency Division Multiple Access Advantages 210
17.3 Orthogonality of Periodic Signals 211
17.4 Frequency Spectrum of Orthogonal Subcarrier 213
17.5 Orthogonal Frequency Division Multiplexing Modulation 214
17.6 Intersymbol Interference and the Cyclic Prefix 217
17.7 Multiple Input and Multiple Output Equalization 220
17.8 Orthogonal Frequency Division Multiple Access System Considerations 221
17.9 Orthogonal Frequency Division Multiple Access Spectral Efficiency 222
17.10 Orthogonal Frequency Division Multiple Access Doppler Frequency Shift 223
17.11 Peak to Average Ratio 223
17.12 Crest Factor Reduction 225
17.13 Digital Predistortion 228
17.14 Remote Radio Head 229

Chapter 18: Radar Basics 231

18.1 Radar Frequency Bands 231
18.2 Radar Antennas 232
18.3 Radar Range Equation 234
18.4 Stealth Aircraft 235
18.5 Pulsed Radar Operation 236
18.6 Pulse Compression 236
18.7 Pulse Repetition Frequency 237
18.8 Detection Processing 239

Chapter 19: Pulse Doppler Radar 241

19.1 Doppler Effect 241
19.2 Pulsed Frequency Spectrum 243
19.3 Doppler Ambiguities 245
19.4 Radar Clutter 246
19.5 Pulse Repetition Frequency Trade-Offs 248
19.6 Target Tracking 250

Chapter 20: Automotive Radar 253

20.1 Frequency-Modulated Continuous-Wave Theory 253
20.2 Frequency-Modulated Continuous-Wave Range Detection 254
20.3 Frequency-Modulated Continuous-Wave Doppler Detection 256

20.4 Frequency-Modulated Continuous-Wave Radar Link Budget 259
20.5 Frequency-Modulated Continuous-Wave Implementation Considerations 260
20.6 Frequency-Modulated Continuous-Wave Interference 261
20.7 Frequency-Modulated Continuous-Wave Beamforming 261
20.8 Frequency-Modulated Continuous-Wave Range-Doppler Processing 263
20.9 Frequency-Modulated Continuous-Wave Radar Front-End Processing 264
20.10 Frequency-Modulated Continuous-Wave Pulse-Doppler Processing 266
20.11 Frequency-Modulated Continuous-Wave Radar Back-End Processing 269
20.12 Noncoherent Antenna Magnitude Summation 269
20.13 Cell Averaging—Constant False Alarm Rate 271
20.14 Ordered Sort—Constant False Alarm Rate 272
20.15 Angle of Arrival Estimation 275

Chapter 21: Space Time Adaptive Processing (STAP) Radar 277
21.1 Space Time Adaptive Processing Radar Concept 278
21.2 Steering Vector 280
21.3 Interference Covariance Matrix 281
21.4 Space Time Adaptive Processing Optimal Filter 283
21.5 Space Time Adaptive Processing Radar Computational Requirements 285

Chapter 22: Synthetic Array Radar 287
22.1 Introduction 287
22.2 Synthetic Array Radar Resolution 287
22.3 Pulse Compression 288
22.4 Azimuth Resolution 288
22.5 Synthetic Array Radar Processing 291
22.6 Synthetic Array Radar Doppler Processing 293
22.7 Synthetic Array Radar Impairments 295

Chapter 23: Introduction to Video Processing 297
23.1 Color Spaces 297
23.2 Interlacing 299
23.3 Deinterlacing 299
23.4 Image Resolution and Bandwidth 301
23.5 Chroma Scaling 302
23.6 Image Scaling and Cropping 302
23.7 Alpha Blending and Compositing 303
23.8 Video Compression 304
23.9 Digital Video Interfaces 304
23.10 Legacy Analog Video Interfaces 307

Chapter 24: DCT, Entropy, Predictive Coding, and Quantization 311
24.1 Discrete Cosine Transform 311
24.2 Entropy 315
24.3 Huffman Coding 316

24.4 Markov Source ... 317
24.5 Predictive Coding ... 319
24.6 Differential Encoding ... 319
24.7 Lossless Compression ... 321
24.8 Quantization ... 322
24.9 Decibels ... 325

Chapter 25: Image and Video Compression Fundamentals ... 329
25.1 Baseline JPEG ... 329
25.2 DC Scaling ... 329
25.3 Quantization Tables ... 330
25.4 Entropy Coding ... 331
25.5 JPEG Extensions ... 333
25.6 Video Compression Basics ... 335
25.7 Block Size ... 336
25.8 Motion Estimation ... 336
25.9 Frame Processing Order ... 339
25.10 Compressing I Frames ... 340
25.11 Compressing P Frames ... 341
25.12 Compressing B Frames ... 341
25.13 Rate Control and Buffering ... 342
25.14 Quantization Scale Factor ... 343

Chapter 26: Introduction to Machine Learning ... 347
26.1 Convolutional Neural Networks ... 347
26.2 Convolution Layer ... 348
26.3 Rectified Linear Unit Layer ... 350
26.4 Normalization Layer ... 350
26.5 Max-Pooling Layer ... 351
26.6 Fully Connected Layer ... 351
26.7 Training Computational Neural Networks ... 353
26.8 Winograd Transform ... 355
26.9 Convolutional Neural Network Numerical Precision Requirements ... 358

Chapter 27: Implementation Using Digital Signal Processors ... 361
27.1 Digital Signal Processing Processor Architectural Enhancements ... 361
27.2 Scalability ... 366
27.3 Floating Point ... 366
27.4 Design Methodology ... 367
27.5 Managing Resources ... 368
27.6 Ecosystem ... 369

Chapter 28: Implementation Using FPGAs ... 371
28.1 FPGA Design Methodology ... 372
28.2 DSP Processor or FPGA Choice ... 373

28.3 Design Methodology Considerations ... 374
28.4 Dedicated Digital Signal Processing Circuit Blocks in FPGAs ... 375
28.5 Floating Point Implementation Using FPGAs ... 382
28.6 Ecosystem ... 383
28.7 Future Trends ... 383

Chapter 29: Implementation With GPUs ... 387

29.1 Characteristics of Graphics Processing Unit Architecture ... 387
29.2 Graphics Processing Unit Programming Environment ... 388
29.3 Memory Hierarchy ... 390
29.4 Interfaces ... 391
29.5 Numerical Precision ... 392
29.6 Future Trends ... 392

Appendix A: Q Format Shift With Fractional Multiplication ... 395

Appendix B: Evaluation of Finite Impulse Response Design Error Minimization ... 397

Appendix C: Laplace Transform ... 401

Appendix D: Z-Transform ... 405

Appendix E: Binary Field Arithmetic ... 409

Index ... 411

Acknowledgments

This book grew out of a need for Altera (now Intel) marketing and technical sales people to have an intuitive-level understanding of digital signal processing (DSP) fundamentals and applications, to better work on issues that our customers face as they implement DSP systems. I am grateful to the Altera management for the support this book has received, in particular from Steve Mensor and Chris Balough.

My understanding of the topics in this book is based on many years of engineering implementation work and collaboration and explanations from many of my colleagues at multiple firms over the years. More recently, within Altera, many people have contributed to my knowledge in these areas. I would like to especially acknowledge a few people who have been helpful both in DSP domain and relevant applications and implementations. Within Altera engineering, this includes Volker Mauer, Martin Langhammer, and Mike Fitton. Within the Altera technical sales organization, people who have been especially helpful to my understanding of some of the relevant DSP applications include Colman Cheung, Ben Esposito, Brian Kurtz, and Mark Santoro.

The automotive radar material grew out of an automotive radar proposal created jointly with Ben Esposito, and some of his figures have been reused here. I am also indebted to Gordon Chiu, Utku Aydonat, Davor Capalija, Andrew Ling, and Shane O'Connell for their patient explanations on machine learning and use of some of their figures that I have used as well.

Within Altera publications, James Adams has been instrumental in getting this project off the ground and working with the publisher.

Finally, the support of my wife, Zaida, and daughter, Ariel, has been most important. This book has been primarily an "evenings and weekends" project, and their patience has been essential.

Introduction

This book is intended for those who work in or provide components for industries that are made possible by digital signal processing, or DSP. Sample industries are wireless mobile phone and infrastructure equipment, broadcast and cable video, DSL modems, satellite communications, medical imaging, audio, radar, sonar, surveillance, electrical motor control—this list goes on. While the engineers who implement these systems must be very familiar with DSP, there are many others—executive and midlevel management, marketing, technical sales and field engineers, business development, and others—who can benefit from a basic knowledge of the fundamental principles of DSP.

Others who are a potential audience include those interested in studying or working in any of these areas. High-school seniors or undeclared college majors considering a future in the industries made possible by DSP technology may gain sufficient understanding that enables them to decide whether to continue further.

That, then, is the purpose of this book: to provide a basic tutorial on DSP. This topic seems to have a dearth of easy-to-read and understand explanations. Unlike most technical resources, this is a treatment in which mathematics is minimized and intuitive understanding maximized. This book attempts to explain many difficult concepts like sampling, aliasing, imaginary numbers, and frequency response using easy-to-understand examples. In addition, there is an overview of the DSP functions and implementation used in several DSP-intensive fields or applications, from error correction to CDMA mobile communication to airborne radar systems.

So this book is intended for those of you who, like me, are somewhat dismayed when presented with a blackboard (or whiteboard) full of equations as an explanation on DSP.

The intended readers include those who have absolutely no previous experience with DSP but are comfortable with high-school–level math skills. Many technical details have been deliberately left out, to emphasize just the key concepts. Although this book is not expected to be used as a university-course-level text, it can initiate readers prior to tackling a proper text on DSP. But it may also be all you need to talk intelligently to other people involved in a DSP-centric industry and understand many of the fundamental concepts.

To start with, just what is DSP? Well, DSP is performing operations on a digital signal of some sort and using a digital semiconductor device. Most commonly, multipliers and adders are used. If you can multiply and add, you can probably understand DSP. Actually, signal processing was around long before digital electronics. Examples of this are radios and TVs. Early tuners used analog circuits with variable capacitors to dial a station. Resistors, capacitors, and vacuum tubes were used to either attenuate or amplify different frequencies or to provide frequency shifting. These are examples of basic signal-processing applications. The signals were analog signals, and the circuits doing the processing were analog, as was the final output.

Today, most signal processing is performed digitally. The reason is that digital circuits have progressively become cheaper and faster, as well as due to the inherent advantages of repeatability, tolerance, and consistency that digital circuits enjoy compared to analog circuits.

If the signal is not in a digital form, then it must be first converted, or digitized. A device called an analog-to-digital converter (ADC) is used. If the output signal needs to be analog, then it is converted back using a digital-to-analog converter (DAC). Of course, many signals are already digitized and can be processed by digital circuits directly.

DSP is at the heart of a wide range of everyday devices in our lives, although many people are unaware of this. A few everyday examples are cellular phones, DSL modems, digital hearing aids, MRI and ultrasound equipment, audio equipment, set top boxes, flat-screen televisions, satellite communications, and DVD players.

As promised, the mathematics will be minimized, but it cannot be eliminated altogether. Some basic trigonometry and the use of complex numbers are unavoidable, so an early chapter is included to introduce these concepts, using as simple examples as possible. There is also one appendix section where very basic calculus is used, but this is not essential to the overall understanding.

CHAPTER 1

Numerical Representation

To process a signal digitally, it must be represented in a digital format. This may seem obvious, but it turns out that there are a number of different ways to represent numbers, and this representation can greatly affect both the result and the amount of circuit resources required to perform a given operation. This particular chapter is focused more for people who are implementing DSP and is not really required to understand fundamental DSP concepts.

Digital electronics operate on bits of course, which are used to form binary words. The bits can be represented as binary, decimal, octal or hexadecimal, or other form. These binary numbers can be used to represent "real" numbers. There are two basic types of arithmetic used in DSP, floating point or fixed point. Fixed point numbers that have a fixed decimal point as part of the number. Examples are 1234 (same as 1234.0), or 12.34 or 0.1234. This is type of number we normally use everyday. Floating point is a number with an exponent. The most common example would be scientific notation, used on many calculators. In floating point, 1,200,000 would be expressed as 1.2×10^6, and 0.0000234 would be expressed as 2.34×10^{-5}. Most our discussion will focus on fixed point numbers, as this is most commonly found in DSP applications. Once we understand DSP arithmetic issues with fixed point numbers, then there is short discussion on floating point numbers.

In DSP, we will pretty much exclusively use signed numbers, meaning that there are both positive and negative numbers. This leads into the next point, which is how to represent the negative numbers.

In signed fixed point arithmetic, the binary number representations will include a sign, a radix, or decimal point, and the magnitude. The sign indicates whether the number is positive or negative, and the radix (also called decimal) point separates the integer and fractional parts of the number.

The sign is normally determined by the left most, or most significant bit (MSB). The convention zero is used for positive, and one for negative. There are several formats to represent negative numbers, but the almost universal method is known as "2s complement". This is the method discussed here.

Furthermore, fixed point numbers are usually represented as either integer or fractional. In integer representation, the decimal point is to the right of the least significant bit (LSB),

Digital Signal Processing 101. http://dx.doi.org/10.1016/B978-0-12-811453-7.00001-9

and there is no fractional part in the number. For an 8-bit number, the range which can be represented is from −128 to +127, with increments of 1.

In fractional representation, the decimal point is often just to the right of the MSB, which is also the sign bit. For an 8-bit number, the range that can be represented is from −1 to +127/128 (almost +1), with increments of 1/128. This may seem a bit strange, but in practice, fractional representation has advantages, as will be explained.

This chapter will present several tables, with each row giving equivalent binary and hexadecimal numbers. The far right column will give the actual value in the chosen representation; for example, 16-bit integer representation. The actual value represented by the hex/binary numbers will depend on which representation format is chosen.

1.1 Integer Fixed Point Representation

Below are some examples showing the 2s complement integer fixed point representation (see Table 1.1).

The 2s complement representation of the negative numbers may seem nonintuitive, but it has several very nice features. There is only one representation of 0 (all 0s), unlike other formats which have a "positive" and "negative" zero. Also, addition and multiplication of positive and negative 2s complement numbers work properly with traditional digital adder and multiplier structures. A 2s complement number range can be extended to the left by simply replicating the MSB (sign bit) as needed, without changing the value.

The way to interpret a 2s complement number is using the following mapping for each bit. A 0 bit in a given location of the binary word means no weight for that bit. A 1 in a given location means to use the weight indicated. Notice the weights double with each bit moving left, and the MSB is the only bit with a negative weight. You should satisfy

Table 1.1: 8-Bit Signed Integer Representation

8-Bit Integer Representation		
Binary	Hexadecimal	Actual Decimal Value
0111 1111	0x7F	127
0111 1110	0x7E	126
0000 0010	0x02	2
0000 0001	0x01	1
0000 0000	0x00	0
1111 1111	0xFF	−1
1111 1110	0xFE	−2
1000 0001	0x81	−127
1000 0000	0x80	−128

Table 1.2: 2s Complement Bit Weighting

8-Bit Signed Integer	MSB							LSB
Bit weight Weight in powers of 2	−128 -2^7	64 2^6	32 2^5	16 2^4	8 2^3	4 2^2	2 2^1	1 2^0

yourself that all negative numbers will have an MSB of 1, and all positive numbers and zero have an MSB of 0 (see Table 1.2).

This can be extended to numbers with larger number of bits. Below is an example with 16 bits. Notice how the numbers represented in a lesser number of bits (say 8 bit for example) can be easily put into 16-bit representation by simply replicating the MSB of the 8-bit number by eight times and tacking onto the left to form a 16-bit number. Similarly, as long as the number represented in the 16-bit representation is small enough to be represented in 8 bits, the left most bits can simply be shaved off to move to the 8-bit representation. In both cases, the decimal point stays to the right of the LSB and does not change location. This can be easily seen by comparing, for example, the representation of −2 in the 8-bit representation table above and again in the 16-bit representation table below (Table 1.3).

Table 1.3: 2s Complement 16-Bit Signed Integer Representation and Weighting

16-Bit Signed Integer Representation		
Binary	**Hexadecimal**	**Actual Decimal Value**
0111 1111 1111 1111	0x7FFF	32,767
0111 1111 1111 1110	0x7FFE	32,766
0000 0000 1000 0000	0x0080	128
0000 0000 0111 1111	0x007F	127
0000 0000 0111 1110	0x007E	126
0000 0000 0000 0010	0x0002	2
0000 0000 0000 0001	0x0001	1
0000 0000 0000 0000	0x0000	0
1111 1111 1111 1111	0xFFFF	−1
1111 1111 1111 1110	0xFFFE	−2
1111 1111 1000 0001	0xFF81	−127
1111 1111 1000 0000	0xFF80	−128
1111 1111 0111 1111	0xFF80	−129
1000 0000 0000 0001	0xFF80	−32,767
1000 0000 0000 0000	0xFF80	−32,768

MSB															LSB
−32,768 -2^{15}	16,384 2^{14}	8192 2^{13}	4096 2^{12}	2048 2^{11}	1024 2^{10}	512 2^9	256 2^8	128 2^7	64 2^6	32 2^5	16 2^4	8 2^3	4 2^2	2 2^1	1 2^0

Now, let us look at some examples of trying to adding combinations of positive and negative 8-bit numbers together, using a traditional unsigned digital adder. We throw away the carry bit from the last (MSB) adder stage.

Case #1: Positive and negative number sum

+15	0000 1111	0x0F
−1	1111 1111	0xFF
= 14	0000 1110	0x0E

Case #2: Positive and negative number sum

−31	1110 0001	0xE1
+16	0001 0000	0x80
= −15	1111 0000	0xF0

Case #3: Two negative numbers being summed

−31	1110 0001	0xE1
−64	1100 0000	0xC0
= −95	1010 0001	0xF0

Case #4: Two positive numbers being summed, result exceeds range

+64	0100 0000	0x40
+64	0100 0000	0x40
= 128	1000 0000[a]	0x80[a]

[a]Notice all the results are correct, except the last case. This is because the result, +128, cannot be represented in the range of an 8 bit 2s complement number. This is a special case, which must be considered by the designer, as this is overflowing the range of numbers which can be represented.

Integer representation is often used in many software applications, because it is familiar and works well. However, in DSP, integer representation has a major drawback. That is because in DSP, there is a lot of multiplication. When you multiply a bunch of integers together, the results start to grow rapidly. It quickly gets out of hand and exceeds the range of values that can be represented. As we saw above, 2s complement arithmetic works well, as long as you do not exceed the numerical range. This has led to the use of fractional fixed point representation.

1.2 Fractional Fixed Point Representation

The basic idea is all values are in the range from +1 to −1, so if they are multiplied, the result will not exceed this range. Notice that to convert from integer to 8-bit signed fractional, the actual values are all divided by 128. This maps the integer range of +127 to −128 to almost +1 to −1 (Tables 1.4 and 1.5).

Table 1.4: 8-Bit Signed Fractional Representation and Bit Weighting

8-bit Fractional Representation								
Binary			**Hexadecimal**				**Actual Decimal Value**	
0111 1111			0x7F				127/128 = 0.99219	
0111 1110			0x7E				126/128 = 0.98438	
0000 0010			0x02				2/128 = 0.01563	
0000 0001			0x01				1/128 = 0.00781	
0000 0000			0x00				0	
1111 1111			0xFF				−1/128 = −0.00781	
1111 1110			0xFE				−2/128 = −0.01563	
1000 0001			0x81				−127/128 = −0.99219	
1000 0000			0x80				−1.00	
8-Bit Signed Fractional	**MSB**							**LSB**
Weight Weight in powers of 2	−1 -2^0	1/2 2^{-1}	1/4 2^{-2}	1/8 2^{-3}	1/16 2^{-4}	1/32 2^{-5}	1/64 2^{-6}	1/128 2^{-7}

Fractional fixed point is often expressed in Q format. The representation shown above is Q15, which means that there are 15 bits to the right of the radix or decimal point. It might also be called Q1.15, meaning that there is 15 bits to right of decimal point, and 1 bit to left.

The key property of fractional representation is that the numbers grow smaller, rather than larger, during multiplication. And in DSP, we commonly sum the results of many multiplication operations. In integer math, the results of multiplication can grow large quickly (see example below). And when we sum many such results, the final sum can be very large, easily exceeding the ability to represent the number in a fixed point integer format.

As an analogy, think about trying to display an analog signal on an oscilloscope. You need to select a voltage range (volts/division) in which the signal amplitude does not exceed the upper and lower range of the display. At the same time, you want the signal to occupy a reasonable part of the screen, so the detail of the signal visible. If the signal amplitude only occupies 1/10 of a division, for example, it is difficult to see the signal.

To illustrate this, imagine using 16-bit fixed point, and the signal has a value of 0.75 decimal or 24,676 in integer (which is 75% of full scale) and is multiplied by a coefficient of value 0.50 decimal or 16,384 integer (which is 50% of full scale).

```
        0.75            24,576
      x 0.50        x   16,384
-----------------   -----------------
        0.375       402,653,184
```

Table 1.5: 16-Bit Signed Fractional Representation and Bit Weighting

16-Bit Signed Fractional Representation		
Binary	Hexadecimal	Actual Decimal Value
0111 1111 1111 1111	0x7FFF	32,767/32768
0111 1111 1111 1110	0x7FFE	32,766/32768
0000 0000 1000 0000	0x0080	128/32768
0000 0000 0111 1111	0x007F	127/32768
0000 0000 0111 1110	0x007E	126/32768
0000 0000 0000 0010	0x0002	2/32768
0000 0000 0000 0001	0x0001	1/32768
0000 0000 0000 0000	0x0000	0
1111 1111 1111 1111	0xFFFF	−1/32768
1111 1111 1111 1110	0xFFFE	−2/32768
1111 1111 1000 0001	0xFF81	−127/32768
1111 1111 1000 0000	0xFF80	−128/32768
1111 1111 0111 1111	0xFF7F	−129/32768
1000 0000 0000 0001	0x8001	−32,767/32768
1000 0000 0000 0000	0x8000	−1

MSB															LSB
−1 -2^{0}	1/2 2^{-1}	1/4 2^{-2}	1/8 2^{-3}	1/16 2^{-4}	1/32 2^{-5}	1/64 2^{-6}	1/128 2^{-7}	1/256 2^{-8}	1/512 2^{-9}	1/1024 2^{-10}	1/2048 2^{-11}	1/4096 2^{-12}	1/8192 2^{-13}	1/16,384 2^{-14}	1/32,768 2^{-15}

Now the larger integer number can be shifted right after every such operation to scale the signal within a range it can be represented, but most DSP designers prefer to represent numbers as fractional, as its a lot easier to keep track of the decimal point.

Now consider multiplication again. If two 16-bit numbers are multiplied, the result is a 32-bit number. As it turns out, if the two numbers being multiplied are Q15, you might expect the result in the 32-bit register to be a Q31 number (MSB to left of decimal point, all other bits to right). Actually, the result is in Q30 format—the decimal point has shifted down to the right. Most DSP processors will automatically left shift the multiplier output to compensate for this, when operating in fractional arithmetic mode. In FPGAs (field programmable gate arrays) or ASIC (application specific integrated circuit) hardware design, the designer may have to take this into account when connecting data buses between different blocks. There is an appendix to explain the need for this extra left shift in detail, as it will be important for those implementing fractional arithmetic on FPGAs or DSPs.

```
      0x4000            value = ½ in Q.15
   x  0x2000            value = ¼ in Q.15
--------------------------------------------------------
 0x0800 0000            value = 1/16 in Q31
```

After left shift by one

0×10000000 value $= 1/8$ in Q31—correct result !

If we use only the top 16-bit word from multiplier output, after the left shift we get

0×1000 value $= 1/8$ in Q15—again correct result !

1.3 Floating Point Representation

Many of these complications can be avoided by using floating point format. Floating point format is basically like scientific notation on your calculator. Because of this, a floating point number can have a much greater dynamic range than a fixed point number with equivalent number of bits. Dynamic range means the ability to represent very small numbers to very large numbers.

The floating point number has both a mantissa (which includes sign) and exponent. The mantissa would be the value in the main field of your calculator, and the exponent would be off to the side, usually as a superscript. Each of these is expressed as a fixed point number (meaning the decimal point is fixed). The mantissa represents the fractional portion, with a range from 0 to 0.999... There is also an implicit one in the mantissa, meaning it is not present in the mantissa. So the range of the fractional portion is actually from 1 to 1.999... The size of the mantissa and exponent in number of bits will vary depending on which floating point format is used (Table 1.6).

Table 1.6: Floating Point Numerical Formats

Floating Point Formats	No of Mantissa Bits	No of Exponent Bits
Half Precision	10, plus 1 bit to determine sign	5, unsigned integer, biased by $2^5-1=31$
Single Precision	23, plus 1 bit to determine sign	8, unsigned integer, biased by $2^8-1=127$
Double Precision	52, plus 1 bit to determine sign	11, unsigned integer, biased by $2^{11}-1=2047$

Common floating point formats require 16 bits, 32 bits, and 64 bits to represent.

Single and double precision are very common in computing, usually designated as "float" and "double," respectively. Half precision has seen little use until recently, when it has become popular in machine learning.

The largest single precision number that can be represented is:

$$2-2^{-23}\times 2^{127}=1.999999881\times 2^{127}=1.70141183\times 10^{38}$$

The smallest single precision number that can be represented without denormalization:

$$1+2^{-23}\times 2^{-127}=1.000000119\times 2^{-126}=\ 1.17549435\times 10^{-38}$$

This is a far great dynamic range that can be expressed using 32-bit fixed point. There are some additional details in floating point to designate conditions such as +/− infinity, or "not a number" NaN, as may occur with a division by zero.

The drawback of floating point calculations is the resources required. When adding or subtracting two floating point numbers, the number with smaller absolute value must be first adjusted so that its exponent is equal to the number with larger absolute value, then the two mantissas can be added. If the mantissa result requires a left or right shift to represent, the exponent is adjusted to account for this shift. When multiplying two floating point numbers, the mantissas are multiplied, and the exponents are summed. Again, if mantissa result requires a left or right shift to represent, the new exponent must be adjusted to account for this shift. All of this requires considerably more circuitry than fixed point computations and have higher power consumption as well. For this reason, many DSP algorithms use fixed point arithmetic, despite the ones on the designer to keep track of where the decimal point is and ensure that the dynamic range of the signal never exceed the fixed point representation, or else become so small that quantization error become significant. We will see more on quantization error in a later chapter.

For those interested in more information on floating point arithmetic, there are many texts that go into this in detail, or else one can google IEEE754 floating point.

CHAPTER 2

Complex Numbers and Exponentials

Complex numbers are one of those things many of us were taught a long time ago and have long since forgotten. Unfortunately, they are important in digital communications and digital signal processing (DSP), so we need to resurrect them.

What we were taught and some of us vaguely remember is that a complex number has a "real" and "imaginary" part, and the imaginary part is the square root of a negative number, which is really a nonexistent number. This right away sounds fishy, and while it's technically true, there is a much more intuitive way of looking at it.

The whole reason for "complex numbers" is that we are going to need a two dimensional number plane to understand DSP. The traditional number line extends from plus infinity to minus infinity, along a single line. To represent many of the concepts in DSP, we need two dimensions. This requires two orthogonal axes, like a North–South line and an East–West line. For the arithmetic to work out, one line, usually depicted as the horizontal line, is the real number line. The other vertical line is the imaginary line. All imaginary numbers are prefaced by "j", which is defined as the square root of -1. Do not get confused by this imaginary number stuff, but rather view "j" as an arbitrary construct we will use to differentiate the horizontal axis (normal) numbers from those on the vertical axis. This is the essence of this whole chapter.

As depicted in Fig. 2.1, any complex number Z has a real and imaginary part, and is expressed as $X + j \cdot Y$, or just $X + jY$. The value of X and Y for any point is determined

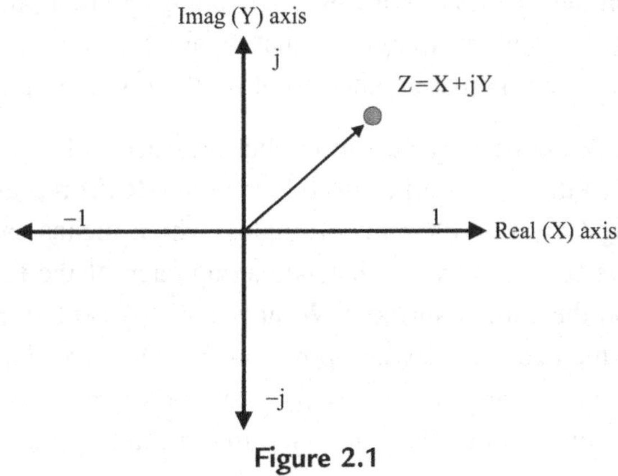

Figure 2.1
I-Q complex number plane.

Digital Signal Processing 101. http://dx.doi.org/10.1016/B978-0-12-811453-7.00002-0

by the distance one must travel in the direction of each axis to arrive at the point. It can also be visualized as a point on the complex number plane, or as a vector originating at the origin and terminating at the point. We need to be able to do arithmetic with complex numbers. There has to be a way to keep track the vertical and horizontal components. That is where the "j" comes in (in some texts, "i" is used instead).

2.1 Complex Addition and Subtraction

Adding and subtracting is simple—just add and subtract the vertical and horizontal components separately. The "j" helps us keep from mixing the vertical and horizontal components. For example:

$$(3 + j4) - (1 - j6) = 2 + j10$$

Scaling a complex number is just simply scaling each component:

$$4 \cdot (3 + j4) = 12 + j16$$

2.2 Complex Multiplication

Multiplication gets a little trickier and is harder to visualize graphically. Here is the way the mechanics of it work:

$$(A + jB) \cdot (C + jD) = A \cdot C + jB \cdot C + A \cdot jD + jB \cdot jD = AC + jBC + jAD + j^2BD$$

Now remember that j^2 is by definition equal to -1. After collecting terms:

$$AC + jBC + jAD + j^2BD = AC + jBC + jAD - BD = (AC - BD) + j(BC + AD)$$

The result is another complex number, with AC−BD being the real part, and BC + AD being the imaginary part (remember, imaginary just means the vertical axis, while real is the horizontal axis). This result is just another point on the complex plane.

This mechanics of this arithmetic may be simple, but we need to be able to visualize what is really happening. To do this, we need to introduce polar (R-Ω) representation. Up until now, we have been using Cartesian (X-Y) coordinates, which means each location on the complex number plane is specified by the distance along each of the two axes (like longitude and latitude on the earth's surface). What polar representation does is to replace these two parameters, which can specify any point on the complex plane, with another set of two parameters, which also can specify any point on the complex plane. The two new parameters are the magnitude and angle. The magnitude is simply the length of the line or

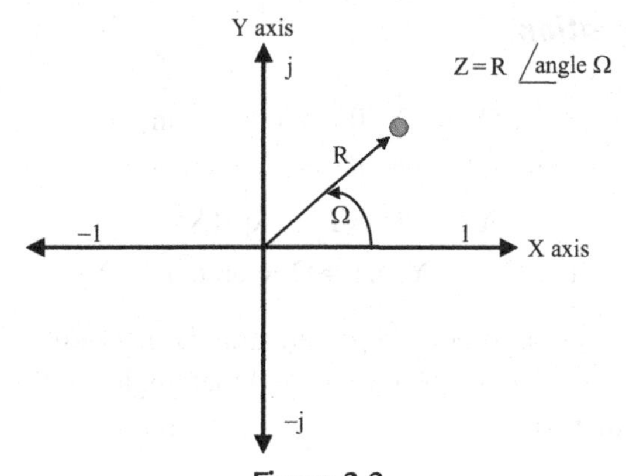

Figure 2.2
Magnitude-angle complex number plane.

vector from the origin to the point. The angle is defined as the angle of this line starting at the positive X axis and arcing counterclockwise.

This is shown in Fig. 2.2, where the same point $Z = X + jY$ is identified by having radius R (length of vector from origin to the point) with angle Ω specified a counterclockwise from positive real axis.

Any point Z on the graph may be specified as $X + jY$ or R with angle Ω.

The relationships between these go back to basic high school math. Consider the right triangle formed in Fig. 2.3, with sides X, Y, R, and angle Ω.

Remember the Pythagorean theorem? For any right triangle, $X^2 + Y^2 = R^2$.

Also, remember that sine is defined as length of opposite side divided by hypotenuse, and cosine is defined as length of adjacent side divided by hypotenuse (but must be a right triangle). Tangent is defined as opposite over adjacent side. So we get the following relationships, which can be used to convert between Cartesian (X, Y) and polar (R, Ω):

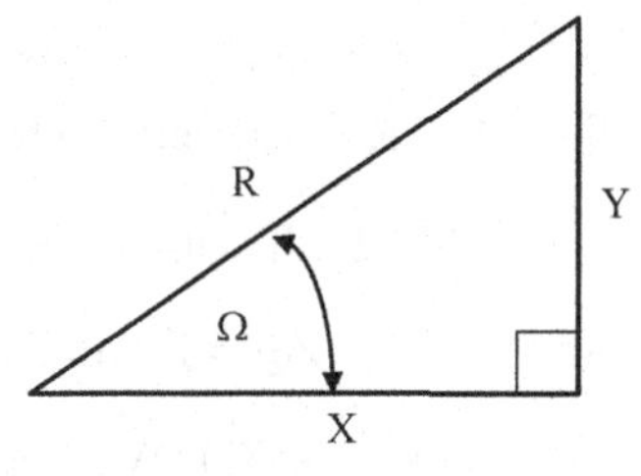

Figure 2.3
Cartisian and polar conversion relationships.

2.3 Polar Representation

$$\sin(\Omega) = (Y/R) \Rightarrow Y = R \cdot \sin(\Omega)$$
$$\cos(\Omega) = (X/R) \Rightarrow X = R \cdot \cos(\Omega)$$
$$X^2 + Y^2 = R^2 \Rightarrow R = \text{sqrt}(X^2 + Y^2)$$
$$\tan(\Omega) = (Y/X) \Rightarrow \Omega = \arctan(Y/X)$$

The reason for this little foray into polar representation is that multiplication (and division) of complex numbers is very easy in polar form, and that angles in the complex plane can be easily visualized in polar form.

2.4 Complex Multiplication Using Polar Representation

We will define two points, Z_1 and Z_2.

$$Z_1 = R_1 \text{ angle}(\Omega_1) \quad Z_2 = R_2 \text{ angle}(\Omega_2)$$
$$Z_1 \cdot Z_2 = (R_1 \cdot R_2) \text{ angle}(\Omega_1 + \Omega_2)$$

What this means is that with any two complex numbers, the magnitude, or distance from the origin to the radius, gets multiplied together to form the new magnitude. This makes sense intuitively. The angles of the two complex numbers get added together to form the new angle. Not so intuitive, so let us try a few examples to get the hang of it.

Let us use real numbers to start—a real number is just a complex number with the "j" part equal to zero. Real numbers are "simplified" version of complex numbers, so any arithmetic rules on a complex number had better work with the real numbers.

$$Z = R \text{ angle}(\Omega) \text{—for real number, the angle must be equal to } 0$$

Then $X = R \cdot \cos(0)$ or $X = R$ and $Y = R \cdot \sin(0)$ or $Y = 0$. This is what we expect—the Y portion must be zero for the number to lie on the X (real number) axis.

Now consider two complex numbers, both with angle zero.

$$Z_1 = R_1 \text{ angle}(0),\ Z_2 = R_2 \text{ angle}(0) \text{with the product}$$
$$Z_1 \cdot Z_2 = R_1 \cdot R_2 \text{ angle}(0 + 0) = R_1 \cdot R_2 \quad \text{(reverts to traditional multiplication)}$$

Now consider another set complex numbers, both with angles of 180°degrees.

$$Z_1 = R_1 \text{ angle}(180),\ Z_2 = R_2 \text{ angle}(180)$$

From relation above, $Y = R \cdot \sin(\Omega)$, $X = R \cdot \cos(\Omega)$. With angle of 180°degrees, $Y = 0$, $X = -R$, meaning the point Z lies on the negative part of real axis. Z is simply a negative real number. If we multiply two real negative numbers, we know that we should get a real positive number. Let us check using complex multiplication.

$$Z_1 \cdot Z_2 = R_1 \cdot R_2 \text{ angle}(180 + 180) = R_1 \cdot R_2 \text{ angle}(360).$$

The angle 360°degrees is all the way around the circle and equal to 0°degree.

$$Z_1 \cdot Z_2 = R_1 \cdot R_2 \text{ angle}(360) = R_1 \cdot R_2 \text{ angle}(0) = R_1 \cdot R_2 = X_1 \cdot X_2$$

The result is a real positive number.

There is another point to all these exercises, which is to explain why we chose something strange like j equals square root of −1 to designate the vertical axis in the complex number plane. Be a little more patient—we are almost there.

There are 4 "special" angles—0, 90, 180, and 270 degrees. Notice that:

Z = R angle (0) = X degrees is a positive real number on positive real axis
Z = R angle (90) = Y degrees is a positive imaginary number on positive imaginary axis
Z = R angle (180) = −X degrees is a negative real number on negative real axis
Z = R angle (270) = −Y degrees is a negative imaginary number on negative imaginary axis

If we add 360 degrees to any complex number, it wraps all the way around the circle. Or we can have a negative angle, which means just going backwards (clockwise) around the circle.

Z = R angle (0) = R angle (360) = R angle (720)...
Z = R angle (120) = R angle (480) = R angle (840)...
Z = R angle (90) = R angle (−270) = R angle (−630)...
Z = R angle (−53) = R angle (307) = R angle (667)...

We now know when multiplying two complex numbers, the magnitudes R are multiplied and the angles Ω are summed. Now let us consider a few example cases to illustrate how this imaginary "j" operator helps us.

Imagine two complex numbers with only an imaginary component. They are both located on the positive imaginary axis.

$$Z_1 = jY_1,\ Z_2 = jY_2 \quad (\text{real parts } X_1 = X_2 = 0)$$
$$Z_1 \cdot Z_2 = jY_1 \cdot jY_2 = j \cdot j \cdot Y_1 \cdot Y_2$$

Recall we defined j = sqrt(−1), so $j^2 = -1$

$$Z_1 \cdot Z_2 = -(Y_1 \cdot Y_2), \text{ a negative real number}$$

Or equivalently,

$$Z_1 = R_1 \text{ angle}(90),\ Z_2 = R_2 \text{ angle}(90)$$
$$Z_1 \cdot Z_2 = R_1 \cdot R_1 \text{ angle}(180) = -(R_1 \cdot R_2) = -(Y_1 \cdot Y_2) \quad \text{since } X_1 = X_2 = 0$$

You can experiment with other combinations, but what you will find is that the arithmetic of adding angles around the circle when multiplying complex numbers works out perfectly when we designate the positive imaginary axis with j, and the negative imaginary axis with−j.

By visualizing this business of going round the circle, you can see by inspection that:

Multiply two positive real numbers, both angles = 0, result has angle of zero.

$$3 \cdot 5 = 15$$

Multiply two negative real numbers, both angles = 180 (or −180), result has angle of zero (or 360).

$$-3 \cdot -5 = 15$$

Multiply a positive real number (angle 0) with a negative real number (angle 180), result has angle of 180—a negative real number.

$$3 \cdot -5 = -15$$

Multiply a positive real number (angle 0) with a positive imaginary number (angle 90), result has angle of 90—an imaginary number.

$$j3 \cdot 5 = j15$$

Multiply a positive imaginary number (angle 90) with a positive imaginary number (angle 90), result has angle of 180—a negative real number.

$$j3 \cdot j5 = j^2 \cdot 15 = -15$$

Multiply a negative imaginary number (angle −90) with a negative imaginary number (angle −90), result has angle of 180—a negative real number.

$$-j3 \cdot -j5 = (-j) \cdot (-j) \cdot 15 = -(-(-(15))) = -15$$

Multiply a positive imaginary number (angle 90) with a negative imaginary number (angle −90), result has angle of 0—a positive real number.

$$j3 \cdot -j5 = j \cdot (-j) \cdot 15 = -(-(15)) = 15$$

2.5 Complex Conjugate

The last example illustrates a special case. Every number Z = R angle (Ω) has, what is called, a complex conjugate, Z* = R angle (−Ω). In the example above, Ω = 90, but Ω can be any angle. The "*" symbol is the complex conjugate symbol and means to take the point Z and mirror it across the X axis as shown in Fig. 2.4.

So Z = X + jY has conjugate Z* = X−jY. We just negate or reverse sign of imaginary part of a number to get its conjugate, or if in polar form, just negate the sign of the angle.

$$Z = \mathrm{R\ angle}(\Omega),\ Z^* = \mathrm{R\ angle}(\Omega)$$

A special property of the complex conjugate is that for any complex number:

$$Z \cdot Z^* = \mathrm{R\ angle}(\Omega) \cdot \mathrm{R\ angle}(-\Omega) = \mathrm{R}^2\ \mathrm{angle}(0).$$

In other words, when you multiply a number by its conjugate, the product is a real number, equal to the magnitude squared. This will become important in digital communication, because it can be used to compute the power of a complex signal.

To summarize, we have tried to show that the imaginary numbers which are used to form things called complex numbers are really not so complex, and imaginary is really a very misleading description. What we have really been after is to create a two dimensional number plane, and define a set of expanded arithmetic rules to manipulate the numbers in it. Now we are ready to move onto the next topic, the complex exponential.

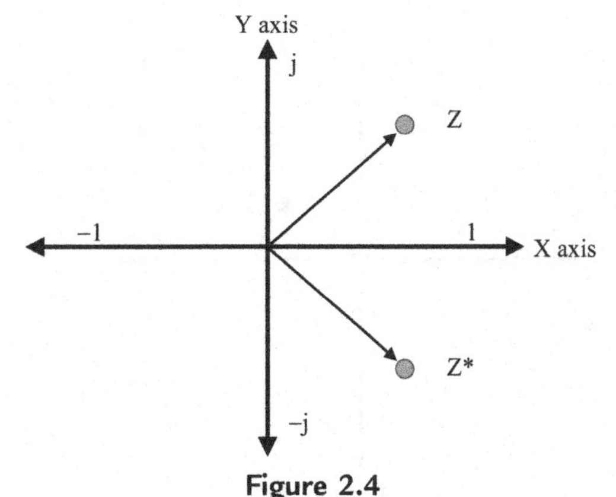

Figure 2.4
Complex conjugate diagram.

2.6 The Complex Exponential

The complex exponential has an intimidating sound to it, but in reality, it is very simple to visualize. It is simply the unit circle (radius = 1) on the complex number plane (Fig. 2.5).

Any point on the unit circle can be represented by "e" or raised to the power (j·angle) or more also expressed $e^{j\Omega}$, which is called a complex exponential function. A few examples should help.

Let the angle $\Omega = 0°$degree. Anything raised to the power 0 is equal to 1. This checks out, since this is the Point 1 on the positive real axis.

Let angle $\Omega = 90°$degrees. The complex exponential is e^{j90}. This is the point j on the positive imaginary axis. We need a way to evaluate the complex exponential to show this. This leads to the Euler equation. This equation can easily be derived using series Taylor expansion for exponential, but we have promised to minimize the math. But the result is:

$$e^{j\Omega} = \cos(\Omega) + j\,\sin(\Omega)$$

Let us try exp(j90) again. Using Euler equation

$$e^{j90} = \cos(90) + j\,\sin(90) = 0 + j \cdot 1 = j$$

Imagine the point $Z = e^{j\Omega}$ with the angle Ω starting at 0°degree and gradually increasing to 360°degrees. This will start at the point +1 on real axis, and move counter-clockwise around the circle until it ends up where it started, at 1 again. If the angle starts at 0 and gradually decreases until it reached −360, the point will do exactly the same thing, except rotate in a clockwise fashion.

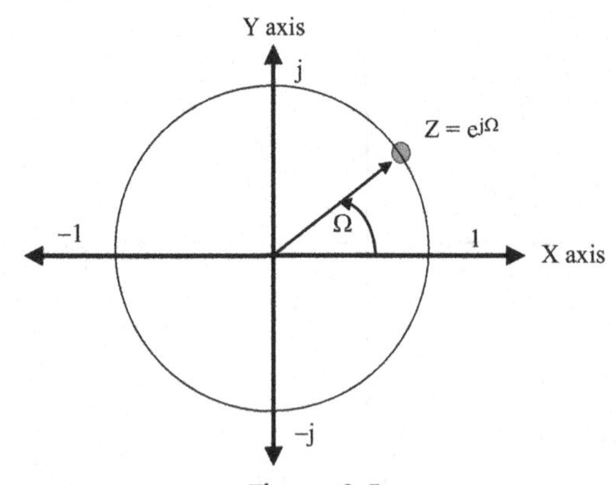

Figure 2.5
Complex exponetial diagram.

Now we know that from the Euler equation that the complex exponential has a real and imaginary component. Try to imagine the movement of the point on the unit circle as reflected on the real axis (imagine a second point, allowed to move only on the real axis, trying to follow the first point as it moves about the circle). The movement of the second point on the real axis will equal to $\cos(\Omega)$. So if we continually rotate in either direction about the unit circle, the real component will move back and forth between $+1$ and -1 using the motion of the cosine function. Similarly, the movement of the point on the unit circle as reflected on the imaginary axis will be similar, except instead of starting at value of $+1$, it will start with value 0. The pattern of motion will lag by 90°degrees. The imaginary axis movement is equal to $j \cdot \sin(\Omega)$, and the imaginary component will move back and forth between j and $-$j using the sine function.

This is shown in Fig. 2.6, where the dashed line represents the imaginary axis movement $j \cdot \sin(\Omega)$ and the dotted line represents the real axis movement of $\cos(\Omega)$.

$$X = \cos(\Omega) \quad \text{(real axis movement)}$$
$$Y = \sin(\Omega) \quad \text{(imaginary axis movement)}$$

This gives us better way to express a complex number in polar coordinates.

$$\text{Recall } Z = X + jY = R \text{ angle } (\Omega)$$

As we saw before,

$$X = R\cos(\Omega)$$
$$Y = R\sin(\Omega)$$

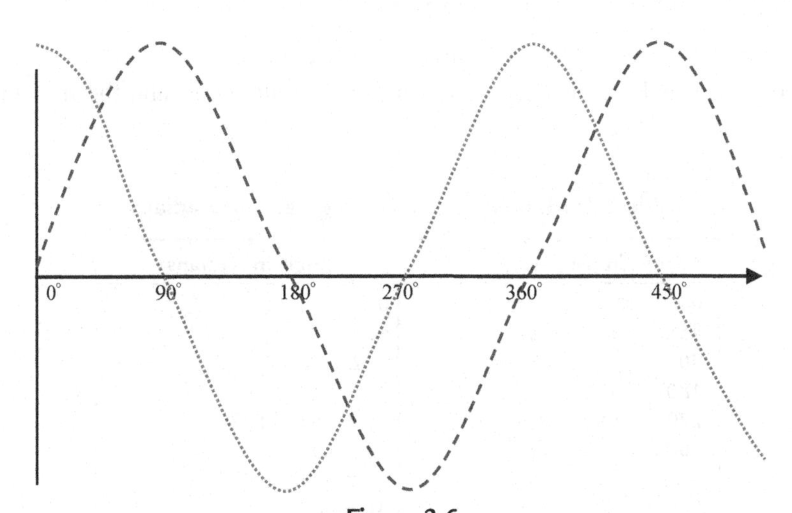

Figure 2.6
Graphing of complex exponetial.

So we can see that angle (Ω) has the same meaning as exp($j\Omega$). Also, for the unit circle, R = 1 by definition. So our new way to express a number in polar form using the complex exponential is:

$$Z = R\ \text{angle}(\Omega) = Re^{j\Omega} \quad \text{(any point in complex plane)}$$
$$Z = \text{angle}(\Omega) = e^{j\Omega} \quad \text{(for R = 1, any point on the unit circle)}$$

This is the way you will often see it in the textbooks and industry literature.

2.7 Measuring Angles in Radians

The last curve ball in this chapter involves measuring angles in radians. You will have to get used to this, because you will see it everywhere in DSP. In our discussion, to make things more familiar, we started measuring angles in degrees, where 360°degrees describes a full circle. More commonly, the angle measurement in radians is based upon π, which is a number defined to have a value of about 3.141592 (it actually is an irrational number, with infinite number of digits, like 1/3 = 0.3333....). It takes exactly 2π radians to describe a full circle (Table 2.1).

Just like angle measurements are periodic in 360°degrees, they are also periodic in 2π radians. Using π is really no different than getting used to meters rather than using feet for measuring distances (or the reverse if you did not grow up in the United States).

We are going to see this same concept later in sampling theory, where everything tends to wraps around or behave periodically. We can visualize this as traveling either clockwise (negative rotation) or counter-clockwise (positive rotation) around the circle.

There is one more DSP convention to be aware of. The real component (we used X in discussion above) is usually called the "I" or in-phase component, and the imaginary

Table 2.1: Mapping Between Degrees and Radians

Angle in Degrees	Angle in Radians
0	$0\ \pi$
45	$\pi/4$
90	$\pi/2$
180 = −180	$\pi = -\pi$
270 = −90	$(3/2)\ \pi = -\pi/2$
360 = 0	$2\pi = 0\pi$
−360 = 0	$-2\pi = 0\pi$
540 = 180	$3\pi = \pi$
−540 = −180 = 180	$-3\pi = -\pi = \pi$

component (we used Y in discussion above) is usually referred to as the "Q" or quadrature phase component. In many DSP algorithms, the digital signal processing must be performed simultaneously on both I and Q data streams, which we now know simply represents the signal's movement, over time, within the two dimensions of our complex number plane.

CHAPTER 3

Sampling, Aliasing, and Quantization

Now that we have the basic background material covered, let us start talking about digital signal processing (DSP). The starting point to understand is sampling and its affect on the signal of interest. To take an analog signal and convert it to a digital signal, we need to sample the signal using a device called an analog to digital converter (ADC). The ADC will measure the signal at rapid intervals, and these measurements are called samples. It will output a digital signal proportional to the amplitude of the analog signal at that instant. This can be compared to looking at an object with only a strobe light for illumination. You can see the object only when the strobe light flashes. If the object is not moving, then everything looks pretty much the same as if we used a normal, continuous light source. Where things get interesting is when we look at a moving object with the strobe light. If the object is moving rapidly, then the appearance of the motion can be quite different that when viewed under normal light. We can also see strange effects even if the object is moving fairly slowly, and we reduce the rate of the strobe light enough. Intuitively, we can see that what is important is the rate of the strobe light compared to the rate of movement of the illuminated object. As long as the light is strobing fast compared to the movement of the object, the movement of the object looks very fluid and normal. When the light is strobing slowly compared to the rate of object movement, the movement of the object looks funny, often like slow motion, as we can see the object is moving, but we miss the sense of continuous fluid movement.

Let us mention one more example many of us have experienced as a child. Imagine trying to make your own animated movie, and sketching a character on index cards. We want to depict this character moving, perhaps jumping and falling. We might sketch 20 or 40 cards, each showing the same character in sequential stages of this motion, with just small movement changes from one card to next. Once we are finished, we can show it to our friends by holding one edge of the deck of index cards, and flipping through it quickly by thumbing the other edge. We see our character in this continuous motion of jumping and falling by flipping through the deck of index cards in a second or two.

Actually, whenever we watch TV, this is what is occurring. But the TV is updating the screen at about 60 times per second, which is rapid enough that we do not notice the separate frames, and we think that we are seeing continuous motion.

Digital Signal Processing 101. http://dx.doi.org/10.1016/B978-0-12-811453-7.00003-2

So it makes sense that if we sample a signal very fast compared to how rapidly the signal is changing, we get a pretty accurate sampled representation of the signal, but if we sample too slow, we will see a distorted version of the signal.

3.1 Sampling Effects

Below are graphs of two different sinusoidal signals being sampled. The slower moving signal (lower frequency) in Fig. 3.1 can be represented accurately with the indicated sample rate, but the faster moving signal (higher frequency) in Fig. 3.2 is not accurately represented by our sample rate. In fact, it actually appears to be a slow moving (low frequency) signal, as indicated by the dashed line. This shows the importance of sampling "fast" enough for a given frequency signal.

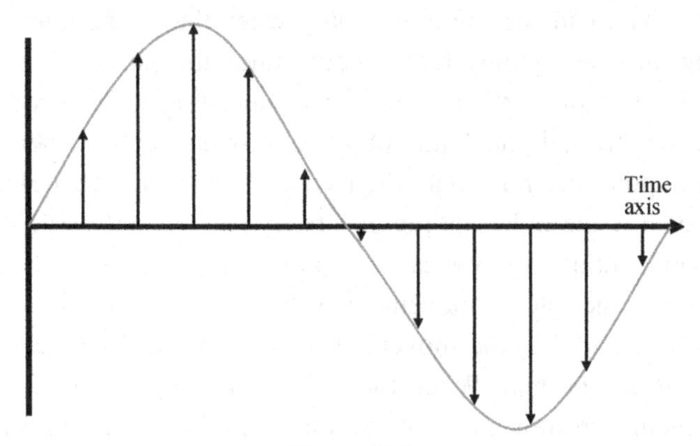

Figure 3.1
Sampling a low-frequency signal (arrows indicate sample instants).

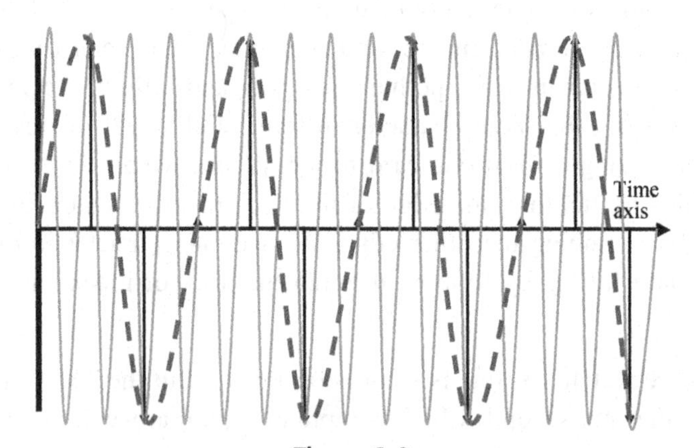

Figure 3.2
Sampling a high-frequency signal (same sample instants).

The dashed line shows how the sampled signal will appear if we connect the sample dots and smooth out the signal. Notice that since the actual (solid line) signal is changing so rapidly between sampling instants, this movement is not apparent in the sampled version of the signal. The sampled version actually appears to be a lower frequency signal than the actual signal. This effect is known as aliasing.

We need a way to quantify how fast we must sample to accurately represent a given signal. We also need to better understand exactly what is happening when aliasing occurs. It may seem strange, but there are some instances when aliasing can be useful.

Let us go back to the analogy of the strobe light, and try another thought experiment. Imagine a spinning wheel, with a single dot near the edge. Let us set the strobe light to flash every 1/8 of a second, or 8 times per second. Below is shown what we see over six flashes, depending on how fast the wheel is rotating. Time increments from left to right in all figures (Figs. 3.3–3.10).

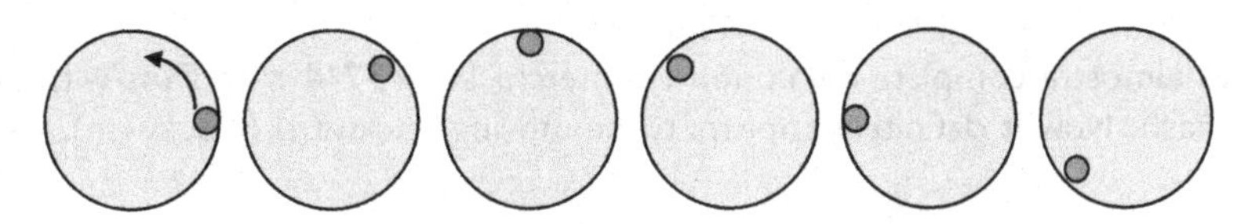

Wheel rotating counterclockwise once per second, or 1 Hz

Figure 3.3
The dot moves 1/8 of a revolution, or $\pi/4$ radians with each strobe flash.

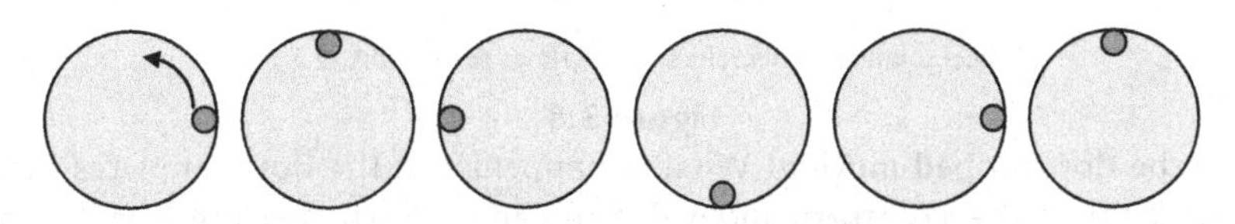

Wheel rotating counterclockwise twice per second, or 2 Hz

Figure 3.4
Now the dot moves twice as fast, ¼ of a revolution, or $\pi/2$ radians with each strobe flash.

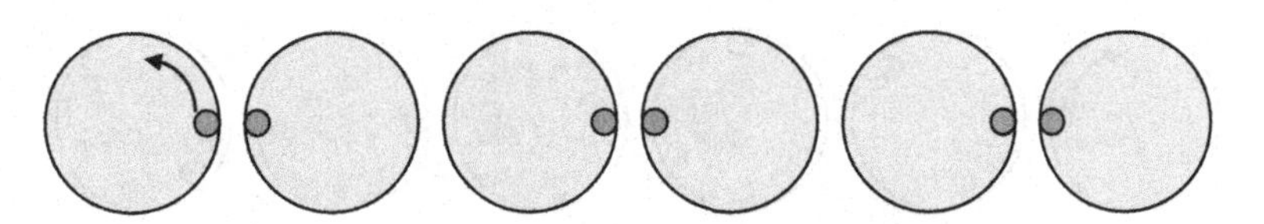

Wheel rotating counterclockwise 4 times per second, or 4 Hz

Figure 3.5
The dot moves ½ of a revolution, or π radians with each strobe flash. It appears to alternate on each side of the circle with each flash. Can you be sure which direction the wheel is rotating?

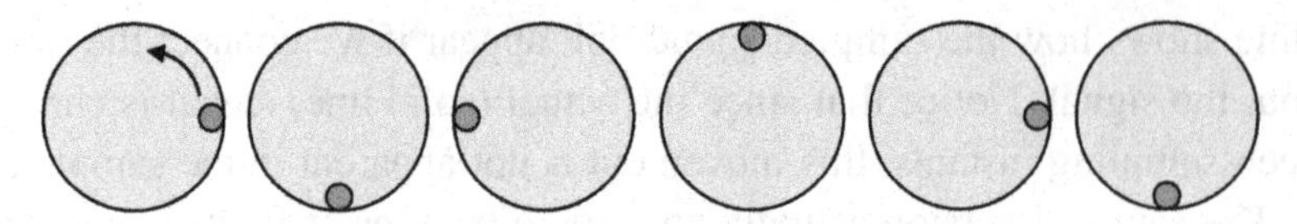

Wheel rotating counterclockwise 6 times per second, or 6 Hz

Figure 3.6
The dot moves counterclockwise 3/4 of a revolution, or 3/2 π radians with each strobe flash. But it appears to be moving backward (clockwise).

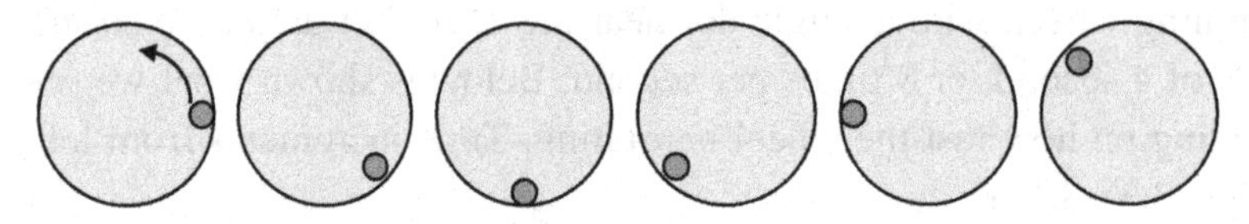

Wheel rotating counterclockwise 7 times per second, or 7 Hz

Figure 3.7
The dot moves almost a complete revolution counterclockwise, 7/4 π radians with each strobe flash. Now it definitely appears to be moving backward (clockwise).

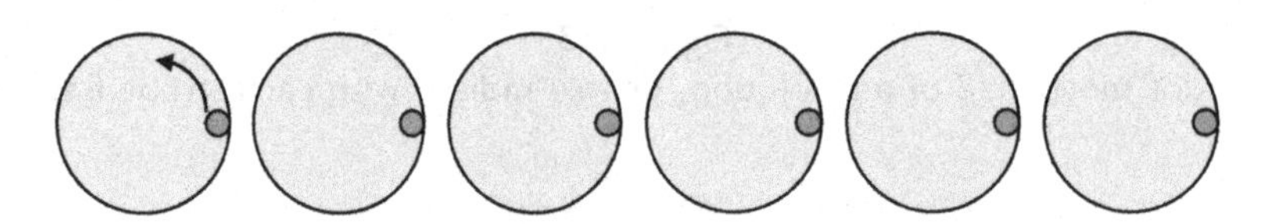

Wheel rotating counterclockwise 8 times per second, or 8 Hz

Figure 3.8
It looks like the dot stopped moving! What is happening is the dot completes exactly one revolution (2 π radians) every strobe interval. You can't tell whether the wheel is moving at all, spinning forwards or backwards. In fact, could the wheel be rotating twice per strobe interval (4 π radians)?

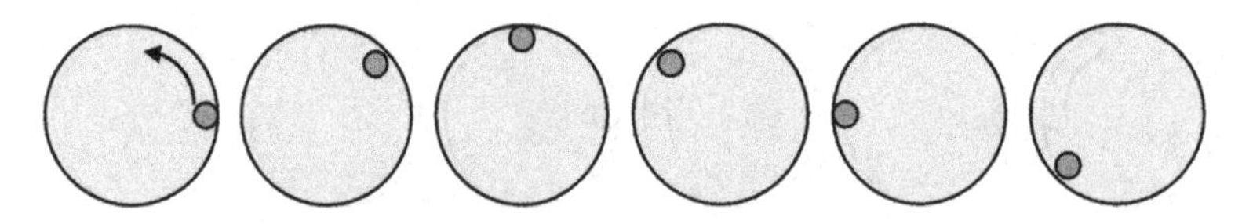

Wheel rotating counterclockwise 9 times per second, or 9 Hz

Figure 3.9
It sure looks the same as when the wheel was rotating once per second, or 1 Hz. In fact, the wheel is moving 9/4 π radians with each strobe flash.

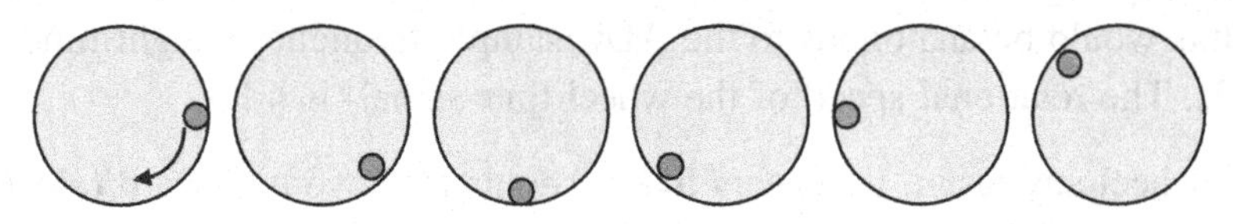

Wheel rotating backward (clockwise) once per second, or –1 Hz

Figure 3.10
Now we stopped the wheel and started rotating backward at $-\pi/4$ radians with each strobe flash. Notice that this appears exactly the same as when we were rotating forward at $7/4\ \pi$ radians with each strobe flash.

Does not all this look familiar from the previous chapter? The positive rotation is wrapping around and appears like a negative rotation as the wheel speed increases. And the rotation perception appears periodic in 2π radians per strobe, just like the angle measurements.

The reality is that once we sample a signal (this is what we are doing by flashing the strobe light), we cannot be sure what has happened in-between flashes. Our natural instinct is to assume that the signal (or dot in our example) took the shortest path from where it appears in one flash and then in the subsequent flash. But as we can see in the examples above, this can be misleading. The dot could be moving around the circle in the opposite direction (taking the longer path) to get to the point where we see it on the next flash. Or imagine that the circle is rotating in the assumed direction, but it rotates one full revolution plus "a little bit extra" every flash (Fig 3.9, or 9 Hz rate). What we see is only the "a little bit extra" on every flash. For that matter, it could go around 10 times plus the same "a little bit extra" and we could not tell the difference.

3.2 Nyquist Sampling Rule

To prevent all this, we have to come up with a sampling rule or convention. What we are going to agree is that we will always sample (or strobe) at least twice as fast as the frequency of the signal we are interested in. And in reality, we need to have some margin, so we better make sure we are sampling *more* than twice as fast as our signal. Going back to our example above, at what point do things start to get fishy?

Consider what happens when we start the wheel moving slowly in a counterclockwise direction. Everything looks fine until we reach a rotational speed of 4 Hz. At this point the dot will appear to be alternating on either side of the circle with each strobe flash. Once we have reached this point, we can no longer tell which direction the wheel is rotating—it will look the same rotating both directions. This is the critical point, where we are sampling at exactly twice as fast as the signal. The sampling speed is the frequency of the strobe light (this would be analogous to the ADC sample frequency), eight times per second or 8 Hz. The rotational speed of the wheel (our signal) is 4 Hz.

If we spin the wheel any faster, it appears like it begins to move backward (clockwise), and by the time we reach a rotational speed of 8 Hz, it appears to stop altogether. Spinning still faster will appear like the wheel moves forward again, until it again appears to start going backward again and the cycle repeats.

To summarize, whenever you have a sampled signal, you cannot really be sure of its frequency. But if you assume that the rule was followed—that the signal was sampled at more than twice the frequency of the signal, then the sampled signal will really represent the same frequency as the actual signal prior to sampling. The critical frequency which the signal must not ever exceed, which is one half of the sampling frequency, is called the *Nyquist* frequency.

If we follow this rule, then we can avoid the aliasing phenomenon we demonstrated with the moving wheel example above. Normally, what is done is that the ADC converter frequency is selected to be high enough to sample the signal in which we want to perform digital signal processing on. To make sure that unwanted signals above the Nyquist frequency do not enter the ADC and cause aliasing, the desired signal in usually passed through an analog low pass filter, which attenuates any unwanted high frequency content signals, just prior to the ADC.

A common example is the telephone system. Our voices are assumed to have a maximum frequency of about 3600 Hz. At the microphone, our voice is filtered by an analog filter to eliminate, or at least substantially reduce, any frequencies above 3600 Hz. Then the microphone signal is sampled at 8000 Hz or 8 kHz. All subsequent digital signal processing occurs on this 8 kSPS (kilo-samples per second) signal. That is why if you hear music in the background while on the telephone, the music will sound flat or distorted. Our ears can detect up to about 15 kHz frequencies, and music generally has frequency content exceeding 3600 Hz. But little of this higher frequency content will be passed through the telephone system.

In the next chapter, we are going to start representing our signals and sampling in the frequency (or spectral) domain. What this means is that when we plot the signal spectrum, the X axis will represent increasing frequency (Fig. 3.11).

So far, we have covered the most important effects of sampling, but there remains one last issue related to sampling, which is quantization. We saw earlier how a digital signal is represented numerically, and how we must be sure that the numbers representing the sampled signal do not exceed the range of our binary or hexadecimal number range.

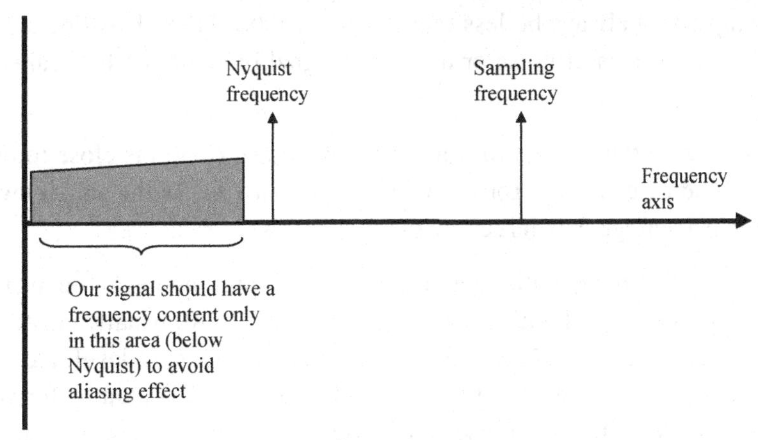

Figure 3.11
Nyquist frequency band.

3.3 Quantization

But what happens if the signals are very small? Remember in our discussion of signed fractional 8 bit fixed point numbers, the range of values we could represent was from −1 to +1 (well, almost +1). The step size was 1/128, which works out to 0.0078125. Let us say the signal has an actual value of 0.5030 at the instant in time we sample it, using an 8 bit ADC. How closely can the ADC represent this value? Let us compare this to a signal that is 1/10 the level of first sample or 0.0503. And again, consider a signal with value 1/10 the level as the second sample, at 0.00503. Below is a table showing the possible outputs from an 8 bit ADC at each of these signal levels, and the error that will result in the conversion of the actual signal to sampled signal value. We say "possible outputs", as we are assuming that the ADC will output either the value immediately above or below the actual input signal level (Table 3.1).

The actual error level remains more or less in the same range over the different signal ranges. This error level will fluctuate, depending upon the exact signal value, but with our

Table 3.1: 8 Bit Quantization Error

Signal Level	Closest 8 Bit Representation	Hexadecimal Value	Actual Error	Error as a Percent of Signal Level (%)
0.50300	0.5000000	0x40	0.00300	0.596
0.50300	0.5078125	0x41	−0.0048128	0.957
0.05030	0.0468750	0x06	0.003425	6.809
0.05030	0.0546875	0x07	−0.0043875	8.722
0.00503	0.000000	0x00	0.00503	100
0.00503	0.0078125	0x01	−0.0027825	55.32

8 bit signed example will always be less than 1/128, or 0.0087125. This fluctuating error signal will be seen as a form of noise or unwanted signal by the DSP. It is called quantization noise.

When the signal level is fairly large for the allowable range, (0.503 is close to one half the maximum value), the percentage error is small = less than 1%. As the signal level gets smaller, the error percentage gets larger, as the table shows.

What is happening is that the quantization noise is always present and is, on average, the same level (any noiselike signal will rise and fall randomly, so we usually concern ourselves with the average level). But as the input signal decreases in level, the quantization noise becomes more significant in a relative sense. Eventually, for very small input signal levels, the quantization noise can become so significant that it degrades the quality of whatever signal processing is to be performed. Think of it as like static on a car radio. As you get further from the radio station, the radio signal gets weaker, and eventually the static noise makes it difficult or unpleasant to listen to, even if you increase the volume.

So what can we do if our signal sometimes is strong (0.503 for example), and other times is weak (0.00503 for example)? Another way of saying this is that the signal has a large dynamic range. The dynamic range describes the ratio between the largest and smallest value of the signal, in this case 100.

Suppose we exchange our 8 bit ADC for a 16 bit ADC? Then our maximum range is still from −1 to +1, but our step size is now 1/32,768, which works out to 0.000030518. Let us make a 16 bit table similar to the 8 bit example (Table 3.2).

What a difference! The actual error is always less than our step size, 1/32,768. But the error as a percent of signal level is dramatically improved. This is what we usually care about in signal processing. Because of the much smaller step size of the 16 bit ADC, the quantization noise is much less, allowing even small signals to be represented with very good precision (<1%). Notice that even for our small signal level, 0.00503, the error about 0.1%.

Table 3.2: 16 Bit Quantization Error

Signal Level	Closest 16 Bit Representation	Hexadecimal Value	Actual Error	Error as a Percent of Signal Level (%)
0.50300	0.5029907	0x4062	0.000009277	0.00185
0.50300	0.5030212	0x4063	−0.00002124	0.00422
0.05030	0.0502930	0x0670	0.00000703	0.0140
0.05030	0.0503235	0x0671	−0.0000235	0.0467
0.00503	0.005005	0xA4	0.0000251	0.499
0.00503	0.0050354	0xA5	−0.0000054	0.107

3.4 Signal to Noise Ratio

Another way of describing this is to introduce the concept of signal to noise power ratio (SNR). This describes the power of the largest signal compared to the background noise. This can be very easily seen on a frequency domain or spectral plot of a signal. There can be many sources of noise, but for now, we are only considering the quantization noise introduced by the ADC sampling.

SNR is usually expressed in decibels (denoted dB), using a logarithmic scale. The SNR of an ideal ADC can be determined by the following equation:

$$\text{SNR}_{\text{quantization}}(\text{dB}) = 6.02 * (\text{Number of bits}) + 1.76$$

Basically, for each additional bit of the ADC, 6 dB of SNR is gained. An 8 bit ADC is capable of representing a signal with an SNR of about 48 dB, a 12 bit ADC can do better at 72 dB, and a 16 bit will give up to 96 dB. This only accounts for the effect of quantization noise; in practice there are other effects that also will degrade SNR in a system.

There is one last important point on decibels. These are very commonly used in many areas of digital signal processing subsystems. A decibel is simply a signal power ratio, similar to percentage. But because of the extremely high ratios commonly used (a billion is not uncommon), it is convenient to express this logarithmically. The logarithmic expression also allow chains of circuits or signal processing operations each with its own ratio (say of output power to input power) to simply be added up to find the final ratio.

Where people commonly get confused is in differentiating between signal levels or amplitude (voltage if an analog circuit) and signal power. Power measurements are virtual in the digital world, but can be directly measured in analog circuits in which DSP systems interface with.

There are two definitions of dB commonly used.

$$\text{dB}_{\text{voltage}} = \text{dB}_{\text{digital value}} = 20 \cdot \log(\text{voltage signal } 1/\text{voltage signal } 2)$$
$$\text{dB}_{\text{power}} = 10 \cdot \log(\text{power signal } 1/\text{power signal } 2)$$

The designations of "signal 1" and "signal 2" depends on the situation. For an RF power amplifier, the dB of gain will be the 10 log (output power/input power). For an ADC, the dB of SNR will be the 20 log (maximum input signal/quantization noise signal level). For a DAC, the dB of spurious free dynamic range will be 20 log (maximum output signal level/largest unwanted frequency component level generated by DAC circuits).

So dB can refer to many different ratios. But it is easy to get confused whether to use to multiplicative factor of 10 or 20, without understanding the reasoning behind this.

Voltage squared is proportional to power. If a given voltage is doubled in a circuit, it requires four times as much power. This goes back to a basic Ohm's law equation.

$$\text{Power} = \text{Voltage}^2/\text{Resistance}$$

In many analog circuits, signal power is used, because that is what the lab instruments work with, and while different systems may use different resistance levels, power is universal (however, 50 Ω is the most common standard in most analog systems).

The important point is that since voltage is squared, this effect needs to be taken into account in the computation of logarithmic decibel relation. Remember, $\log x^y = y \log x$. Hence, the multiply factor of "2" is required for voltage ratios, changing the "10" to a "20".

In the digital world, the concept of resistance and power do not exist. A given signal has specific amplitude, expressed in a digital numerical system (such as signed fractional or integer for example).

Understanding dB increases using the two measurement methods is important. Let us look at doubling of the amplitude ratio and doubling of the power ratio.

$$6.02\ \text{dB}_{\text{voltage}} = 6.02\ \text{dB}_{\text{digital value}} = 20\cdot\log(2/1)$$
$$3.01\ \text{dB}_{\text{power}} = 10\cdot\log(2/1)$$

This is why shifting a digital signal left 1 bit (multiply by 2) will cause a 6 dB signal power increase, and why so often the term 6 dB/bit is used in conjunction with ADCs, DACs, or digital systems in general.

By the same reasoning, doubling in power to an RF engineer means a 3 dB increase. This will also impact the entire system. Coding gain, as used with error correcting code methods, is based upon power. All signals at antenna interfaces are defined in terms of power, and the decibels used will be power ratios.

In both systems, ratios of equal power or voltage are 0 dB. For example, a unity gain amplifier has again of 0 dB.

$$0\ \text{dB}_{\text{power}} = 10\cdot\log(1/1)$$

A loss would be expressed as a negative dB. For example, a circuit whose output is equal to ½ the input power.

$$-3.01\ \text{dB}_{\text{power}} = 10\cdot\log(1/2)$$

CHAPTER 4

Frequency Response

Once we have a sampled digital signal, we are ready to perform digital signal processing (DSP) on this signal. In the last chapter, we briefly touched on representing the signal in the frequency domain. Usually, our goal is to modify the signal's frequency representation. This is normally performed using a filtering function. This is probably the most fundamental of DSP functions. In the previous example of the telephone system, we talked about sampling the voice signal at 8 kHz. Let us say we want to build an automated system to detect touch tones (on a touch tone phone, whenever you press a number key, the phone creates two specific tones or frequencies in the audio signal band—which is what you hear when pressing the button). We could build a digital filter for each of the possible tones, feed the sampled audio signal into each filter, and monitor the outputs of the filters. In this way, we could detect when the telephone user presses the buttons, and which button was pressed, which is exactly what touch tone phone systems do.

4.1 Frequency Response and the Complex Exponential

In this chapter, we need to understand the concept of frequency response. Then in the next chapter, we will develop a way to relate a filter's frequency and time response. First, let us begin with an intuitive way to understand the frequency response of a filter. From the last chapter, we learned that we can create a complex exponential signal, which is just a positive or negative frequency rotation about the unit circle (radius = 1). Furthermore, when the frequency of the complex exponential reaches the Nyquist frequency, we have reached the maximum frequency that can be represented for a given sampling rate.

A complex exponential signal has the following forms:

$$e^{j\omega t} = \cos(\omega t) + j\,\sin(\omega t) \quad \text{or}$$
$$e^{j2f\pi} = \cos(2\pi ft) + j\,\sin(2\pi ft)$$

This is very similar to what we saw before. The earlier angle Ω has been replaced by ωt or by 2πft. The significance of this is that we are no longer representing a point or vector on the unit circle of the complex number plane (as determined by angle Ω). Now we are representing a time-varying signal. This signal will move about the unit circle with a rotational speed of ω radians per second. For a given ω (rotational speed), we can determine the position of the signal at any given point in time (denoted by "t").

Digital Signal Processing 101. http://dx.doi.org/10.1016/B978-0-12-811453-7.00004-4

The second equation is equivalent to the first, except the rotational speed is expressed in cycles (revolutions) per second, denoted by "f". Try to make sure this is clear before moving on, because both forms will appear interchangeably in the DSP world.

t represents time, in seconds.
ω represents rotational speed, in radians per second (2π radians = one revolution).
f represents rotational speed, in cycles (or revolutions) per second.

The variables ω and f simply describe the same thing, using different units, like inches and centimeters. Recall from before that it takes exactly 2π radians to complete a full circle. So 1 cycle per second equals 2π radians per second, or $\omega = 2\pi f$.

Both ω and f denote the rotational speed in the counterclockwise direction. A negative value for ω or f means we are rotating in the opposite direction (clockwise). Let us now consider an example. Let the rotational speed equal to 1/8 of a circle per second (takes 8 s to complete one revolution). Therefore, $f = 1/8$. Also, $\omega = 2\pi f$, or $\omega = \pi/4$. This signal (let us call it "s") can be described as:

$$s(t) = e^{j\pi t/4} = \cos(\pi t/4) + j\,\sin(\pi t/4)$$

We can evaluate s(t) at any given time t. For example, refer to Table 4.1:

We can also plot s(t). Shown in Fig. 4.1 are of separate plots of the real (dotted line) part of s(t) and the imaginary (dashed line) part of s(t). Also shown in Fig. 4.2, in on a separate plot is the complete signal s(t) on the unit circle of the complex number plane with time labels at each point.

4.2 Normalizing Frequency Response

Notice that we are measuring the value of our signal once per second. It turns out that it is often convenient to set the sampling frequency F_s equal to 1 s. This is called normalization. The frequency response of a filter can be expressed as a normalized

Table 4.1: Rotational Representations of s(t)

Time t in Seconds	$s(t) = e^{j\pi t/4}$	Angle of s(t) in Degrees	Angle of s(t) in Radians
0	1	0	0
1	0.707 + 0.707 j	45	$\pi/4$
2	J	90	$\pi/2$
3	−0.707 + 0.707 j	135	$3\pi/4$
4	−1	180	π
5	−0.707−0.707 j	225	$5\pi/4 = -3\pi/4$
6	−j	270	$3\pi/2 = -\pi/2$
7	0.707−0.707 j	315	$7\pi/8 = -\pi/4$
8	1	360 = 0	$2\pi = 0$

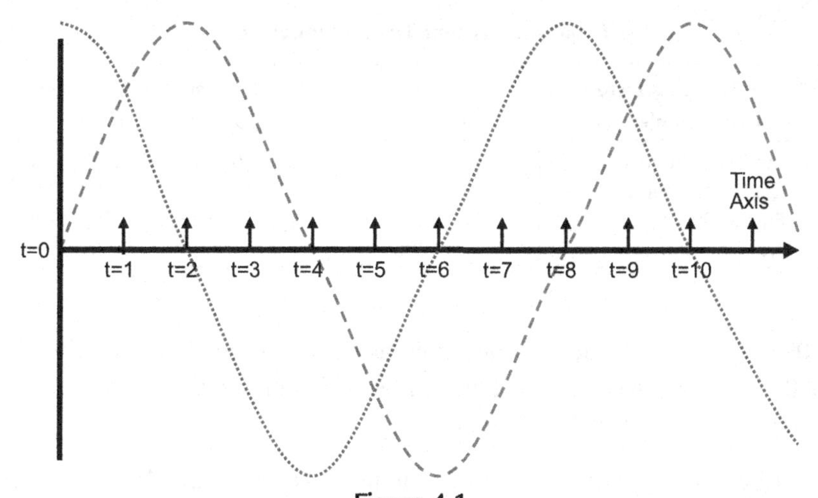

Figure 4.1
Real and imaginary components of s(t).

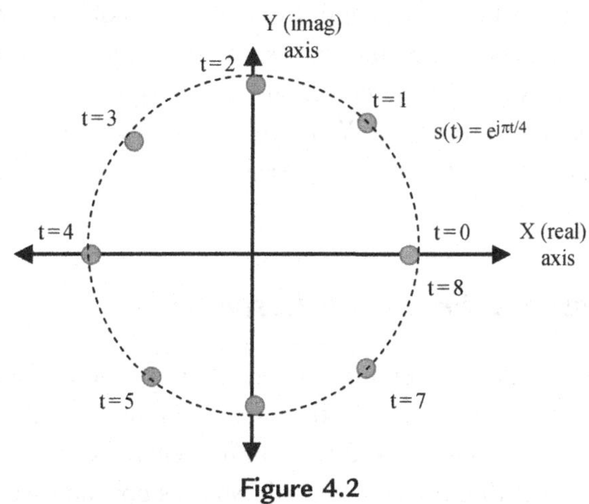

Figure 4.2
Complex plane representation of s(t).

response, where the input frequency is expressed as a fraction of sampling frequency F_s. Let us illustrate this using the telephone system as an example. In this case, $F_s = 8000$ Hz, $F_{Nyquist} = 4000$ Hz, and one of the touch tones we need to detect is at 770 Hz. We could build a filter with a passband (portion of frequency band to "pass through" the signal, with minimum attenuation) to detect 770 Hz. Since we need a little tolerance, lets decide to make the passband from 760 to 780 Hz and our desired stopbands (portion of frequency band to "stop" or block the signal, with maximum attenuation) is from 0 to 760 and from 780 to 4000 Hz (recall $F_{Nyquist} = F_s/2$). Do not forget that we are assuming no signal at frequencies above $F_{Nyquist}$, this has been already taken care of before sampling. This could also be expressed as a passband from 0.0950 to 0.0975 F_s, and stopbands from 0 to 0.0950 F_s

Table 4.2: Touch Tone Frequencies

Touch Tone Bandpass Detection Filter Example	Actual Frequency (Hz)	Normalized Frequency (F_s = 8000 Hz)
Touch Tone Frequency	770	0.09625 F_s (770/8000)
Start of Passband	760	0.0950 F_s (760/8000)
End of Passband	780	0.0975 F_s (780/8000)
Nyquist Frequency	4000	0.50 F_s (4000/8000)

and from 0.0975 to $F_s/2$. All these frequencies have been normalized, by dividing with our $F_s = 8000$ Hz. When designing a digital filter, it is common to normalize the frequency response in terms of F_s (Table 4.2).

Now let us get back to frequency response. Suppose we take our filter, and input a series of complex exponential signals into the filter. Each exponential will be a little higher frequency than the previous. We check the output of each frequency. What we will find is if we input a complex exponential of a given frequency, we will see at the output a complex exponential signal of the same frequency. That is because a digital filter is a linear device, so it cannot create new frequencies, or change the frequency of a signal passing through it. What it can do is to change the amplitude and phase of the input signal. Let us leave the phase part out of it for now, and focus on the amplitude.

4.3 Sweeping Across the Frequency Response

Suppose we build a digital signal generator that can create a complex exponential signal of any frequency we desire, from 0 to $F_{Nyquist}$ (this is called a numerically controlled oscillator or NCO, and it is common in digital communication systems). We drive our filter with the NCO, incrementing the frequency in small steps, and we measure the amplitude of the filter output signal. If we plot this data, we will have the magnitude response of our filter with all frequencies from 0 to $F_{Nyquist}$.

If we do this for the touch tone filter above, we will get zero output until we input a complex exponential at frequency 0.0950 F_s. From frequency 0.0950 F_s until 0.0975 F_s, the complex exponential will pass through our filter, and could be detected. Above 0.0975 F_s up until $F_{Nyquist}$, we will get zero output. This type of filter is called a bandpass filter, because it allows only a portion of the frequency band to pass and blocks frequencies above and below this band.

Now remember, we can also have negative frequency complex exponentials, so we could also do a similar plot from 0 to $-F_{Nyquist}$. For the vast majority of digital filters, the frequency response will be the same whether the input is positive or negative frequency. All filters with real (as opposed to complex) coefficients will have this property.

Referencing everything to F_s is also convenient when using our NCO. The NCO does not know that it is being clocked at 8 kHz sampling frequency. Instead when we program it, we need to set the desired frequency output in terms of F_s. For example, to test our filter with an input of 770 Hz, we need the NCO to produce a complex exponential at 0.09,625 F_s or (770/8000) F_s.

4.4 Example Frequency Responses

After this long winded explanation, let us show some examples of frequency responses to help clarify things. In this chapter, we will depict the frequency response of ideal filters. First we will show low pass, then high pass, and lastly a bandpass, like our touch tone 770 Hz detector.

Please note that the filter response repeats at intervals of F_s. This is an artifact of sampling, as explained in the last chapter. The valid portion of the sampled frequency response is from $-F_{Nyquist}$ to $+F_{Nyquist}$. Just as we saw the apparent rate of rotation of the wheel with the red dot reach a maximum at a rate of ½ the sample rate, and then slowly decrease until it finally stopped when the rate of rotation equaled the sample rate, the frequency response of the filter will behave similarly between zero and F_s. So the filter response near zero will be the same as that near F_s and the filter response just below $F_{Nyquist}$ will be the same as just above $F_{Nyquist}$. In fact, since this phenomenon is well understood, there is really no reason to plot a digital filter's response above $F_{Nyquist}$ and below $-F_{Nyquist}$. We are doing it here in Fig. 4.3 mainly to show this point.

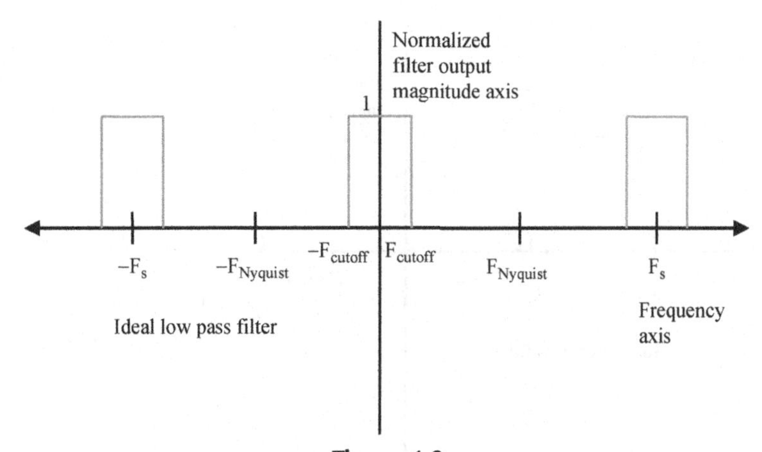

Figure 4.3
Ideal low pass filter frequency response.

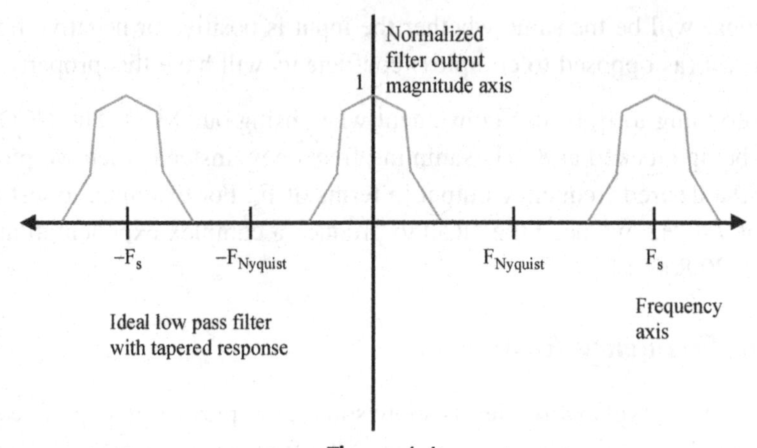

Figure 4.4
Realizable low pass filter frequency response.

The low pass filter passes frequencies near zero. Both positive and negative frequencies are passed. When the frequency reaches the chosen F_{cutoff}, the filter no longer passes the signal. As we approach any multiple of F_s, we see the aliasing of the filter response, in Fig. 4.4.

Here in Fig. 4.4 we gradually reduced the low pass filter response as it approaches F_{cutoff}. This is to show that it is possible to design for an arbitrary passband response, and a gradual transition to the stopband, not just a flat response in the filter passband with an abrupt transition at F_{cutoff}.

The high pass filter in Fig. 4.5 passes frequencies near $F_{Nyquist}$. Both positive and negative frequencies are passed. When the frequency falls below F_{cutoff}, the filter no longer passes the signal. Aliasing is again seen by the symmetry about $F_{Nyquist}$.

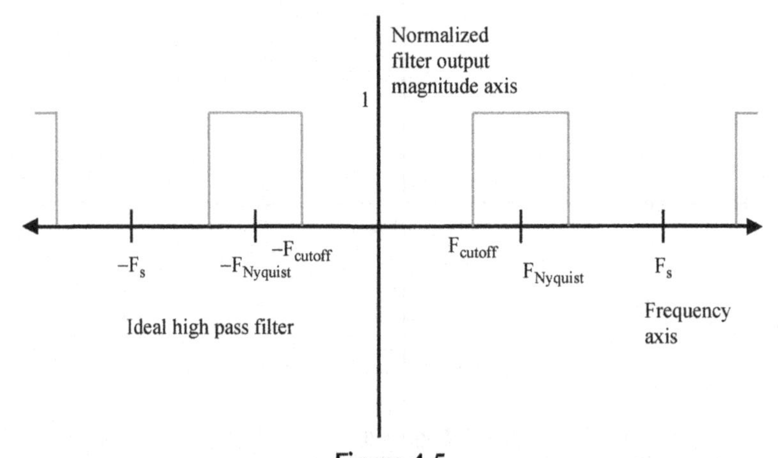

Figure 4.5
Ideal high pass filter frequency response.

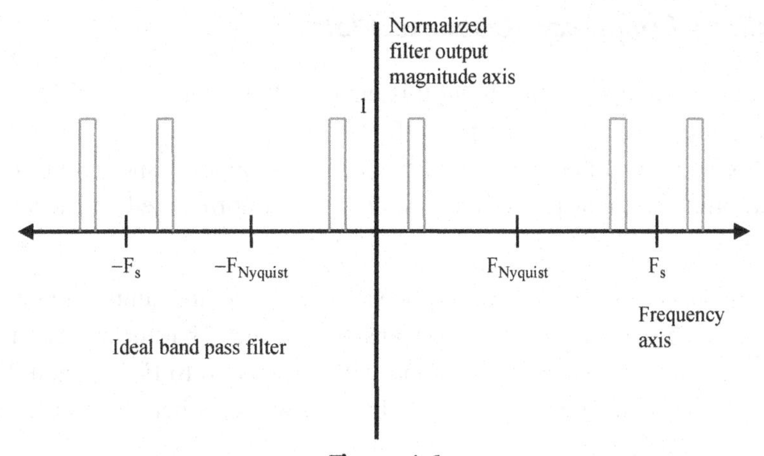

Figure 4.6
Ideal passband filter frequency response.

Bandpass (do not confuse this with the early term "passband") filters pass only a specific band, or portion of frequencies, which does not include either DC (0 Hz) or $F_{Nyquist}$. The band-pass filter in Fig. 4.6 again shows the alias effect. This repeats to infinity at every multiple of F_s. Going back to Fig. 4.3, the low pass filter frequency plot, we see the alias centered at F_s. So according to our plot, the filter will pass frequencies around F_s, even though this is above our F_{cutoff}. So if we input a signal near F_s to an ADC, it will appear as a frequency near zero, and be passed by our filter.

This is a difficult concept, which is why there is so much repetition on this theme. Do not be worried if you have to go back and forth a few times to the diagrams or earlier chapters to really satisfy yourself that you understand it. It is a key concept for much of what follows, and unfortunately, there are some experienced people working in DSP industry who still have trouble with these fundamentals.

4.5 Linear Phase Response

Earlier, we decided to ignore the phase response. We are just concerned with the magnitude response (remember that complex numbers have both a magnitude and phase). This is because, for most filters, the phase response is not something we need to worry about. The most common type of digital filter is called the finite impulse response, or FIR, and it has what is called a linear phase response. This means that every frequency passing through the filter experiences the same delay, which works out to a linearly increasing phase as the frequency increases. This sounds complicated, but the short story is that for FIR filters, we do not need to worry about phase response. This is the type of filter we will discuss in detail in the next chapter and is the most common DSP function implemented.

4.6 Normalized Frequency Response Plots

Now we have one more topic to finish out this chapter. We now know that it is sufficient to describe a digital filter response from $-F_{Nyquist}$ to $+F_{Nyquist}$. But frequently, in textbooks or filter design programs, the frequency response will be given in terms of normalized radians per second, ω, rather than normalized frequency, as shown in Fig. 4.7.

Again, let us refer back to our complex exponential, $e^{j\omega t}$. We are sampling at a normalized frequency of 1 sample per second, or 2π radians per second. That means that a complex exponential of frequency π radians per second will correspond to $F_{Nyquist}$ and 2π radians per second will correspond to F_S. This is exactly what we see when we plot the frequency response using the ω axis.

As a review, see Table 4.3 providing example low pass response cutoff both in terms of ω and F_S. It is important to be comfortable going between these two representations, because

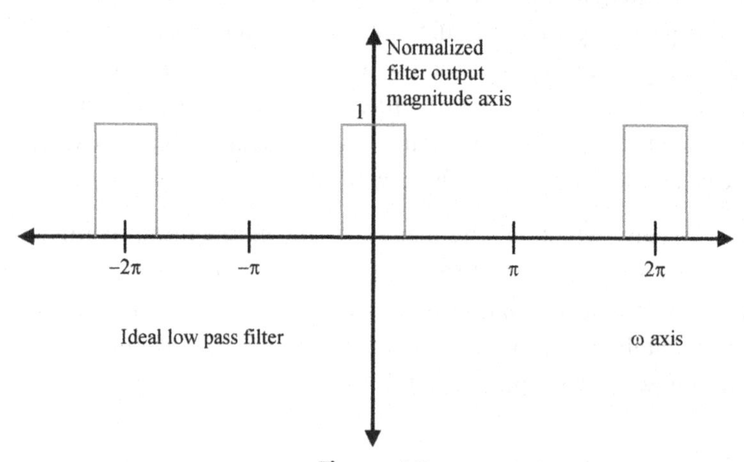

Figure 4.7
Normalized ideal low pass filter frequency response.

Table 4.3: Converting Radians to Hertz

% of Maximum Possible Low Pass Filter Bandwidth (%)	Ω (Radians per Second)	F_s (Hz or Cycles per Second)
10	π/10	$F_s/20$
10	−π/10	$-F_s/20$
25	π/4	$F_s/8$
50	π/2	$F_s/4$
80	4π/5	$2F_s/5$
100	π	$F_s/2$ or $F_{Nyquist}$

most filter design programs are defined in terms of ω, but to design your filter, you need to be able to translate to the real frequencies used in your application to the filter response in terms of ω.

The cutoff point is defined as a percentage of the maximum bandwidth possible for a low pass filter, with 100% becoming an all-pass filter (every frequency passes through). Normally, the filter frequency response will be symmetric, so $F_{cutoff} = -F_{cutoff}$.

CHAPTER 5

Finite Impulse Response (FIR) Filters

This chapter will focus on the workhorse of digital signal processing (DSP)—the finite impulse response (FIR) filter. We are going to discuss three main topics. First we will talk about the structure of an FIR filter, how to build one, and some of its properties. Next we are going to show how, given the filter coefficients, to compute the frequency response of the filter (normally, this is done with software, but we can gain insight by understanding how to do this). Last, we will show a method to compute coefficients to meet a given frequency response. This last step is what is commonly required of a DSP designer—to find the number and value of filter coefficients required for frequency response required by the application. Again, this is normally done with software, but we need to gain insight as to what is involved. Filter design is a process where we cannot have everything, and compromise is necessary. To use software tools to design filters, we will need to understand what the different trade-offs are, and how they interact.

5.1 Finite Impulse Response Filter Construction

Let us begin with how to construct an FIR filter. An FIR filter is built of multipliers and adders. It can be implemented in hardware or software, and run in a serial fashion, parallel fashion, or some combination. We will focus on the parallel implementation, because it is the most straightforward to understand.

FIR filters, and DSP in general, often use delay elements. A delay element is simply a clocked register, and a series of delay elements is simply a shift register. However, in DSP, a one-sample delay element is often represented as a box with a z^{-1} symbol. This comes from the mathematical properties of the z-transform, which we have not covered here in the interests of minimizing mathematics (although there is a discussion of z-transform in Appendix D). But we cannot skip the use of z^{-1} representing a single clock or register delay, as it is prevalent in DSP diagrams and literature. We will use this here, so that you can get used to seeing this representation.

A key property of an FIR filter is the number of taps or multipliers required to compute each output. In a parallel implementation, the number of taps equals the number of multipliers. In a serial implementation, one multiplier is used to perform all of the multipliers sequentially for each output. Assuming single clock cycle multipliers, a parallel

Digital Signal Processing 101. http://dx.doi.org/10.1016/B978-0-12-811453-7.00005-6

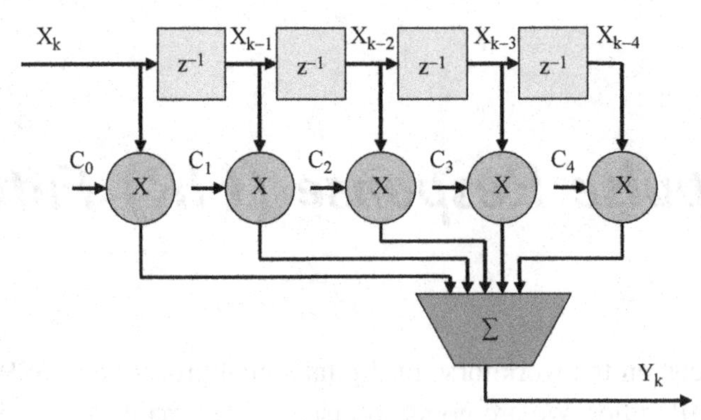

Figure 5.1
Finite impulse response filter structure.

FIR filter can produce one output each clock cycle, and a serial FIR filter would require N clock cycles to produce each output, where N is the number of filter taps. Filters can sometimes have hundreds of taps. Below shown is a small 5-tap parallel filter (Fig. 5.1).

The inputs and outputs of the FIR filter are sampled data. For simplicity, we will assume that the inputs, outputs, and filter coefficients C_m are all real numbers. The input data stream will be denoted as x_k and the output y_k. The "k" subscript is used to identify the sequence of data. For example, x_{k+1} follows x_k, and x_{k-1} precedes x_k. Often for the purpose of defining a steady state response, we assume that the data streams are infinitely long in time, or that k extends from $-\infty$ to $+\infty$.

The coefficients are usually static (meaning they do not change over time), and determine the filter's frequency response.

In equation form, the filter could be represented as:

$$y_k = C_0 \cdot x_k + C_1 \cdot x_{k-1} + C_2 \cdot x_{k-2} + C_3 \cdot x_{k-3} + C_4 \cdot x_{k-4}$$

It is just the sum of multipliers. This could get pretty tedious to write as the number of taps gets larger, so the following short hand summation is often used:

$$y_k = \sum_{i=0 \text{ to } 4} C_i x_{k-i}$$

We can also make the equation for any length of filter. To make our filter of length "N," we simply replace the 4 (5−1 taps) with N−1.

$$y_k = \sum_{i=0 \text{ to } N-1} C_i x_{k-i}$$

Another way to look at this is that the data stream $\ldots x_{k+2}, x_{k+1}, x_k, x_{k-1}, x_{k-2} \ldots$ is sliding past a fixed array of coefficients. At each clock cycle, the data and coefficients are cross-multiplied and the outputs of all multipliers for that clock cycle are summed, to form a single output (this process also known as dot product). Then on the next clock cycle, the data is shifted one place relative to the coefficients (which are fixed), and the process repeated. This process is known as convolution.

The FIR structure is very simple, yet it has the ability to create almost any frequency response, given sufficient number of taps. This is very powerful, but unfortunately not at all intuitive. It's somewhat analogous to the brain—a very simple structure of interconnected neurons, yet the combination can produce amazing results. During the rest of the chapter, we will try to gain some understanding of how this happens.

Below is an example using actual numbers, to illustrate this process called convolution.

We will define a filter of 5 coefficients $\{C_0,C_1,C_2,C_3,C_4\} = \{1,3,5,3,1\}$.

Our x_k sequence will be defined as $\{x_0,x_1,x_2,x_3,x_4,x_5\} = \{-1,1,2,1,4,-1\}$ and

$x_k = 0$ for $k < 0$ and for $k > 6$ (everywhere else).

Let us start by computing $y_{-1} = \sum_{i=0 \text{ to } N-1} C_i x_{-1-i}$

We can see that the subscript on x will be negative for all $i = 0$ to 4. In this example $y_k = 0$ for $k < 0$. What this is saying, is that until there is a nonzero input x_k, the output y_k will also be zero. Things start to happen at $k = 0$, because x_0 is the first nonzero input.

$$y_{-1} = \sum_{i=0 \text{ to } N-1} C_i x_{-1-i} = (1)(0) + (3)(0) + (5)(0) + (3)(0) + (1)(0) = 0$$

$$y_0 = \sum_{i=0 \text{ to } N-1} C_i x_{0-i} = (1)(-1) + (3)(0) + (5)(0) + (3)(0) + (1)(0) = -1$$

$$y_1 = \sum_{i=0 \text{ to } N-1} C_i x_{1-i} = (1)(1) + (3)(-1) + (5)(0) + (3)(0) + (1)(0) = -2$$

$$y_2 = \sum_{i=0 \text{ to } N-1} C_i x_{2-i} = (1)(2) + (3)(1) + (5)(-1) + (3)(0) + (1)(0) = 0$$

$$y_3 = \sum_{i=0 \text{ to } N-1} C_i x_{3-i} = (1)(1) + (3)(2) + (5)(1) + (3)(-1) + (1)(0) = 9$$

$$y_4 = \sum_{i=0 \text{ to } N-1} C_i x_{4-i} = (1)(\mathbf{4}) + (3)(1) + (5)(2) + (3)(1) + (1)(-1) = 19$$

$$y_5 = \sum_{i=0 \text{ to } N-1} C_i x_{5-i} = (1)(-1) + (3)(\mathbf{4}) + (5)(1) + (3)(2) + (1)(1) = 23$$

$$y_6 = \sum_{i=0 \text{ to } N-1} C_i x_{6-i} = (1)(0) + (3)(-1) + (5)(\mathbf{4}) + (3)(1) + (1)(2) = 22$$

$$y_7 = \sum_{i=0 \text{ to } N-1} C_i x_{7-i} = (1)(0) + (3)(0) + (5)(-1) + (3)(\mathbf{4}) + (1)(1) = 8$$

$$y_8 = \sum_{i=0 \text{ to } N-1} C_i x_{8-i} = (1)(0) + (3)(0) + (5)(0) + (3)(-1) + (1)(\mathbf{4}) = 1$$

$$y_9 = \sum_{i=0 \text{ to } N-1} C_i x_{9-i} = (1)(0) + (3)(0) + (5)(0) + (3)(0) + (1)(-1) = -1$$

$$y_{10} = \sum_{i=0 \text{ to } N-1} C_i x_{10-i} = (1)(0) + (3)(0) + (5)(0) + (3)(0) + (1)(0) = 0$$

$$y_{11} = \sum_{i=0 \text{ to } N-1} C_i x_{11-i} = (1)(0) + (3)(0) + (5)(0) + (3)(0) + (1)(0) = 0$$

$$y_{12} = \sum_{i=0 \text{ to } N-1} C_i x_{12-i} = (1)(0) + (3)(0) + (5)(0) + (3)(0) + (1)(0) = 0$$

This is definitely tedious. There are a couple of things to notice. Follow the input $x_4 = 4$ (highlighted) in our example. See how it moves across, from one multiplier to the next. Each input sample x_k will be multiplied by each tap in turn. Once it passes through the filter, that input sample is discarded and has no further influence on the output. In our example, x_4 is discarded after computing y_8.

Once the last nonzero input data x_k has shifted it's way through the filter taps, the output data y_k will go to zero (this starts at $k = 10$ in our example).

Now let us consider a special case, where $x_k = 1$ for $k = 0$, and $x_k = 0$ for $k \neq 0$. This means that we only have one nonzero input sample, and it is equal to 1. Now if we again compute the output, which is simpler this time, we get:

$$y_{-1} = \sum_{i=0 \text{ to } N-1} C_i x_{-1-i} = (1)(0) + (3)(0) + (5)(0) + (3)(0) + (1)(0) = 0$$

$$y_0 = \sum_{i=0 \text{ to } N-1} C_i x_{0-i} = (1)(1) + (3)(0) + (5)(0) + (3)(0) + (1)(0) = 1$$

$$y_1 = \sum_{i=0 \text{ to } N-1} C_i x_{1-i} = (1)(0) + (3)(1) + (5)(0) + (3)(0) + (1)(0) = 3$$

$$y_2 = \sum_{i=0 \text{ to } N-1} C_i x_{2-i} = (1)(0) + (3)(0) + (5)(1) + (3)(0) + (1)(0) = 5$$

$$y_3 = \sum_{i=0 \text{ to } N-1} C_i x_{3-i} = (1)(0) + (3)(0) + (5)(0) + (3)(1) + (1)(0) = 3$$

$$y_4 = \sum_{i=0 \text{ to } N-1} C_i x_{4-i} = (1)(0) + (3)(0) + (5)(0) + (3)(0) + (1)(1) = 1$$

$$y_5 = \sum_{i=0 \text{ to } N-1} C_i x_{5-i} = (1)(0) + (3)(0) + (5)(0) + (3)(0) + (1)(0) = 0$$

$$y_6 = \sum_{i=0 \text{ to } N-1} C_i x_{6-i} = (1)(0) + (3)(0) + (5)(0) + (3)(0) + (1)(0) = 0$$

Notice that output is the same sequence as the coefficients. This should come as no surprise once you think about it. This output is defined as the filter's impulse response, named as it occurs when the filter input is an impulse, or a single nonzero input equal to 1. This gives the FIR filter its name. By "finite impulse response" or FIR, this indicates that if this type of filter is driven with an impulse, we will see a response (the output) has a finite length, after which it becomes zero. This may seem trivial, but it is a very good property to have, as we will see in the chapter on infinite impulse response filters.

5.2 Computing Frequency Response

What we have covered so far is the mechanics of building the filter, and how to compute the output data, given the coefficients and input data. But we do not have any intuitive feeling as to how this operation can allow some frequencies to pass through, and block other frequencies. A very basic understanding of a low-pass filter can be gained by the concept of averaging. We all know that if we average multiply results, we get a smoother, more consistent output, and rapid fluctuations are damped out. A moving average filter is simply a filter with all the coefficients set to 1. The more filter taps, the longer the averaging, and the more smoothing takes place. This gives an idea of how a filter structure can remove high frequencies or rapid fluctuations. Now imagine if the filter taps were alternating $+1$, -1, $+1$ -1... and so on. A slowly varying input signal will have adjacent samples nearly the same, and these will cancel in the filter, resulting in a nearly zero output. This filter is blocking low frequencies. On the other hand, an input signal near the Nyquist rate will have big changes from sample to sample and will result in a much larger output. However, to get a more precise handle on how to configure the coefficient values to get the desired frequency response, we are going to need using a bit of math.

We will start by computing the frequency response of the filter from the coefficients. Remember, the frequency response of the filter is determined by the coefficients (also called the impulse response).

Let us begin by trying to determine the frequency response of a filter by measurement. Imagine if we take a complex exponential signal of a given frequency and use this as the input to our filter. Then we measure the output. If the frequency of the exponential signal is in the passband of the filter, it will appear at the output. But if the frequency of the exponential signal is in the stopband of the filter, it will appear at the output with a much lower level than the input, or not at all. Imagine we start with a very low frequency exponential input, and do this measurement, then slightly increase the frequency of the exponential input, measure again, and keep going until the exponential frequency is equal to the Nyquist frequency. If we plot the level of the output signal across the frequency from 0 to $F_{Nyquist}$, we will have the frequency response of the filter. It turns out that we do not have to do all these measurements, instead we can compute this fairly easily. This is shown below.

$y_k = \sum_{i=0 \text{ to } 4} C_i x_{k-i}$	Output of our 5-tap example filter
$y_k = \sum_{i=-\infty \text{ to } \infty} C_i x_{k-i}$	Same equation, except that we are allowing infinite number of coefficients (no limits on filter length)
$x_m = e^{j\omega m} = \cos(\omega m) + j \sin(\omega m)$	This is our complex exponential input at ω radians per sample.

Let us take a close look at the last equation, and review a bit.

It is just a sampled version of a signal rotating around the unit circle. We sample at time = m, and then sample again at time = m + 1. So from one sample to the next, our sampled signal will move ω radians around the unit circle. If we are sampling at 10 times faster than we are moving around the unit circle, then it will take 10 samples to get around the circle and move 2π/10 radians each sample.

$$x_m = e^{j2\pi m/10} = \cos(2\pi m/10) + j\sin(2\pi m/10) \quad \text{when } \omega = 2\pi/10$$

To clarify, an example table below shows $x_m = e^{j2\pi m/10}$ evaluated at various m. If you want to check using a calculator, remember that the angles are in units of radians, not degrees (Table 5.1).

We could also increment x_m so that we rotate the unit circle every five samples. This is twice as fast as before. Hopefully you are getting more comfortable with complex exponentials.

$$x_m = e^{j2\pi m/5} = \cos(2\pi/5) + j\sin(2\pi m/5) \quad \text{when } \pi = 2\pi/5$$

Table 5.1: Evaluating Complex Exponentials

m = 0	$x_0 = e^{j0} = \cos(0) + j\sin(0)$	1 + j0
m = 1	$x_1 = e^{j\pi/5} = \cos(\pi/5) + j\sin(\pi/5)$	0.8090 + j0.5878
m = 2	$x_2 = e^{j2\pi/5} = \cos(2\pi/5) + j\sin(2\pi/5)$	0.3090 + j0.9511
m = 3	$x_3 = e^{j3\pi/5} = \cos(3\pi/5) + j\sin(3\pi/5)$	−0.3090 + j0.9511
m = 4	$x_4 = e^{j4\pi/5} = \cos(4\pi/5) + j\sin(4\pi/5)$	−0.8090 + j0.5878
m = 5	$x_5 = e^{j\pi} = \cos(\pi) + j\sin(\pi)$	−1 + j0
m = 6	$x_6 = e^{j6\pi/5} = \cos(6\pi/5) + j\sin(6\pi/5)$	−0.8090 −j0.5878
m = 7	$x_7 = e^{j7\pi/5} = \cos(7\pi/5) + j\sin(7\pi/5)$	−0.3090 −j0.9511
m = 8	$x_8 = e^{j8\pi/5} = \cos(8\pi/5) + j\sin(8\pi/5)$	0.3090 − j0.9511
m = 9	$x_9 = e^{j9\pi/5} = \cos(9\pi/5) + j\sin(9\pi/5)$	0.8090 − j0.5878
m = 10	$x_{10} = x_0 = e^{j2\pi} = \cos(2\pi) + j\sin(2\pi)$	1 + j0

Now we go back to the filter equation and substitute the complex exponential input for x_{k-i}.

$y_k = \sum_{i=-\infty \text{ to } \infty} C_i x_{k-i}$	
$x_m = e^{j\omega m} = \cos(\omega m) + j\sin(\omega m)$	insert k−i for m
$x_{k-i} = e^{j\omega(k-i)} = \cos(\omega(k-i)) + j\sin(\omega(k-i))$	next, replace in x_{k-i} in filter equation
$y_k = \sum_{i=-\infty \text{ to } \infty} C_i e^{j\omega(k-i)}$	

There is a property of exponentials that we frequently need to use.

$$e^{(a+b)} = e^a \cdot e^b \quad \text{and} \quad e^{(a-b)} = e^a \cdot e^{-b}$$

If you remember your scientific notation, this makes sense. For example,

$$10^2 \cdot 10^3 = 100 \cdot 1000 = 100,000 = 10^5 = 10^{(2+3)}$$

OK, now back to the filter equation.

$$y_k = \sum_{i=-\infty \text{ to } \infty} C_i e^{j\omega(k-i)} = \sum_{i=-\infty \text{ to } \infty} C_i e^{j\omega k} \cdot e^{-j\omega i}$$

Let us do a little algebra trick. Notice that the term $e^{j\omega k}$ does not contain the term i used in the summation. So we can pull this term out in front of the summation.

$$y_k = e^{j\omega k} \cdot \sum_{i=-\infty \text{ to } \infty} C_i e^{-j\omega i}$$

Notice that the term $e^{j\omega k}$ is just the complex exponential we used as an input.

$$y_k = x_k \cdot \sum_{i=-\infty \text{ to } \infty} C_i e^{-j\omega i}$$

Voila! The expression $\sum_{i=-\infty \text{ to } \infty} C_i e^{-j\omega i}$ gives us the value of the frequency response of the filter at frequency ω. It is solely a function of ω and the filter coefficients.

This expression applies a gain factor to the input, x_k, to produce the filter output. Where this expression is large, we are in the passband of the filter. If this expression is close to zero, we are in stopband of the filter.

Let us give this expression a less cumbersome representation. Again, it is a function of ω, which we expect, because the characteristics of the filter vary with frequency. It is also a function of the coefficients, C_i, but these are assumed fixed for a given filter.

$$\text{Frequency response} = H(\omega) = \sum_{i=-\infty \text{ to } \infty} C_i e^{-j\omega i}$$

Now in reality, it is not as bad as it looks. This is the generic version of the equation, where we must allow for an infinite number of coefficients (or taps). But suppose we are determining the frequency response of our 5-tap example filter.

$$H(\omega) = \sum_{i=0 \text{ to } 4} C_i e^{-j\omega i} \quad \text{and} \quad \{C0, C1, C2, C3, C4\} = \{1, 3, 5, 3, 1\}$$

Let us find the response of the filter at a couple different frequencies. First, let $\omega = 0$. This corresponds to DC input—we are putting a constant level signal into the filter. This would be $x_k = 1$ for all values k.

$$H(0) = C_0 + C_1 + C_2 + C_3 + C_4 = 1 + 3 + 5 + 3 + 1 = 13$$

This one was simple, since $e^0 = 1$. The DC or zero frequency response of the filter is called the gain of the filter. Often, it may be convenient to force the gain = 1, which would involve dividing all the individual filter coefficients by $H(0)$. The passbands and stopbands characteristics are not altered by this process, since all the coefficients are scaled equally. It just normalizes the frequency response so the passband has a gain equal to 1.

Now compute the frequency response for $\omega = \pi/2$.

$$\begin{aligned} H(\pi/2) &= C_0 e^0 + C_1 e^{-\pi/2} + C_2 e^{-\pi} + C_3 e^{-3\pi/2} + C_4 e^{-4\pi/2} \\ &= 1 \cdot 1 + 3 \cdot (-j) + 5 \cdot (-1) + 3 \cdot (j) + 1 \cdot 1 = -3 \end{aligned}$$

So the magnitude of the frequency response has gone from 13 (at $\omega = 0$) to 3 (at $\omega = \pi/2$). The phase has gone from 0°degree (at $\omega = 0$) to 180°degrees (at $\omega = \pi/2$), although generally we are not concerned about the phase response of FIR filters. Just from these two points of the frequency response, we can guess that the filter is probably some type of low-pass filter.

Recall that the magnitude is calculated as follows:

$$\text{Magnitude } Z = X + jY = (X^2 + Y^2)^{1/2} = |Z|$$

Our example calculation above turned out to have only real numbers, but that is because the imaginary components of $H(\pi/2)$ canceled out to zero. The magnitude of $H(\pi/2)$ is

$$\text{Magnitude}|-3 + 0\,j| = 3$$

A computer program can easily evaluate $H(\omega)$ from $-\pi$ to π and plot it for us. Of course, this is almost never done by hand. Fig. 5.2 is a frequency plot of this filter using an FIR Filter program.

Not the best filter, but it still is a low-pass filter. The frequency axis is normalized to F_s, and the magnitude of the amplitude is plotted on a logarithmic scale, referenced to a passband frequency response of 1.

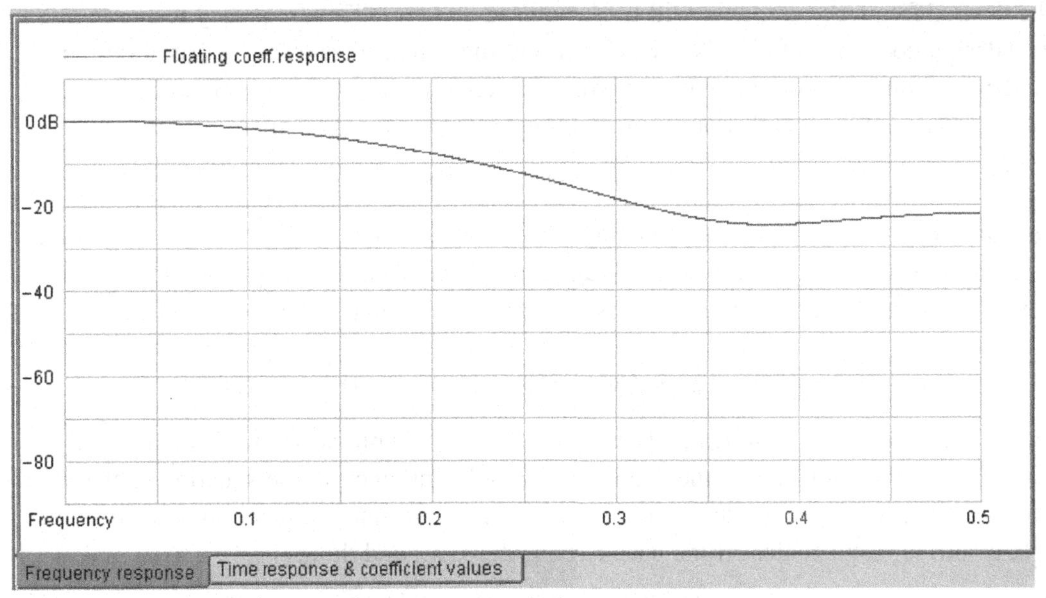

Figure 5.2
Frequency response plot.

We can verify our hand calculation was correct. We calculated the magnitude of $|H(\pi/2)|$ at 3, and $|H(0)|$ at 13. The logarithmic difference is:

$$20 \log_{10}(3/13) = -12.7 \text{ dB}$$

If you check the frequency response plot above, you will see at frequency $F_s/4$ (or 0.25 on the normalized frequency scale), which corresponds to $\pi/2$ in radians, the filter does indeed seem to attenuate the input signal by about 12–13 dB relative to $|H(0)|$. Other filter programs might plot the frequency axis referenced from 0 to π, or from $-\pi$ to π.

5.3 Computing Filter Coefficients

Now suppose you are given a drawing of a frequency response and told to find the coefficients of a digital filter that best matches this response. Basically, you are designing the digital filter. Again, you would use a filter design program to do this, but to do this, it is helpful to have some understanding of what the program is doing. To optimally configure the program options, you should understand the basics of filter design. We will explain a technique known as the Fourier design method. This requires more math than we have used so far, so if you are a bit rusty, just try and bear with it. Even if you do not follow everything in the rest of the chapter, the ideas should still be very helpful when using a digital filter design program.

The desired frequency response will be designated as $D(\omega)$. This frequency response is your design goal. As before, $H(\omega)$ will represent the actual filter response based on the number and value of your coefficients. We now define the error, $\xi(\omega)$, as the difference between what we want and what we actually get from a particular filter.

$$\xi(\omega) = D(\omega) - H(\omega)$$

Now all three of these functions are complex—when evaluated, they will have magnitude and phase. We are concerned with the magnitude of the error, not its phase. One simple way to eliminate the phase in the $\xi(\omega)$ is to work with the magnitude squared of $\xi(\omega)$.

$$\left|\xi(\omega)\right|^2 = \{\text{Real part } \xi(\omega)\}^2 + \{\text{Imag part } \xi(\omega)\}^2 = \xi(\omega)\xi(\omega)*$$

where * is the complex conjugate operator (recall that magnitude squared is a number multiplied by its conjugate, in the chapter on complex numbers). The squaring of the error function differentially amplifies errors. It makes the error function much more responsive to large errors than smaller errors, usually considered a good thing.

To get the cumulative error, we need to evaluate the magnitude squared error function over the entire frequency response.

$$\text{Error } = \xi = \int_{-\pi}^{\pi} |\xi(\omega)|^2 d\omega$$

The classic method to minimize a function is to evaluate the derivative with respect to the parameter, which we have control of. In this case, we will try to evaluate the derivative of ξ with respect to the coefficients, C_i. This will lead us to an expression that will allow us to compute the coefficients that result in the minimum error, or minimize the difference between our desired frequency response and the actual frequency response. As I promised to minimize the math (and many of you probably would not have started reading this otherwise), this derivation is located in Appendix A. If you have trouble with this derivation, do not let it bother you. It is the result that is important anyway.

$$C_i = (1/2\pi) \cdot \int_{-\pi}^{\pi} D(\omega) e^{j\omega i} \, d\omega$$

This provides a design equation to compute the filter coefficients, which give a response best matching the desired filter response, $D(\omega)$.

Let us try an example. Let $D(\omega)$ be defined as a low-pass filter with cutoff at $\omega = \pi/2$ (Fig. 5.3).

We can change the limits on the integral to $-\pi/2$ to $\pi/2$, as $D(\omega)$ is zero in the remainder of the integration interval.

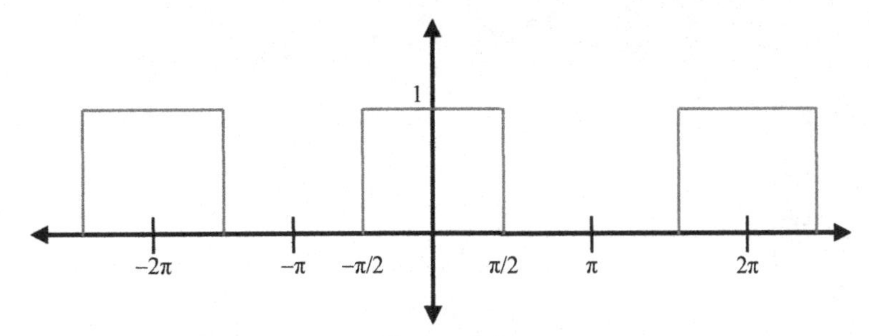

Figure 5.3
Ideal low-pass filter frequency response.

$$C_i = (1/2\pi)\cdot \int_{-\pi}^{\pi} D(\omega)e^{j\omega i}\, d\omega = (1/2\pi)\cdot \int_{-\pi/2}^{\pi/2} 1\cdot e^{j\omega i}\, d\omega$$

From an integration table in a calculus book we will find

$$\int e^{x} dx = (1/k)\cdot e^{kx}$$

so that we get

$$\int e^{j\omega i} d\omega = (1/ji)\cdot e^{j\omega i}$$

The filter coefficients are therefore this expression, evaluated at $-\pi/2$ and $\pi/2$.

$$C_i = (1/2\pi)\cdot \int_{-\pi/2}^{\pi/2} 1\cdot e^{j\omega i} d\omega = (1/2\pi)\cdot(1/ji)\cdot e^{j\omega i}\Big|_{-\pi/2}^{\pi/2}$$

Next, plug in the integral limits.

$$C_i = (1/2\pi)\cdot(1/ji)\cdot\left[e^{j\pi i/2} - e^{-\pi i/2}\right]$$

By using the Euler equation for $e^{j\pi i/2}$ and $e^{-j\pi i/2}$, we will find the cosine parts cancel.

$$C_i = (1/2\pi)\cdot(1/ji)\cdot 2j\,\sin(\pi i/2) = (1/2\pi i)\cdot\sin(\pi i/2)$$

This expression gives the ideal response for a digital low-pass filter. The coefficients, which also represent the impulse response, decrease as 1/i as the coefficient index i gets larger. It is like a sine wave, with the amplitude gradually diminishing on each side. This function is called the sinc function, also known as sin(x)/x. It is a special function in DSP, because it gives an ideal low pass frequency response. The sinc function is plotted in Fig. 5.4, both as a sampled function and as a continuous function.

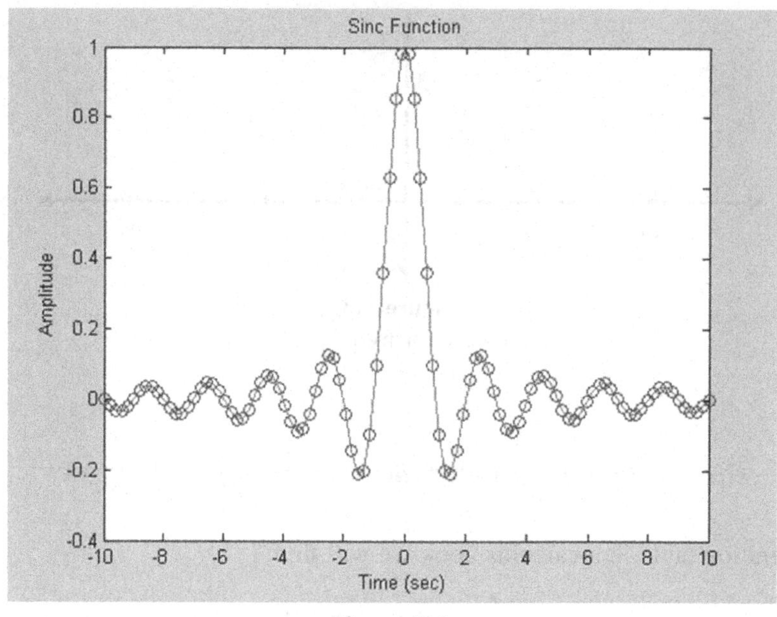

Figure 5.4
Sinc function.

Our filter instantly transitions from passband to stopband. But before getting too excited about this filter, we should note that it requires infinite number of coefficients to realize this frequency response. As we truncate the number of coefficients (which is also the impulse response, which cannot be infinitely long in any real filter), we will get a progressively sloppier and sloppier transition from passband to stopband as well as less attenuation in the stopband. This can be seen in the figures below. What is important to realize is that to get a sharper filter response requires more coefficients or filter taps, which in turn requires more multiplier resources to compute. There will always be a trade-off between quality of filter response and number of filter taps, or coefficients.

5.4 Effect of Number of Taps on Filter Response

A picture is worth a thousand words or so it is said. Below are multiple plots of this filter with indicated number of coefficients. By inspection, you can see that as the number of coefficients grows, you will see actual $|H(\omega)|$ approaching desired $|D(\omega)|$ frequency response.

Filter plots are always given on a logarithmic amplitude scale. This allows us to see passband flatness, as well as see how much rejection, or attenuation, the filter provides to signals in its stopband.

All the following filter and coefficient plots are done using FIR Filter program. These filters can be easily implemented in FPGA or DSP processor.

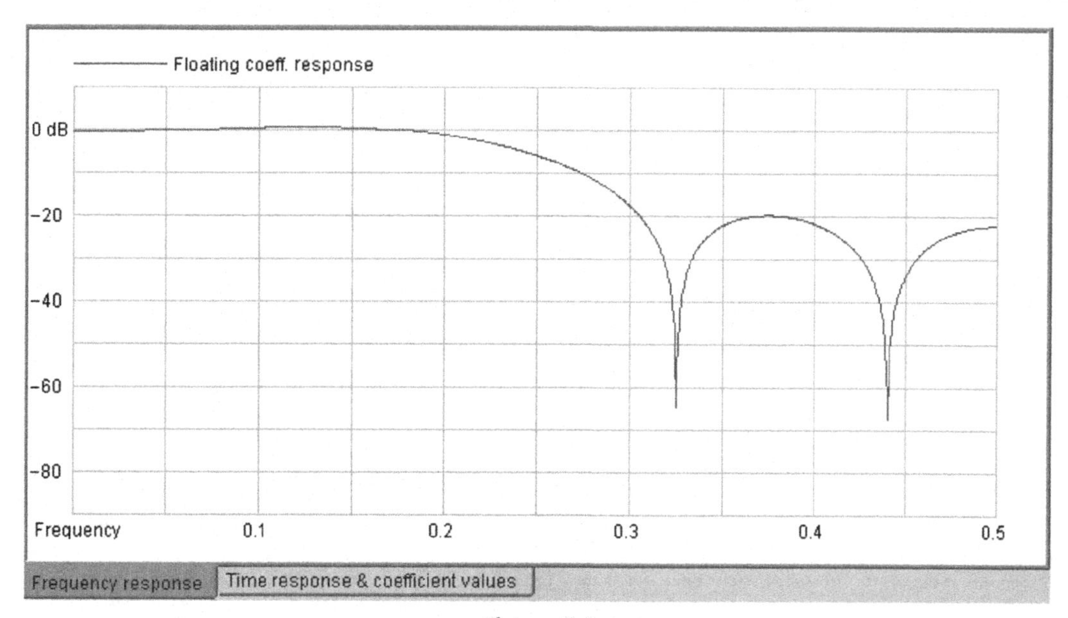

Figure 5.5
7-tap ideal low-pass filter frequency response.

Above in Fig. 5.5 is a frequency plot of our ideal low pass filter. It does not look very ideal. The problem is that it is only 7 coefficients long. Ideally, it should be unlimited coefficients.

We can see some improvement in Fig. 5.6, with about twice the number of taps.

Now this is starting to look like a proper low-pass filter (Fig. 5.7).

You should notice how the transition from passband to stopband gets steeper as the number of taps increases. The stopband rejection also increases (Fig. 5.8).

This is a very long filter, with closer to ideal response. Notice that how as the number of filter taps grow, the stopband rejection is increasing (it is doing a better job attenuating unwanted frequencies). For example, using 255 taps, by inspection $|H(\omega = 0.3)| \sim -42$ dB. With 1023 taps, $|H(\omega = 0.3)| \sim -54$ dB. Suppose a signal of frequency $\omega = 0.3$ radians per second, with peak to peak amplitude equal to 1, is our input. The 255-tap filter would produce an output with peak to peak amplitude of $\sim$0.008. The 1023-tap filter would give 4x better rejection, producing an output with peak to peak amplitude of $\sim$0.002 (Fig. 5.9).

Above in Fig. 5.10 is the plot of the 63-tap sinc filter coefficients (this is not a frequency response plot). The sinc shape of the coefficient sequence can be easily seen. Notice that the coefficients fall on the zero crossing of the sinc function at every other coefficient.

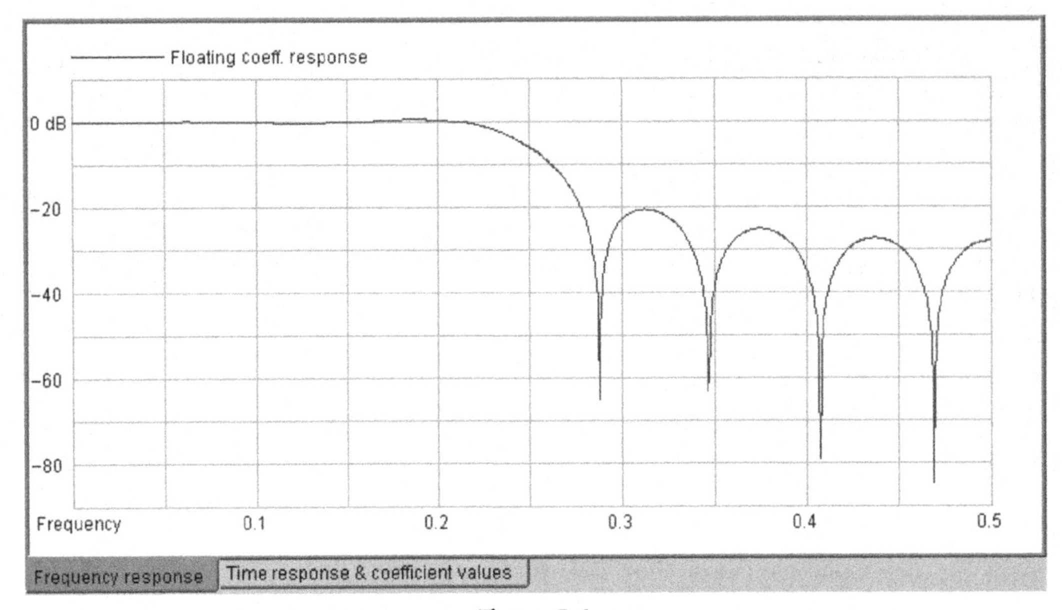

Figure 5.6
15-tap ideal low-pass filter frequency response.

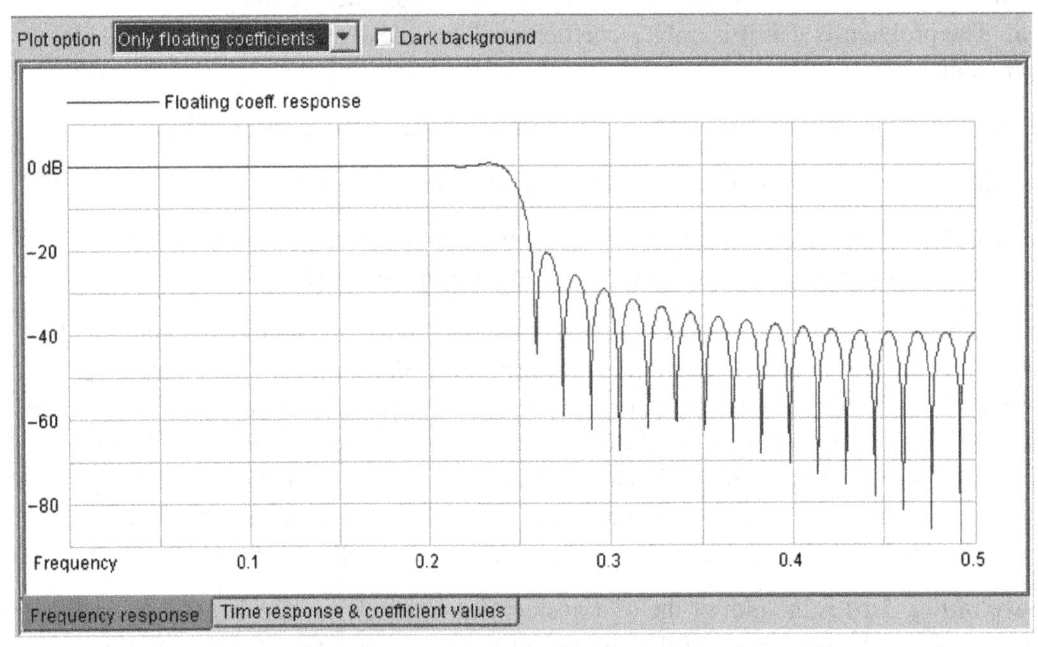

Figure 5.7
63-tap ideal low-pass filter frequency response.

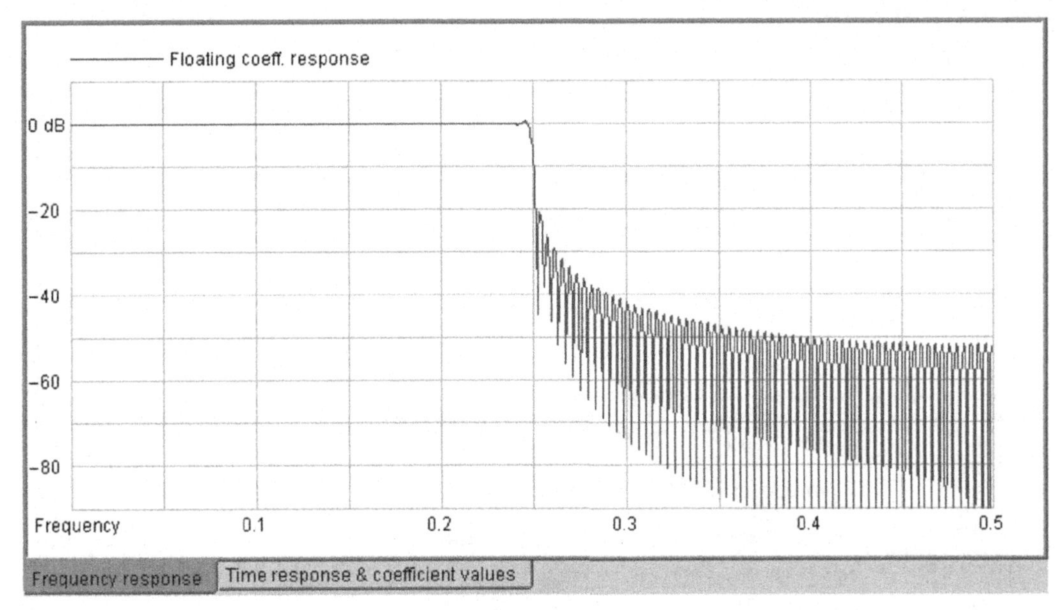

Figure 5.8
255-tap ideal low-pass filter frequency response.

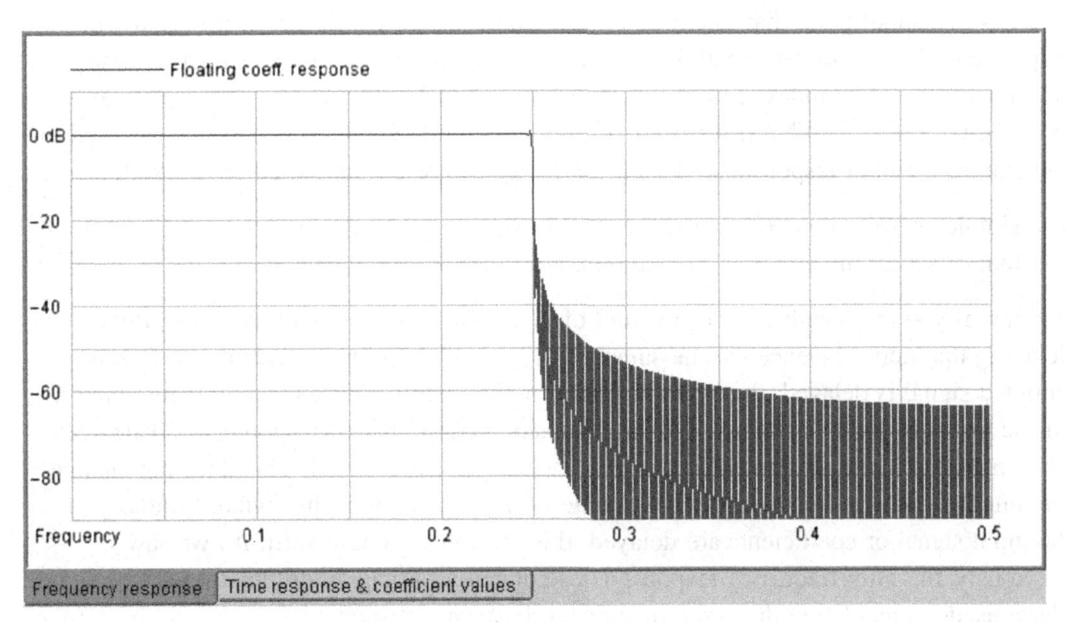

Figure 5.9
1023-tap ideal low-pass filter frequency response.

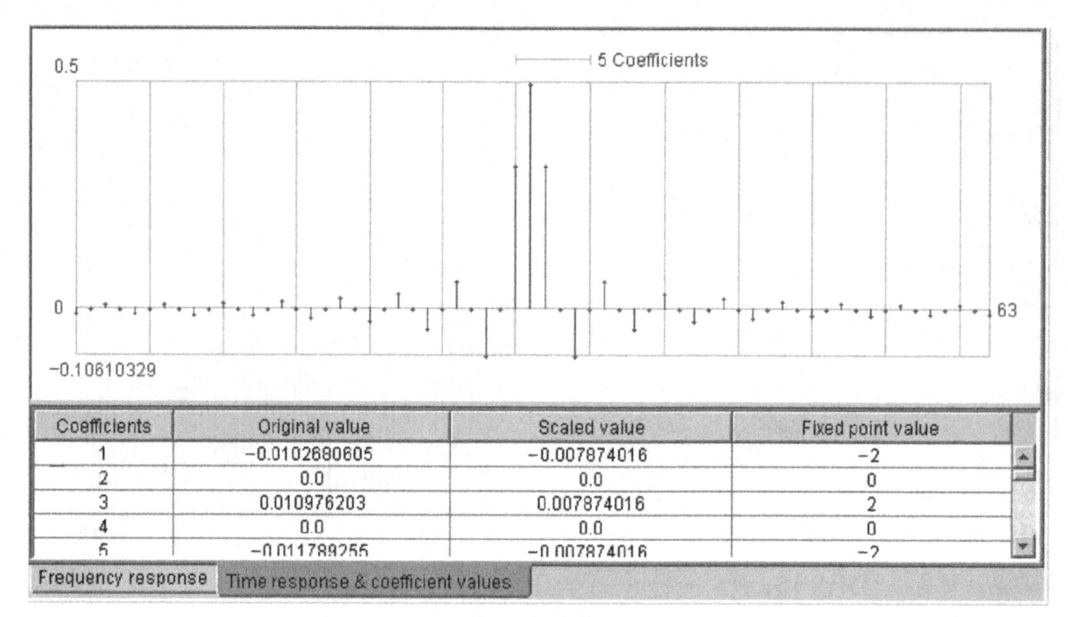

Figure 5.10
63-tap ideal low-pass filter coefficients.

There is one small point that might confuse an alert reader. Often a filter program will display the filter coefficients with indexes from –N/2 to N/2. In other cases, the same coefficients might be indexed from 0 to N, or from 1 to N+1. Our 5-tap example was $\{C_0, C_1, C_2, C_3, C_4\}$. What if we used $\{C_{-2}, C_{-1}, C_0, C_1, C_2\}$? These indexes are used in calculating the filter response, so does it matter how they are numbered or indexed?

The simple answer is no. The reason for this is that changing all the indexes by some constant offset has no effect on the *magnitude* of the frequency response.

Imagine if you put a shift register in front of your FIR filter. It would have the effect of delaying the input sequence by one sample. This is simply a one-sample delay. When a sampled signal is delayed, the result is simply a phase shift. Think for a moment about a cosine wave. If there is a delay of ¼ of the cosine period, this corresponds to 90°degrees. The result is a phase-shifted cosine, in this case a sine wave signal. This does not change the amplitude or the frequency, rather changes only the phase of the signal. Similarly, if the input signal or coefficients are delayed, this is simply a phase shift. As we saw previously, the filter frequency response is calculated as a magnitude function. There is no phase used in calculating the magnitude of the frequency response.

FIR filters will have a delay or latency associated with them. Usually, this delay is measured by comparing an impulse input to the filter output. We will see the output, starting one clock after the impulse input. This output or filter impulse response spans the length of the filter. So we measure the delay from the impulse input to the largest part of the filter output. In most cases, this will be the middle or center tap. For example, with our 5-tap example filter, the delay would be three samples. This is from the input impulse to the largest component of the output (impulse response), which is 5. In general, for an N-tap filter, where N is odd, the delay will be N/2 + 1. When N is even, it will be N/2 + ½. Since an FIR filter adds equal delay to all frequencies of the signal, it introduces no phase distortion, and it has a property called linear phase.

This is not as clear an explanation as you might like, but the essence is that since no phase distortion occurs when the signal passes through an FIR filter, we do not need to consider filter phase response in our design process. This is one reason why FIR filters are preferred over other types of digital filters.

The next chapter will cover with a topic called "windowing," which is a method to optimize FIR filter frequency response without increasing the number of coefficients.

CHAPTER 6

Windowing

In the previous chapter, we developed a technique for calculating filter coefficients to generate a desired frequency response. Our desired filter response will generate an infinite number of coefficients, and we have to decide how many coefficients to keep and throw away or truncate the rest. As the previous plots show, when the number of coefficients grows larger, the transition region from passband to stopband grows smaller, and the stopband rejection or attenuation increases.

$$\text{Frequency response} = H(\omega) = \sum_{i=-\infty \text{ to } \infty} C_i e^{-j\omega i}$$

The smaller transition region, or steepness of the transition from pass to stopband, is due to higher frequency components in the frequency response. The higher indexes of "i" in the frequency response correspond to higher frequency complex exponentials. Complex exponentials have a sinusoidal characteristic, so to get a quick response, one must use higher frequency exponentials.

6.1 Truncation of Coefficients

The process of truncating the infinite number of coefficients is called windowing. We can imagine $C_{i=-\infty \text{ to } \infty}$, and multiplying it term by term with a window function, $W_{i=-\infty \text{ to } \infty}$. Let us consider the examples of the plots in the previous chapter, with our low-pass filter with a cutoff frequency of $\pi/2$.

$$C_i = (1/\pi i)\cdot\sin(\pi i/2) \quad \text{for} -\infty < i < \infty$$

For our 7-tap filter,

$$W_i = 1 \quad \text{for} -3 \le i \le 3 \text{ and } W_i = 0 \text{ otherwise.}$$

Similarly, for our 15-tap filter

$$W_i = 1 \quad \text{for} -7 \le i \le 7 \text{ and } W_i = 0 \text{ otherwise.}$$

In both cases, W_i is a rectangular window. This means that the coefficients are unaltered within the window and are zeroed or truncated outside the window. The length of the rectangular window determined the length of the impulse response of the filter. This is called rectangular window because the window coefficients W_i are all "1" within the window and "0" outside.

Digital Signal Processing 101. http://dx.doi.org/10.1016/B978-0-12-811453-7.00006-8

With the rectangular window, we are abruptly truncating the impulse response of the filter. Obviously, for realistic filter implementations, we have to limit the impulse response at some point, as each tap or coefficient requires a multiply operation to compute each filter output. But perhaps we can get a more desirable response by reducing the coefficient values gradually at either end of the impulse response before we reach the point of impulse response truncation.

6.2 Tapering of Coefficients

This has led to efforts to develop other window functions besides the default rectangular window. Window design and analysis involved a fair bit of mathematics. But after the rigors of the last chapter, many readers will probably not mind too much if we skip over this. Actually, many filter designers do not know the details of the various window functions offered by their filter design software but work iteratively instead. That is, the designer will experiment with moving the frequency cutoff point slightly and playing the allowable number of taps, the various window options, and sometimes the numerical precision (number of bits) of the input data and coefficients. By observing the computer-generated frequency plots, the designer can iterate to find an optimum combination of these parameters that meet the application requirements.

Often, the requirements will be a certain degree of filter rejection or attenuation at one or more specific frequency points, a maximum amount of ripple or variance in the passband region of the frequency response, and a specified region of the frequency response (Fig. 6.1).

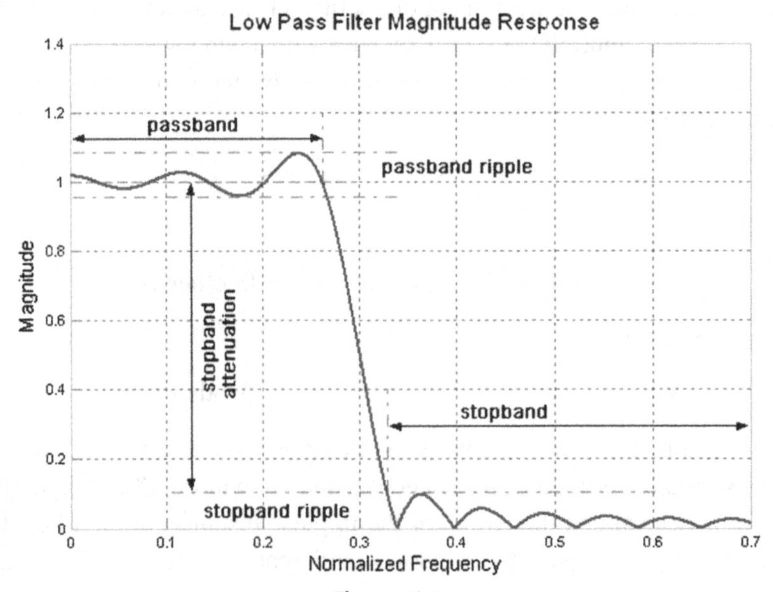

Figure 6.1
Low-pass filter frequency response.

Most windows are named after their inventors. These include Hamming, von Hann, Kaiser, Blackman, Bartlett, and others. The window coefficients will not be equal to 1 as in the rectangular window but will gradually change from 1 to 0 in some fashion near the edges of the window. The form of this transition or tapering off of the coefficients defines the window properties. Note that this is a different function than filter design, which produced the original and ideal set of coefficients. Windowing is used to avoid the abrupt truncation of the filter coefficient set, required to allow implementation of the filter with a finite number of multiply–add operations.

In general, a window cannot increase the steepness of the transition region, but it can be used to reduce either the passband or stopband ripple in the frequency response. Most filter design programs offer several window options. Below are the frequency responses with different windows for comparison. All the filters shown are for a 61-tap band-pass filter. The major trade-off between the different windows is the width of the transition band and the amount of attenuation in the stopband region of the frequency response. Often, the filter designer will make trade-offs in transition width, passband ripple, stopband attenuation (including rippler of lobes in stopband), number of coefficients, and the chosen window to achieve the application requirements.

6.3 Sample Coefficient Windows

Below in Figs. 6.2–6.5, several windows are shown with the rectangular window as the baseline. Notice that the rectangular window (Fig. 6.2) provides the steepest transition

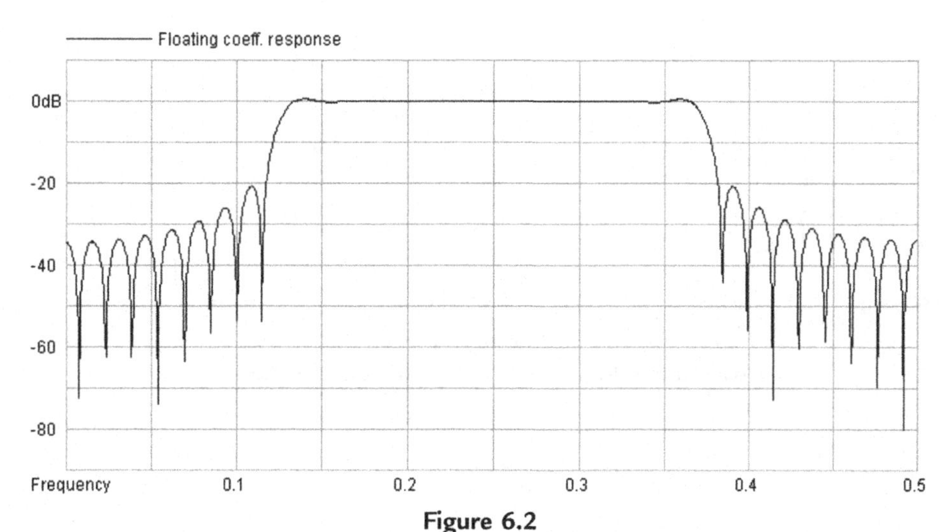

Figure 6.2
61-tap band-pass filter with rectangular window.

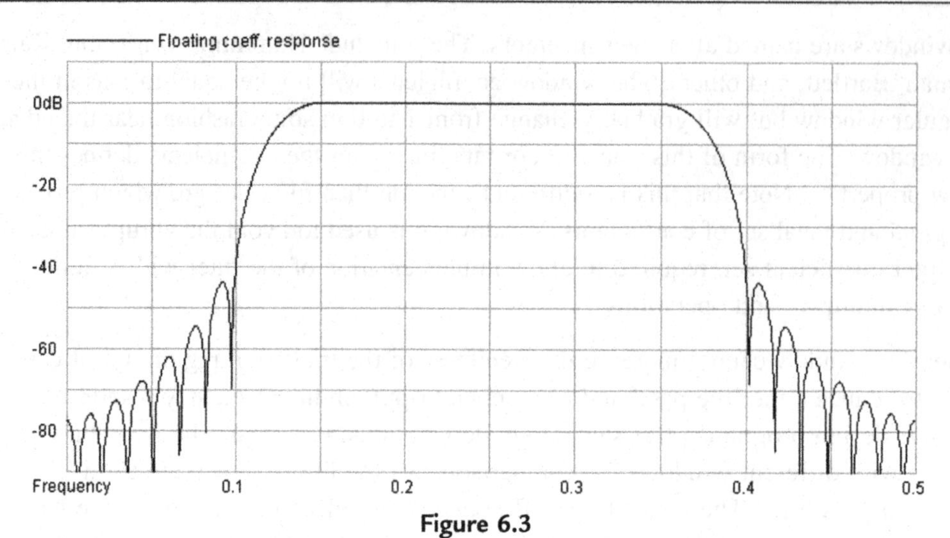

Figure 6.3
61-tap band-pass filter with Hanning window.

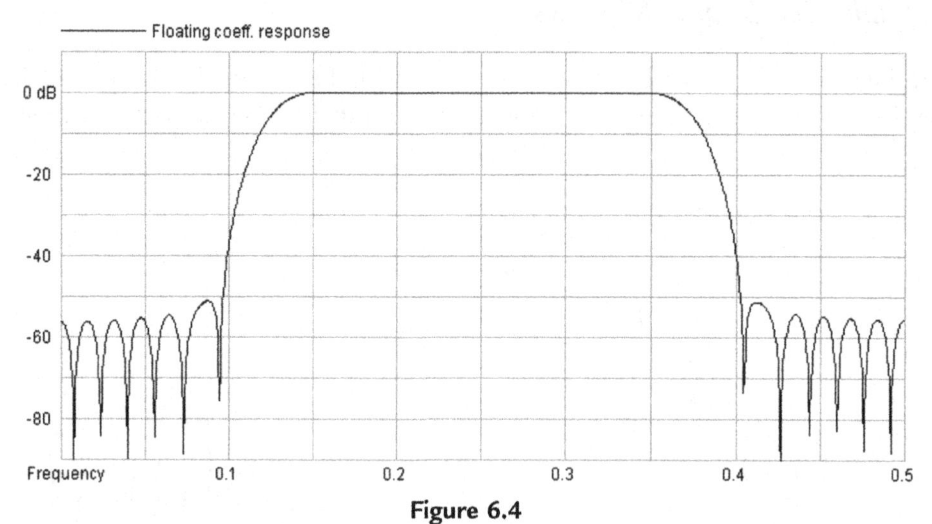

Figure 6.4
61-tap band-pass filter with Hamming window.

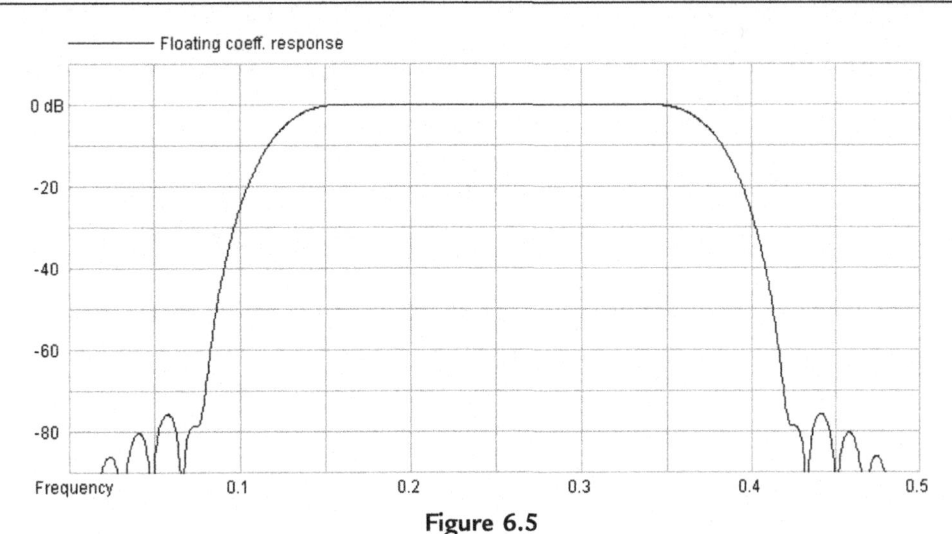

Figure 6.5
61-tap band-pass filter with Blackman window.

band. However, the stopband sidelobes are very high, reducing the amount of stopband attenuation. The Hanning, Hamming, and Blackman windows provide increasing stopband attenuation, at the expense of a wider transition band. Windowing is supported in all finite impulse response filter design software programs.

These windows have a similar effect on the transition band and sidelobes whether applied to low-pass filters, high-pass filters, or band-pass filters.

CHAPTER 7

Decimation and Interpolation

In this chapter, we are going to discuss decimation and interpolation. Decimation is the process of reducing the sample rate F_s in a signal processing system, and interpolation is the opposite, increasing the sample rate F_s in a signal processing system. This process is very common in signal processing systems and is nearly always performed using a finite impulse response (FIR) filter.

First of all, why are sampling rates changed? The most common reason is that higher sample rates make it easier in the analog domain, whereas lower rates reduce the digital computational requirements. The ADCs and DACs often operate a higher frequency, while most filtering and other functions can be operated at a lower frequency. As we have seen in previous chapters, signals have a frequency representation, and this frequency representation must be less than the Nyquist frequency, which is defined as $F_s/2$. This sets a lower bound on F_s. The amount of hardware or software processing resources is normally proportional to F_s, so we usually want to keep F_s as small as practical. So while there is no upper bound on F_s, it is usually less than 10× the frequency representation of the signal. A minimum F_s is needed to ensure the highest frequency portion of the signal does not approach the $F_{Nyquist}$ frequency.

In some cases, there are advantages to highly oversampling a signal, where $F_s >> F_{signal}$. One such example might be when sampling a signal using an analog to digital converter (ADC). If F_s is high, then so is $F_{Nyquist}$. Recall that we can accurately represent all signals below $F_{Nyquist}$ in the sampled domain without aliasing occurring. By making $F_{Nyquist}$ large, any unwanted signals in that frequency space can be eliminated, or at least highly attenuated, by using a digital filter with its passband matching our desired frequency and with its stopband for the rest of frequency spectrum up to $F_{Nyquist}$. The analog filter needs to filter out frequencies above $F_{Nyquist}$. In this manner, by increasing F_s and therefore $F_{Nyquist}$, we can simplify the requirements for analog filtering prior to the ADC. Now once we have successfully filtered out these unwanted signals through both the analog and digital filters, we have no further need of keeping the $F_s >> F_{signal}$ and should consider lowering F_s to reduce resources required for subsequent processing.

7.1 Decimation

To discuss decimation, or down sampling as it is also called, we need to consider the frequency representation of the signal as well as the time domain. First we will look at the

Digital Signal Processing 101. http://dx.doi.org/10.1016/B978-0-12-811453-7.00007-X

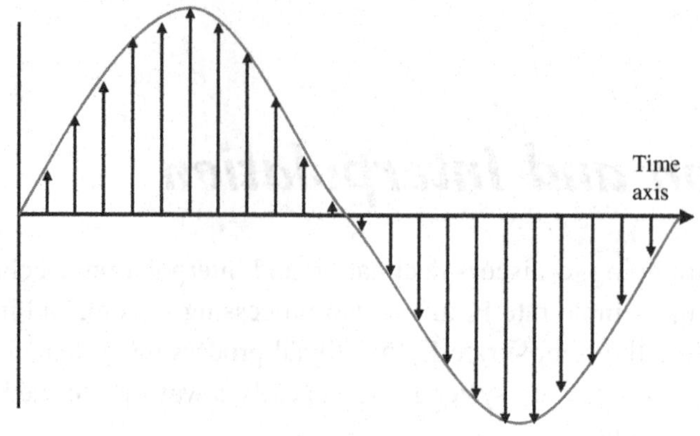

Figure 7.1
Original sampling rate = F_s.

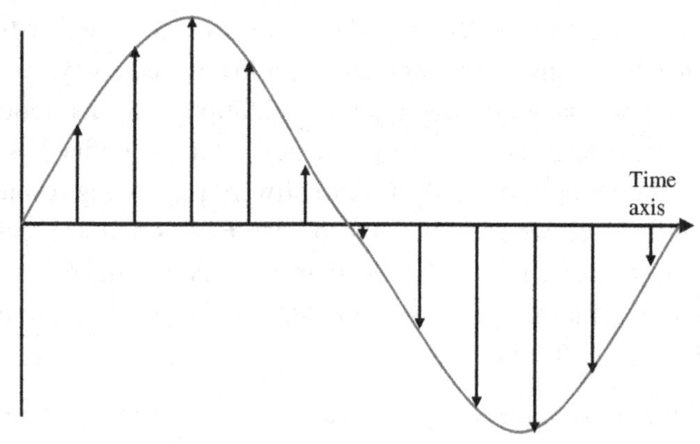

Figure 7.2
New sampling rate = $F_s' = F_s/2$.

sampled signal sampled at F_s (Fig. 7.1) and at F_s', where $F_s' = F_s/2$ (Fig. 7.2). The new sampling rate F_s' is the sampling rate after decimating by 2.

Below in Fig. 7.3, is the frequency domain perspective of decimation.

By examining the frequency plots, it is evident the signal itself has not changed, only the sampling rate frequency and corresponding Nyquist rate frequency. As long as the new Nyquist frequency is larger than the signal frequency, no aliasing will occur.

The signal images, which are periodic in F_s, and a natural consequence of sampling, are also shown. When decimating, the images become periodic in the new sample rate F_s'.

One could decimate by simply throwing away samples. In our example, by throwing away every other sample, the sample rate will be reduced by a factor of ½, and result will be as shown above.

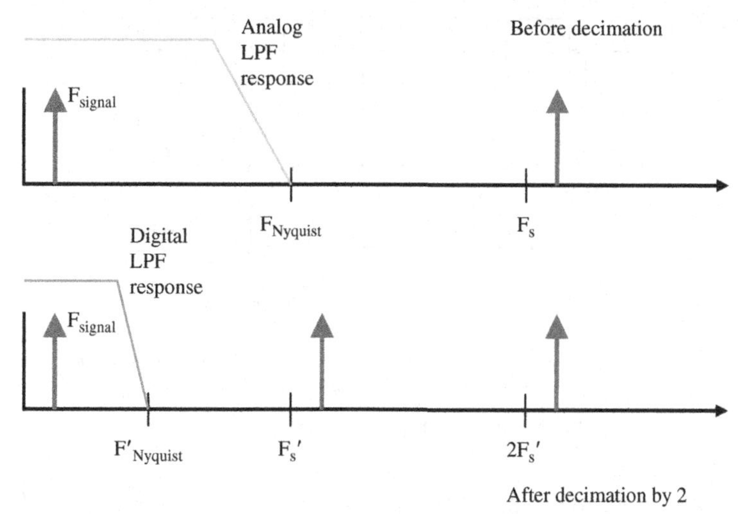

Figure 7.3
Decimation in frequency domain. *LPF*, low-pass filter.

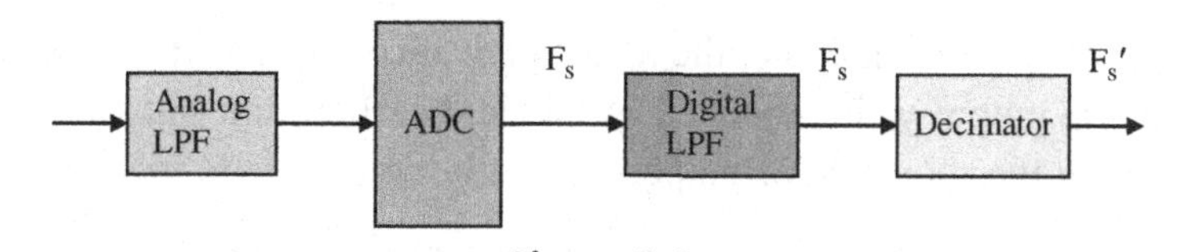

Figure 7.4
Sampling chain functions. *ADC*, analog to digital converter; *LPF*, low-pass filter.

In practice, however, the decimation process usually has a low-pass filter (LPF) incorporated. Let us go back to our ADC example. This could be implemented using the block diagram in Fig. 7.4. In this example, the analog LPF is responsible for removing all unwanted signals above $F_{Nyquist}$, which would otherwise alias into the frequency region where our signal is. With this analog LPF prior to the ADC, we can then be confident that any signals we see below $F_{Nyquist}$ are legitimate signals, and not aliased versions of some higher frequency unwanted signal or noise. We must similarly provide an LPF prior to the decimator, to remove any frequency components between $F'_{Nyquist}$ and F_s', else any frequency components in this frequency band will alias, or fold back, into the frequency band below our new Nyquist frequency $F'_{Nyquist}$. The approximate frequency response of the analog and digital LPF is depicted in the previous frequency domain (Fig. 7.3).

Now we can ask an interesting question. Why bother to compute samples at rate F_s in the digital LPF if we are going to discard ½ of them in the decimator? This is a waste of processing resources. Instead, we can build a digital filter which only computes every other sample, and therefore accomplishes the function of the decimator (shown separately

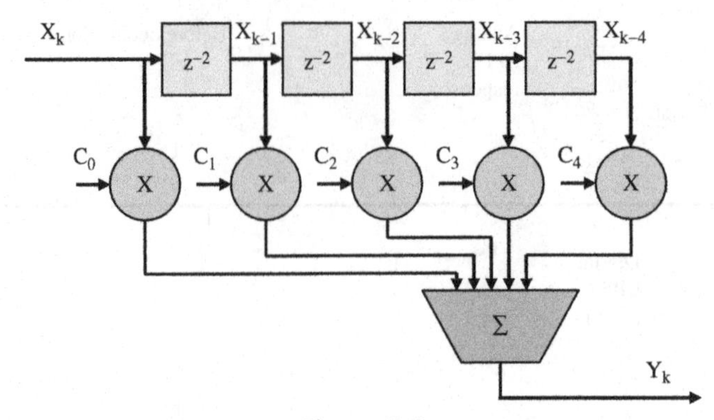

Figure 7.5
Finite impulse response filter structure.

in our block diagram). And in the process, we find that we only need one half the multiplier rate to implement the digital LPF.

To see how we might do this, we will reconsider the FIR block diagram Fig. 7.5.

In the normal case, at each clock cycle, the x_k data advance into the shift register of the filter structure, and an output y_k is produced. As we did earlier, one can build a sequence of inputs and compute the sequence of outputs. The filter equation is repeated below:

$$y_k = C_0 \cdot x_k + C_1 \cdot x_{k-1} + C_2 \cdot x_{k-2} + C_3 \cdot x_{k-3} + C_4 \cdot x_{k-4}$$

Now suppose that on every even clock cycle (k is even), we operate the filter normally, but on the odd clock cycles (k is odd), we merely shift the input data, but disable the operation of multipliers, summation, and update of output register (this register not explicitly shown). The output register will just hold the previous value, since there has been no update.

We will show how to compute a few outputs,

$$y_0 = C_0 \cdot x_0 + C_1 \cdot x_{-1} + C_2 \cdot x_{-2} + C_3 \cdot x_{-3} + C_4 \cdot x_{-4}$$

(y_1 output not computed, only x_k input data shifted through)

$$y_2 = C_0 \cdot x_2 + C_1 \cdot x_1 + C_2 \cdot x_0 + C_3 \cdot x_{-1} + C_4 \cdot x_{-2}$$

(y_3 output not computed, only x_k input data shifted through)

$$y_4 = C_0 \cdot x_4 + C_1 \cdot x_3 + C_2 \cdot x_2 + C_3 \cdot x_1 + C_4 \cdot x_0$$

And so forth.

The output sequence is the decimated, filtered sequence $\{\ldots y_0, y_2, y_4, \ldots\}$. Only the shift registers operate at the input rate. The rest of the circuitry can be clocked at the output clock rate. If implemented using DSP processors, or if you are clever in your hardware filter design, you can utilize less multipliers by operating them at the faster input speed. This is beyond our scope here, but in general, whether implementing DSP algorithms in hardware (FPGA or ASIC) or in software (DSP processor), the multipliers can be multiplexed such that it does not matter whether you have a few very fast multipliers, or a large number of slow multipliers or anything in between, so long as the cumulative multiply-accumulate capacity is sufficient for the DSP algorithm requirement.

Decimation is limited to integer values. So this concept can be extended to decimate by 3, 4, 10, and so forth. In each case, the decimation filter only computes the required samples. The input data is shifted right by the decimation rate between each computation. In our decimate-by-2 example above, the input data is shifted right by two places between each output computation (check the x indexes in the example computations above).

7.2 Interpolation

As we just saw, decreasing the sample rate F_s by an integer value can be accomplished by a decimation FIR filter. Similarly, the sample rate F_s can be increased by an integer value using a type of FIR filter called an interpolation filter. This is called upsampling and is the opposite of decimation. As long as the signal frequency content is below the Nyquist frequency at the lowest sampling frequency, one can decimate a signal and then turn around and interpolate it and recover the same signal.

Interpolation requires that the sample rate be increased by some integer factor. New samples need to be created and inserted between the existing samples. Let us look at the simplest example of interpolation. Let us go back to our sine wave example and interpolate it by a factor of two in Fig. 7.6.

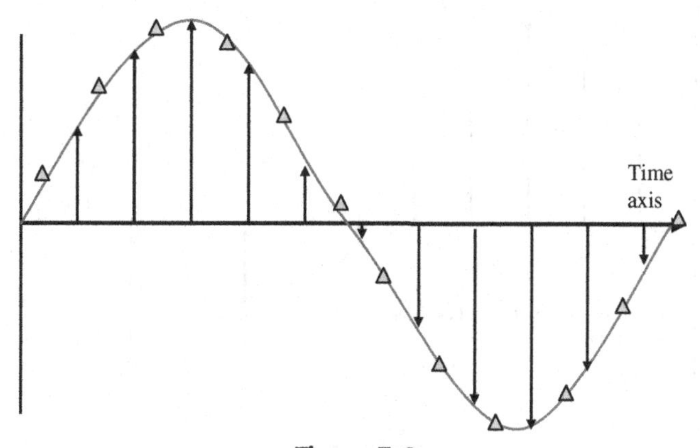

Figure 7.6
Interpolation in time domain.

Shown is the original signal at sample rate F_s where the "triangles" indicate approximately where new samples must be created to interpolate up to sample rate $F_s' = 2 \cdot F_S$. This seems straight forward enough. We do not have to worry about aliasing issues, because we are doubling both the sample frequency and Nyquist frequency.

The simplest and most intuitive way to interpolate is called linear interpolation. In linear interpolation, one simply draws a straight line between the original samples, and calculates the new samples along this line. In our case, if we interpolate by 2, then we need the point located mid-way along the line between the original points. Linear interpolation, whether by factor of 2, 3, 4, ... is equivalent to drawing a line between all the original points, and will look something like the following figure.

Signal at $F_s' = 2 \cdot F_S$ shown in Fig. 7.7 with linear interpolation. Obviously, this does not look quite right. But let us consider how we would build an interpolation filter.

An interpolation filter is actually several filters running in parallel, each with the same data input x_k. Each filter computes a different intermediate sample. One filter has a single tap = 1, which provides the original signal at the output (this could just be a shift register to provide correct delay). Each of the other filters calculates one of the new samples between the original samples. When interpolating by N, there will be N of these filters, including the trivial filter that generates the original samples. For example, when interpolating by four, N = 4, and there will be four of these filters. Every input sample will produce four output samples, one from each filter, which will be interleaved at the output at a combined rate four times larger than the input rate. This concept, of multiple filters creating a single output with higher rate sequence, is called polyphase filtering.

The length or number of taps in each of the N interpolation filters will largely determine the quality of the interpolation. With linear interpolation, the number of filter taps is only

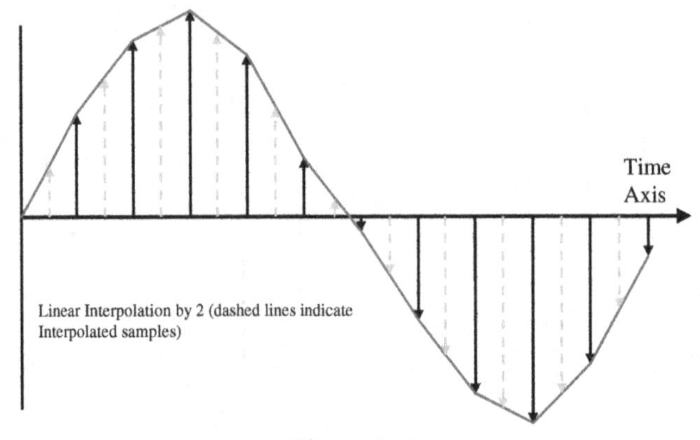

Figure 7.7
Linear interpolation.

2, so the quality of the interpolated signal is rather poor, as can be seen above. The ideal filter is an LPF with cutoff frequency at $F_{Nyquist}$ of the original signal sampling rate. As we learned previously, an ideal filter has infinite number of coefficients. We will have to compromise at less than infinity, and so each individual filter will have M taps.

Shown in Fig. 7.8 is an interpolate-by-4 (N = 4) polyphase filter, with 5 taps (M = 5) used for each phase. The input data stream x_k is sampled at F_s, and the serialized output data stream y_m is sampled $F_s' = N \cdot F_S$. Coefficient representations A_n, B_n, and C_n are used to indicate that each phase of the interpolation filter may have a different set of coefficients.

You should note how the first filter could be eliminated and replaced by a shift register. All of the taps except the center one are multiplying by zero, so the multiplier and adder logic is not required. The original input sample simply passes through with some delay.

Interpolating polyphase filters are a little tricky, so we will also show what is happening in the time or sample domain. Shown below in Fig. 7.9 is the input and output sequences, which are time aligned for clarity. The output data rate is four times the input, and like all FIR filters, there is some processing delay through the filter. The function of the serializer is to input N samples in parallel at rate F_s, and output a serial sample stream at F_s', which is the new interpolated sample rate.

The dashed lines indicate the interpolated samples. The dotted lines show the delay of the original samples through the interpolation filter. This delay is dependent on filter design, but is generally equal to (M − 1)/2 input samples, plus any additional register pipeline delays. In our example above, M = 5, so the delay is (5 − 1)/2 = 2 input sample intervals.

Note that the "m" output index is increasing faster than the "k" input index. The interpolation filter must produce N output samples for every input sample, so the output index "m" needs to run N times faster than "k".

7.3 Resampling by Noninteger Value

Suppose that you need to align the sample rates between sets of digital circuitry running at different sampling rates. For example, the sampling rate of the first circuit is 3 MSPS, and the second circuit has a sampling rate of 2 MSPS. You will need to decimate, or downsample, by 2/3.

This can be done by a combination of interpolating and decimating. In this case, you would interpolate by 2 and then decimate by 3, as shown in Fig. 7.10.

Next, a frequency domain representation of both interpolation and decimation steps is shown in Fig. 7.11.

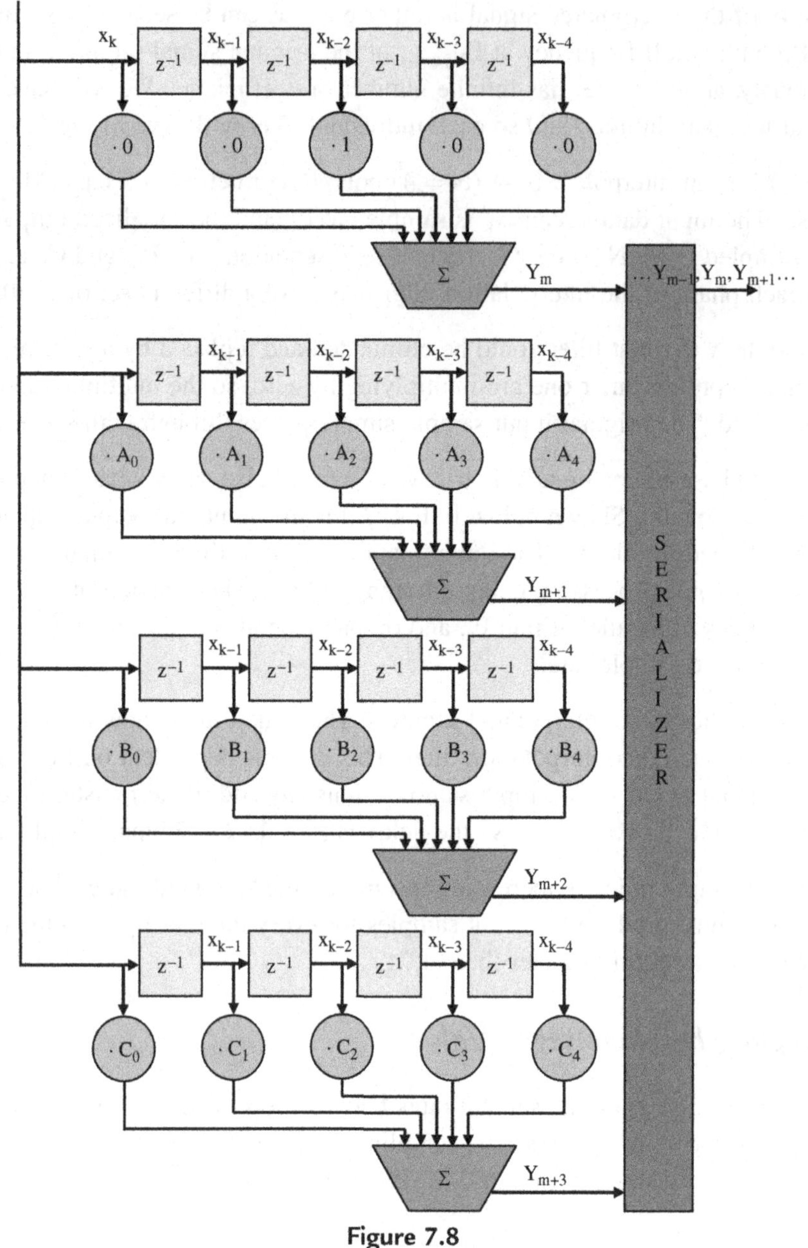

Figure 7.8
Interpolation filter structure.

Both the interpolation and decimation filters incorporate a low-pass filtering function. The reason for this LPF, however, is quite different for each case. For decimation, the LPF serves to eliminate high frequency components in the spectrum. If these components were not filtered out, they would alias when the reduction in sample rate is performed.

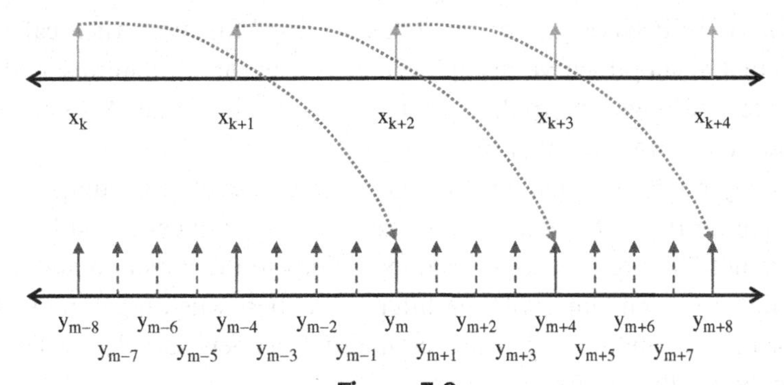

Figure 7.9
Interpolation samples.

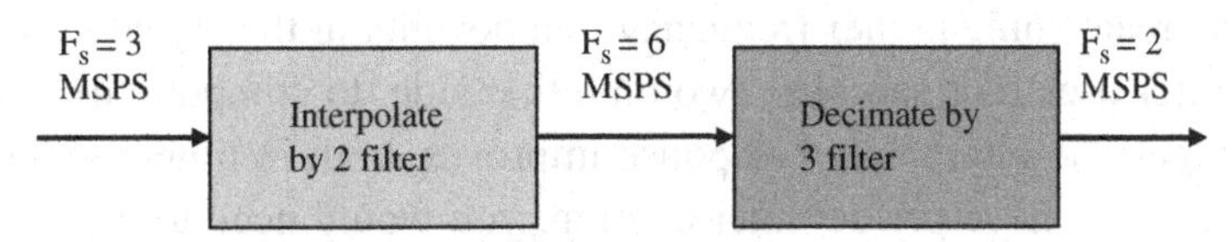

Figure 7.10
Resampling function.

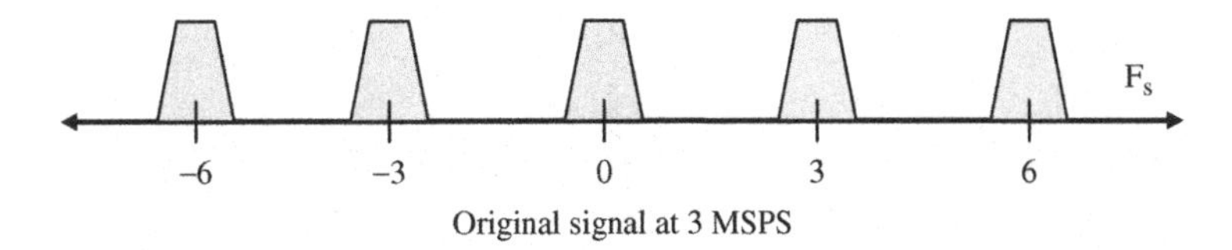

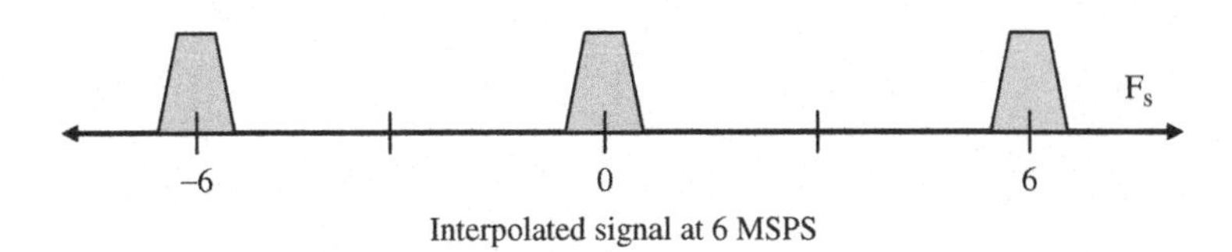

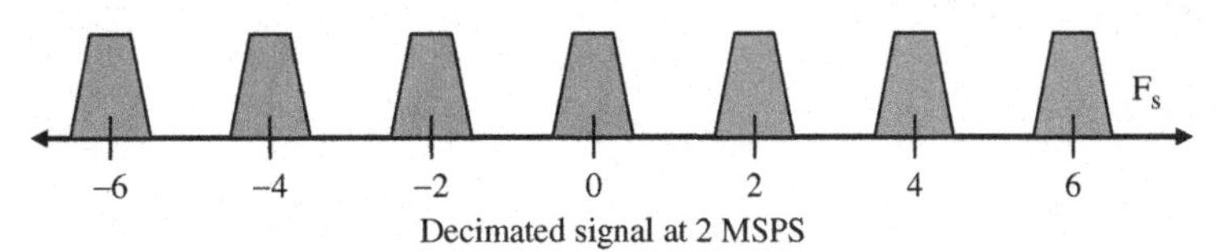

Figure 7.11
Interpolation in the frequency domain.

For interpolation, the LPF serves to provide a "smoothing" function when calculating the new samples, so that a smooth curve results when the new samples are inserted between the original samples. A longer interpolating filter (more taps = larger M) will use a weighted calculation of larger number of the adjacent samples to determine the new samples. An analogy might be when a person is driving a car on a winding road. If you look only a few feet in front of the car, you cannot take the curves smoothly. The reason is that since you are not looking ahead, you cannot anticipate the direction and rate of curves and smoothly adjust to it. The interpolating filter works best when it can look at samples on both sides (or in front and behind) when computing the new samples, which should smoothly fit in between the existing samples.

When $M = 2$, which is linear interpolation, only the two adjacent samples are used, and the filter computes the sample which lies on a straight line between the two points. As we saw in the example, this does not give a very smooth response. This lack of smooth response can also create some higher frequency components in the signal spectrum. If $M = 4$, then the filter uses four samples, two on either side, to compute the new samples. As M becomes larger, the interpolated response improves, both in time domain and frequency domain. To achieve perfect interpolation, you would need to use an infinite number of samples, and build a perfect LPF. This perfect interpolation filter would be in the form of the sinc function, also known as $\sin(X)/X$, introduced in the FIR Filter chapter, and be of infinite length. In practice, using between 8 and 16 samples, or filter coefficients, is usually enough to a reasonable job of interpolating a signal for most applications.

CHAPTER 8

Infinite Impulse Response (IIR) Filters

This chapter discusses infinite impulse response (IIR) filters. The IIR filter is unfortunately a much more complex topic than the FIR filter, and due to its nonlinear behavior, it is very difficult to analyze. It is also more complex to implement, due to the feedback, though it typically does require far less multipliers than a FIR filter. This feedback can cause instability in IIR filters, and overflow or underflow can easily occur using fixed point numbers. For these reasons, IIR is used much less than finite impulse response (FIR) filters. On the plus side, since it is not commonly used, understanding IIR filters is not essential to the fundamental concepts of digital signal processing (DSP). The IIR filter design technique is usually considered a bit of a specialty in the DSP world, so do not feel that you need to master this topic.

All the popular IIR filter designs are based upon analog filter circuits. The nature of analog components—capacitor, inductors, and opamps tend to be naturally suited to recursive filter designs. By contrast, FIR filters are naturally implemented digitally. This chapter will introduction to IIR filters, and then focus primarily on how to convert an analog IIR filter into a sampled digital implementation.

Mathematics for IIR filters tends to be more daunting. If the math proves too much, please note that the discussion in the chapter is not necessary to continue onto the other topics in the remainder of this book. Plus, the chapter following on digital modulation is really interesting; you would not want to miss it.

Most digital IIR filter designs are derived from analog filter designs. As analog filters were around long before digital filters existed, this provides for many types of filters.

The basic design procedure will be to take an analog filter design, and convert it to a digital IIR filter implementation. Since we do not have time or space to go through analog filter fundamentals, some material presented here may be difficult for readers who do not have any familiarity with analog filter design techniques.

Now, enough of the disclaimers, and let us begin.

An IIR filter is basically an FIR filter with feedback added. The FIR filter takes a stream of input data and multiplies it by a series of coefficients to produce each output. The IIR filter also does this, but in addition, it feeds the output data stream back through another series of multipliers and coefficients. This structure is shown in the diagram below in Fig. 8.1.

Digital Signal Processing 101. http://dx.doi.org/10.1016/B978-0-12-811453-7.00008-1

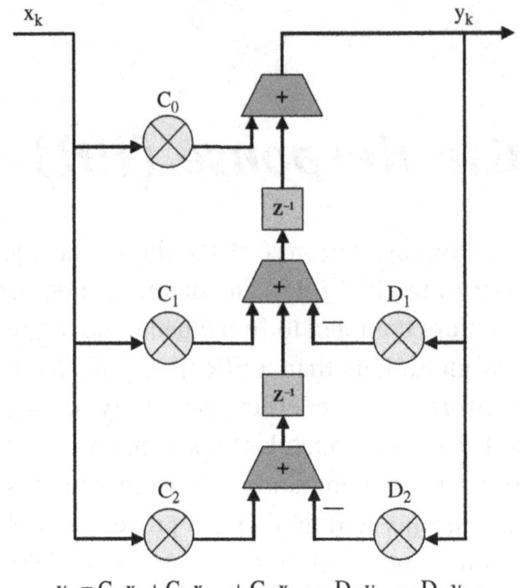

$$y_k = C_0 \cdot x_k + C_1 \cdot x_{k-1} + C_2 \cdot x_{k-2} - D_1 \cdot y_{k-1} - D_2 \cdot y_{k-2}$$

Figure 8.1
Infinite impulse response filter biquad structure.

This feedback eliminates many of the linear properties of the FIR filter and makes the IIR filter much more difficult to analyze. It can also create some undesired behavior, depending on the choice of coefficients, where the impulse response may have an infinite duration or even an infinite magnitude.

8.1 Infinite Impulse Response and Finite Impulse Response Filter Characteristic Comparison

Before we get too far into IIR filters, it is useful to compare IIR and FIR filters. Below is a summary of advantages and disadvantages of IIR and FIR digital filters.

- FIR filters have linear phase, meaning that no phase distortion of the signal occurs. IIR filters will always cause some phase distortion,[1] so the phase as well as magnitude response needs to be considered by the filter designer.
- FIR filters are always stable and have a finite length impulse response. IIR filters generally have an infinite length impulse response and may have infinite magnitude output (become unstable) under some conditions.

[1] Phase distortion happens when the phase response of the filter changes in nonlinearly across the filter's passband frequency response. Recall that filters are complex; they can affect both the magnitude and phase of the signal.

- FIR filters can be designed with a specified amount of quantization noise (remember quantization from Numerical Representation chapter), which can be made as small as necessary. This is not the case with IIR filters.
- FIR filters can be implemented efficiently in multirate systems or systems that have decimation or interpolation steps.
- IIR filters are very sensitive to coefficient values and numerical precision, particularly in designs which require a sharp cutoff frequency response.
- IIR filters are more natural digital form to replace existing analog filters.
- An IIR filter can provide a much sharper cutoff frequency response compared to the same order (number of stages of multipliers) FIR filter. In other words, for sharp response in an FIR filter, more resources (multipliers and adders) are required than in IIR filter.

Based on this, one wonders why anyone would want to use IIR filters. Well, aside from the last bullet, which lists a key advantage of IIR filters, the reason is that IIR filters can best approximate the performance of many analog filter responses. Sometimes when a system with analog filtering is being upgraded to digital implementation, it is important to preserve its performance characteristics, especially in the phase domain. An example of this might be professional audio equipment.

Intuitively, the reason IIR filters can have a much more rapid frequency response is because their frequency response is determined by both "zeros" and "poles". Zeros are caused by cancellations. Remember the FIR discussion, with coefficients of $+1$, -1, $+1$, -1, and so on? This filter causes cancellations at high frequencies, and this can be described as zero at a specific frequency. Filter response is determined by canceling various frequencies.

Poles, on the other hand, involve feedback and result in an effect similar to division by zero at specific frequencies. Actually, this is not exactly divided by exactly zero, as this would produce an infinite response, but the idea is to divide by a very small number at certain frequencies, which can produce a large output in the signal in that frequency region. Think of a rope representing frequency response. The poles would hold the rope up at specific points, and the zeros would be like a lead weight holding it to the ground at other points. This is analogous to how poles and zeros can act on the frequency response.

The additional flexibility of using poles with zeros, while resulting in some unstable filter responses, if not careful (imagine a pole which can provide very high, or even infinite gain), can also provide for more rapid changes in the frequency response. This makes IIR filters more sensitive and complicated than FIR filters.

In this introductory chapter, we will not try to explain the design of IIR filters using pole and zero placement. There are whole books on this topic and is too much for this introductory treatment here. Rather we will show how to take popular analog filter designs, with a defined pole and zero arrangement and thus frequency response, and convert to a digital implementation, as this is a more common task for a DSP system designer.

Even a basic understanding of IIR filters will require some mathematics. The reason is that analog filters are analyzed using something called the "Laplace transform." Digital filters are analyzed using something called the "z-transform." Because of their simplicity, we managed to avoid the z-transform when discussing FIR filters, but we will not have that option here. Please refer to the appendices at the end of book introducing Laplace and Z-transforms if these are new to you.

8.2 Bilinear Transform

The normal design procedure for IIR filters is to specify the filter response and design an analog filter using analog filter techniques (using Laplace transform). Alternately, you might be given an analog filter design, and be asked to convert it to a digital implementation. The analog filter design is based on the location of poles and zeros in the "s" plane. A digital filter response can be characterized by using z-transform. The equivalent pole and zero domain for digital filters is called the "z" plane. The idea is to map the "s" plane poles and zeros to the "z" plane. The mapping technique most often used between the "s" and "z" domains is called the bilinear transform. There are alternative techniques, but we will not cover these here. Further, only a rudimentary coverage will be attempted here, as the topics involved are fairly mathematical.

Again, please note that both the Laplace and z-transforms are reviewed in appendices at the end of this book. If you have not looked at these recently, you may want to spend a little time going through Laplace and Z-transform appendices.

We use $\acute{\omega}$ for the s-plane to distinguish between ω of the sampled domain z-plane. Similarly, we will define the frequency response of the analog filter as H_s and the frequency response of the digital filter as H_z.

In analog filter design, the frequency response of an analog filter is defined by setting $s = j\acute{\omega}$ and evaluating for $-\infty < \acute{\omega} < \infty$. This corresponds to the imaginary axis of the s-plane.

The frequency response of a digital filter is defined by setting $z = e^{j\omega}$, and evaluating for $-\pi < \omega < \pi$. This corresponds to the unit circle of the z-plane.

To go between these two domains, we need a mapping function from the s-plane to the z-plane. Then we can map the zeros and poles across the two domains.

This is performed by replacing s in the expression for H_s.

$$s = 2 \cdot \left(1 - z^{-1}\right) / \left(T \cdot \left(1 + z^{-1}\right)\right)$$

where $T = 1/F_s$ (T = time between sampling interval of the digital filter in seconds).

Let us go through an example of converting and analog filter to a digital IIR filter. There are a great many analog filter types. Here we will discuss only one, as analog filters are not the topic we want to focus on. A very common analog filter is the Butterworth filter. It has the characteristic of not having ripples in the passband or stopband.

We will be taking a third-order Butterworth analog filter and converting to an IIR digital filter using the bilinear transformation technique. Many analog filters are known simply by their pole and zero locations (since this is another way of defining frequency response). We can eliminate quite a bit of algebra by using the relationship above between s and z domains to come up with a relationship between poles and zeros in the s and z domains. These derivations are not discussed here, only the result is discussed.

The IIR digital filter will have poles and zeros at locations:

$$\text{z\_pole}_i = (2 + \text{s\_pole}_i \cdot T)/(2 - \text{s\_pole}_i \cdot T)$$
$$\text{z\_zero}_i = (2 + \text{s\_zero}_i \cdot T)/(2 - \text{s\_zero}_i \cdot T)$$

We will set the cutoff frequency of our third-order Butterworth filter to 100 Hz, or 628 radians/second ($100 \cdot 2 \cdot \pi$). For a third-order Butterworth filter, the three pole locations in the s-plane are located at:

$$\text{s\_pole}_1 = 628\ \text{angle}(120^\circ) = -314 + j544$$
$$\text{s\_pole}_2 = 628\ \text{angle}(180^\circ) = -628$$
$$\text{s\_pole}_3 = 628\ \text{angle}(240^\circ) = -314 - j544$$

We will set our digital sampling frequency at 1000 samples per second, so $T = 0.001$. The s-domain poles will map to the following poles in the z-plane.

$$\text{z\_pole}_1 = [2 + (-314 + j544) \cdot 0.001]/[2 - (-314 + j544) \cdot 0.001] = 0.745\ \text{angle}(72.56^\circ)$$

$$\text{z\_pole}_2 = [2 + (-628) \cdot 0.001]/[2 - (-628) \cdot 0.001] = 0.523$$
$$\text{z\_pole}_3 = [2 + (-314 - j544) \cdot 0.001]/[2 - (-314 - j544) \cdot 0.001] = 0.745\ \text{angle}(-72.56^\circ)$$

The Butterworth filter has three zeros located at infinity. These can be evaluated as:

$$\text{z\_zero}_{1,2,3} = [2 + \infty \cdot 0.001]/[2 - \infty \cdot 0.001] = \infty/-\infty = -1$$

Now that we have the poles and zeros on the z-plane, we can determine the z-transform of the digital IIR filter approximating the response of the analog Butterworth filter.

$$H(z) = \left[\prod\nolimits_{i=0\text{ to }M-1} (z - \text{z\_zero}_i)\right] / \left[\prod\nolimits_{i=0\text{ to }N-1} (z - \text{z\_pole}_i)\right]$$

where M = number of zeros and N = number of poles. In our example, M = N = 3. With a bit of algebra, we can multiply this whole mess out, to get in the more familiar form.

$$H(z) = \left(\sum\nolimits_{i=0 \text{ to } M-1} C_i \cdot z^{-i}\right) \Big/ \left(1 - \sum\nolimits_{i=12 \text{ to } N-1} D_i \cdot z^{-i}\right)$$

In this form, we can pick off our coefficients needed to implement the IIR filter.

Multiplying the numerator and denominator components, we will get:

$$H(z) = \frac{z^3 + 3z^2 + 3z + 1}{z^3 - 0.970z^2 + 0.108z - 0.290} = \frac{1 + 3z^{-1} + 3z^{-2} + z^{-3}}{1 - (0.970z^{-1} - 0.108z^{-2} + 0.290z^{-3})}$$

From inspection,

$C_0 = 1 \quad C_1 = 3 \quad C_2 = 3 \quad C_3 = 1$

and

$D_1 = 0.970 \quad D_2 = -0.108 \quad D_3 = 0.290 \quad D_3 = 0.290$

These C and D coefficients apply to the IIR filter structure depicted earlier in the chapter (although this is a third-order filter, the example diagram is second order).

8.3 Frequency Prewarping

We still have another point to consider. Our technique is to take the pole and zero locations (in the s-plane) of an analog filter, and to map them to pole and zero locations (in the z-plane) of an IIR digital filter.

The problem is that this relationship is not linear. For example, the y-axis of the s-plane, which is infinite in length, is mapped to the unit circle in the z-plane. This relationship is

$$\acute{\omega}(s - \text{plane}) = (2/T) \cdot \tan(\omega/2) \quad \text{and}$$
$$\omega(z - \text{plane}) = (2/T) \cdot \tan^{-1}(\acute{\omega}T/2) \quad \text{where } T = 1/F_s$$

This nonlinear mapping will cause filter response distortion, particularly at higher frequencies. A method to mitigate this is to prewarp the analog filter. The whole analog filter is not prewarped, instead key breakpoints in the analog filter response are prewarped. This will in essence stretch the analog filter in frequency, so that when it is compressed by the bilinear transform, the prewarped breakpoint(s) will be mapped correctly. This ensures that transition points between pass and stopband are accurately converted by the bilinear transform.

The table below shows the distortion cause by mapping between the s and z domains. For this table, we set T = 1. Note how this distortion or warping increases with frequency (Table 8.1).

Table 8.1: s to z Plane Frequency Mapping

$\acute{\omega}$, Analog Frequency (s-plane)	ω, Digital Frequency (z-plane)
$0.0\cdot\pi$	$0.0\cdot\pi$
$0.1\cdot\pi$	$0.0992\cdot\pi$
$0.2\cdot\pi$	$0.1938\cdot\pi$
$0.3\cdot\pi$	$0.2804\cdot\pi$
$0.4\cdot\pi$	$0.3571\cdot\pi$
$0.5\cdot\pi$	$0.4238\cdot\pi$
$0.6\cdot\pi$	$0.4812\cdot\pi$
$0.7\cdot\pi$	$0.5302\cdot\pi$
$0.8\cdot\pi$	$0.5702\cdot\pi$
$0.9\cdot\pi$	$0.6081\cdot\pi$
$1.0\cdot\pi$	$0.6391\cdot\pi$

Let us consider a simple example. We have an analog low-pass filter, with a 3 dB breakpoint at 100 Hz (3 dB breakpoint is the point in transition band where filter response is 3 dB lower than the passband response). We want to implement this filter with a digital IIR filter, which will have an $F_s = 1000$ Hz.

The digital breakpoint frequency should be $(100/1000)\cdot 2\pi = 0.2\cdot\pi$. The warping will cause this breakpoint to occur at digital frequency $0.1938\cdot\pi$ instead.

The analog filter breakpoint needs to be moved, or prewarped, to compensate for this.

Substituting into $\acute{\omega}$ (s-plane) $= (2/T)\cdot\tan(\omega/2)$, we find

$$\acute{\omega} = (2/0.001)\cdot\tan(0.2\cdot\pi/2) = 649.8\ \text{radians/s} = 103.4\ \text{Hz}$$

We should design the analog filter with a 3 dB point at 103.4 Hz, rather than 100 Hz prior to converting to a digital IIR filter using the bilinear transform to map the pole/zero locations.

To illustrate the effect as the frequency rises, let us revise the problem to building a digital filter to replace an analog filter with a 3 dB breakpoint at 250 Hz. We will still use $F_s = 1000$ Hz.

The digital frequency of the breakpoint is $\omega = (250/1000)\cdot 2\pi = 0.5\cdot\pi$.

The analog frequency of the breakpoint needs to be moved from 250 Hz to prewarp the filter. The new analog filter breakpoint is found to be:

$$\acute{\omega} = (2/0.001)\cdot\tan(0.5\cdot\pi/2) = 2000\ \text{radians/s} = 318.3\ \text{Hz}$$

We can see that the importance of prewarping the analog filter prior to applying the bilinear transform as we move our breakpoint closer to the Nyquist frequency of the digital filter (in this example, equal to 500 Hz). If there are multiple transition points, they can all be prewarped, and the analog filter modified to meet these new transition points.

For those of you without much analog filter design experience, you may be wondering how to modify the analog filter, to find prewarped analog poles and zeros which will then be mapped to the digital domain. This is a valid question, but unfortunately is beyond the scope of this chapter. The focus of our discussion is how to learn the basics of converting an analog filter to a digital IIR filter. Since analog filter design a complex and mathematical subject, this just too much to try to cover here. However, as you might expect, there are software programs which can be used to perform analog filter design, and even convert these to a digital IIR design.

CHAPTER 9

Complex Modulation and Demodulation

We are going to try to take an unusual approach here. The normal explanations on this topic are heavily based on mathematics and equations. Here, we will try to take an almost entirely intuitive approach, based on examples. There will be no attempt to establish any mathematical foundation or to calculate performance.

Modulation is the process of taking information bits and mapping them to symbols. The sequence of symbols is then filtered to produce a baseband waveform, with the desired spectral properties. This baseband waveform or signal is then upconverted to a carrier frequency, which can be transmitted over the air, through coaxial cable, through fiber or some other medium. The key idea here is the concept of a symbol.

9.1 Modulation Constellations

We are going to use a common modulation method, known as QPSK (quadrature phase shift keying), as our first example. Do not let the technically sounding names of these different modulations confuse you. We will talk about what these names mean later. With QPSK, every two input bits will map to one of four symbols, as shown below in Fig. 9.1.

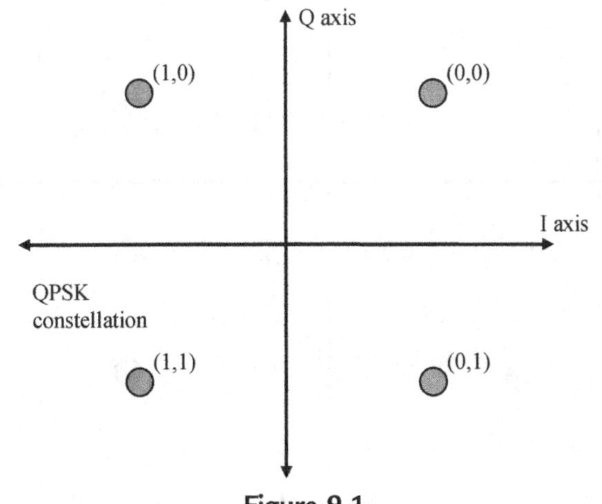

Figure 9.1
Quadrature phase shift keying constellation.

Digital Signal Processing 101. http://dx.doi.org/10.1016/B978-0-12-811453-7.00009-3

The bitstream of zeros and ones input to the modulator is converted into a stream of symbols. Each symbol is represented as the coordinates of a location in the I-Q plane. In QPSK, there are four possible symbols, arranged as shown. Since there are four symbols, then the input data is arranged as groups of 2 bits, which are mapped to the appropriate symbol. The arrangement of symbols in the I-Q plane is also called the constellation (See Table 9.1).

Another common modulation scheme is known as 16-QAM (quadrature amplitude modulation), which has 16 symbols, arranged as shown in Fig. 9.2. Again, do not worry about the name of the modulation. Since we have 16 possible symbols, each symbol will map to 4 bits. To put it another way, in QPSK, each symbol carries 2 bits of information, while in 16-QAM, each symbol carries 4 bits of information.

We can easily see that the 16-QAM is the more efficient modulation method. In this chapter, we are going to pick a convenient symbol rate and reference our discussion to

Table 9.1: Quadrature Phase Shift Keying Symbol to Bit Mapping

Input Bit Pair	Symbol Location on Complex Plane	I Value	Q Value	Symbol Value (Same as Location on Complex Plane
0, 0	$1+j$	1	1	$I+jQ => 1+j$
0, 1	$1-j$	1	−1	$I+jQ => 1-j$
1, 0	$-1+j$	−1	1	$I+jQ => -1+j$
1, 1	$-1-j$	−1	−1	$I+jQ => -1-j$

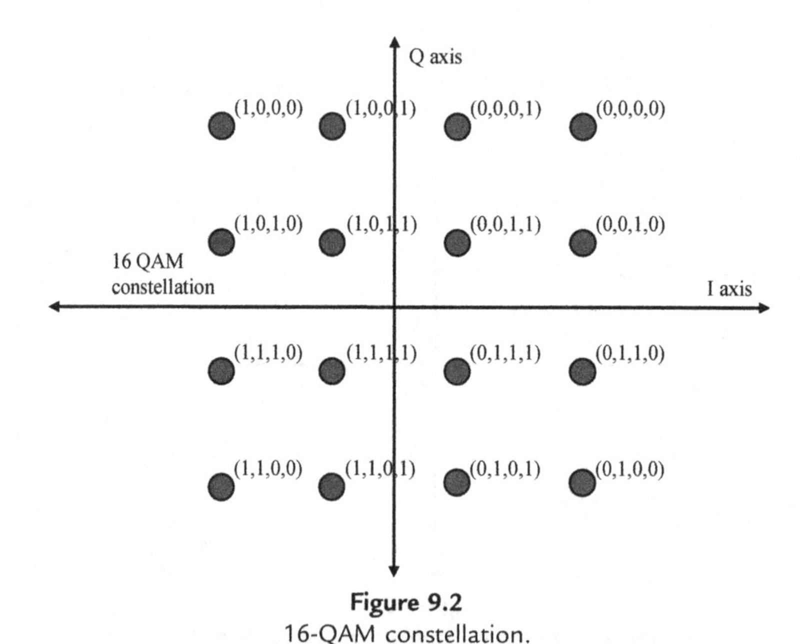

Figure 9.2
16-QAM constellation.

this symbol rate. But this is an arbitrary choice on my part, and systems are designed with symbol rates ranging from a few kHz to hundreds of MHz. We will decide to transmit symbols at a rate of 1 MHz, or 1 MSPS (million symbols per second). Then our system, if using the 16-QAM modulation, will be able to send 4 Mbits/s. If instead QPSK is used in this system, it will able to send only 2 Mbits/s. We could also use an even more efficient constellation, 64-QAM. Since there are 64 possible symbols, arranged as eight rows of eight symbols each, then each symbol carries 6 bits of information, and support a data rate of 6 Mbits/s. This is shown in the table below for a few sample constellation types (See Table 9.2).

The frequency bandwidth is determined mainly by the symbol rate. A QPSK signal at 1 MSPS will require about the same bandwidth as a 16-QAM signal at 1 MSPS. Notice the 16-QAM modulator is able to send twice the data within this bandwidth, compared to the QPSK modulator. There is a trade-off, however. As we increase the number of symbols, it becomes more and more difficult for the receiver to detect which symbol was sent. If receiver needs to choose among 16 possible symbols that could have been transmitted, rather than choose from among four possibilities, it is more likely to make errors. The level of errors will depend on the noise and interference present in the receive signal, the strength of the receive signal, and how many possible symbols the receiver must select from. In cellular systems, there are often high levels of interfering noise or weak signals due to buildings or other objects block the transmission path. In this situation, it is often preferable to use a simple constellation, such as QPSK. Even with a weak signal, the receiver can usually make the correct choice of four possible symbols. Other systems, such as microwave radio systems, usually have directional receive and transmit antennas facing each other from building roofs or mountaintops. Because of this, the interfering noise level is usually very low, and complex constellations such as 64-QAM or 256-QAM can be used. Assuming the receiver is able to make the correct choice from among 64 symbols, which allows three times more bits to be encoded into each symbol, resulting in a 3× higher data rate. Recently, sophisticated communication systems such as LTE and WiMax have been introduced, which allow the transmitter to dynamically switch between constellation types depended on quality of reception the receiver experiences.

Table 9.2: Modulation Bit Rates

Modulation Type	Possible Number of Symbols	Bits Per Symbol	Transmitted Bit Rate
QPSK	$4 = 2^2$	2	2 * symbol rate
8 PSK	$8 = 2^3$	3	3 * symbol rate
16-QAM	$16 = 2^4$	4	4 * symbol rate
64-QAM	$64 = 2^6$	6	6 * symbol rate
256-QAM	$256 = 2^8$	8	8 * symbol rate

9.2 Modulated Signal Bandwidth

Now that we understand what a constellation is, we still need to discuss some of the steps in taking this set of constellation points and transmitting this over some medium to a receiver. Let us take a look at a QPSK constellation, with transmission rate of 1 MSPS. The baseband signal is two dimensional, so must be represented with two orthogonal components, which are by convention denoted I and Q.

Let us consider a sequence of 5 QPSK symbols, at time t = 1,2,3,4, and 5 respectively. First, let us look at the sequence in the two dimensional complex constellation plane. It will appear as a signal trajectory moving from one constellation point to another over time.

We can also look at the I and Q signals individually, plotted against time in Fig. 9.4. We can take this two-dimensional signal, and plot each component separately. In actuality, the I and Q baseband signals are generated as two separate signals, and later combined together with a carrier frequency to form a single passband signal (See Fig 9.3). This is discussed in Chapter 11.

Notice the sharp transitions of the I and Q signals at each symbol. Intuitively, we know that a sharp transition requires the signal to have high-frequency content. A signal that is of low frequency can change only slowly and smoothly.

The high-frequency content of these I and Q signals can cause problems, because in most systems, it is important to minimize the frequency content or bandwidth of the signal. Remember the early discussion on frequency response, where a

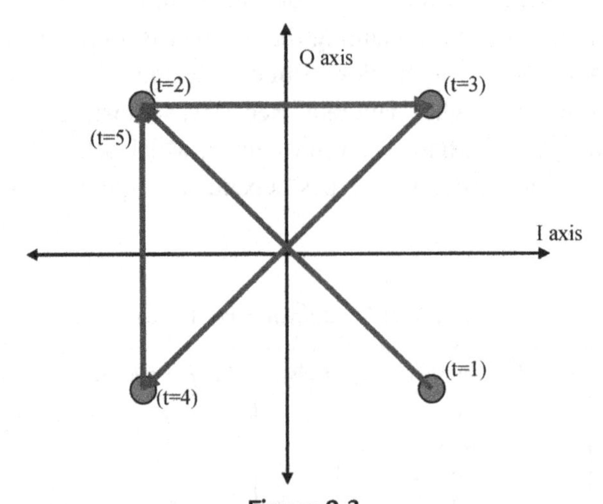

Figure 9.3
Quadrature phase shift keying signal trajectory.

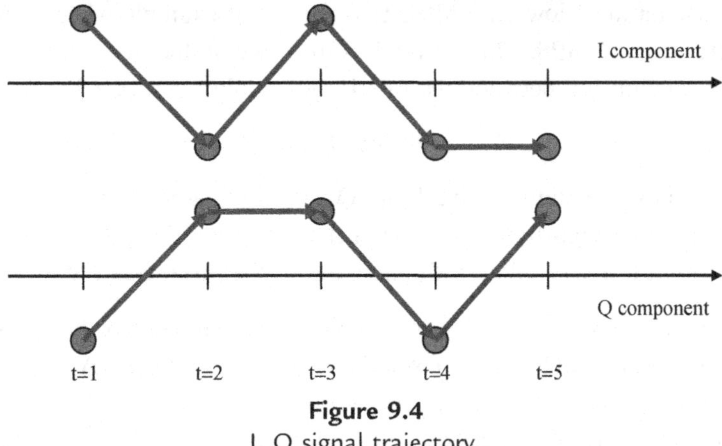

Figure 9.4
I, Q signal trajectory.

low-pass filter removes fast transitions or changes in a signal (or eliminates the high frequency components of that signal). If the frequency response of the signal is reduced, this is the same as reducing its bandwidth. The smaller the bandwidth of the signal, the more signal channels and therefore capacity can be packed into a given amount of frequency spectrum. Thus the channel bandwidth is often carefully controlled.

A simple example is FM radio. Each station, or channel, is located 200 kHz from its neighbor. That means that each station has 200 kHz spectrum or frequency response it can occupy. The station on 101.5 is actually transmitting with a center frequency of 101.5 MHz. The channels on either side transmit with center frequencies of 101.3 and 101.7 MHz. Therefore it is important to restrict the bandwidth of each FM station to within $\pm$ 100 kHz, which ensure it does not overlap or interfere with neighboring stations. As we know by now, one way to restrict the bandwidth of a signal is to use a filter.

In this discussion, we are assuming that a given signal's frequency response, or spectrum, can be moved up or down the frequency axis at will. This is true, and is called up or down conversion, and will be discussed in Chapter 11.

9.3 Pulse-Shaping Filter

To accomplish this frequency limiting of the modulated signal, one needs to pass the I and Q signals through a low pass filter. This filter is often called a pulse-shaping filter, and it determines the bandwidth of the modulated signal. But it is not quite that simple. We need to consider what the filter does to the signal in the time domain as well.

Suppose that we use an ideal low pass filter. Let us use our example where symbols are generated at a rate R of 1 MSPS. The period T is the symbol duration, and equal to 1 μs in our example. The relationship between the rate R and symbol period T is:

$$R = 1/T \text{ and } T = 1/R$$

If we alternate with positive and negative I and Q values at each sample interval (this is the worst case in terms of high-frequency content), the rate of change will be 500 kHz. So we will start with a low-pass filter with passband of 500 kHz (See Fig. 9.5).

This filter will have the sinc impulse or time response. The impulse response is shown in red. It has zero crossings at intervals of T seconds and decays slowly. A very long filter is needed to approximate the sinc response. The impulse response of the symbols immediately preceding and following the center symbol is in Fig. 9.6 shown below. The actual transmitted signal will be the sum of all the symbols impulse response (we are just showing three symbols here). If the I or Q sample has a negative value for a particular symbol, then the impulse response for that symbol will be inverted from what is shown below.

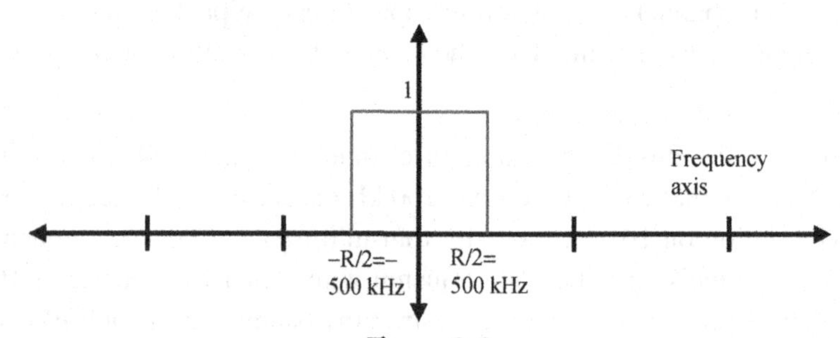

Figure 9.5
Ideal symbol rate filter frequency response.

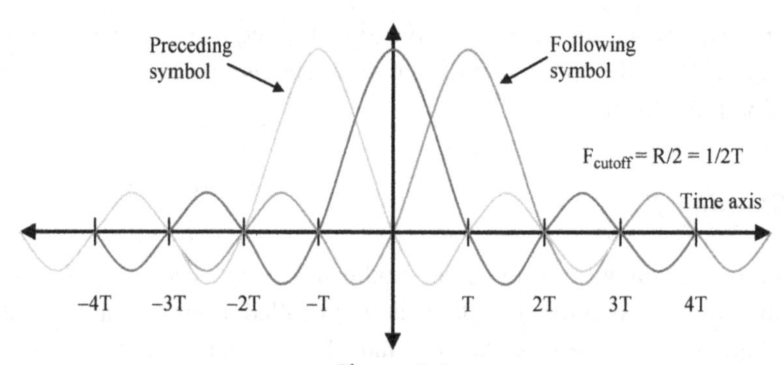

Figure 9.6
Symbol impulse responses.

Think for a moment about the job of the receiver. The receiver is sampling the signal at T intervals in time to detect the symbol values. It must sample at the T intervals shown on the time axis above (leave aside for now the question of how the receiver knows exactly where to sample). At time t = −T, the receiver will be sampling the first symbol. Notice how the two later symbols have zero crossings at t = −T, and so have no contribution at this instant. At t = 0, the receiver will be sampling the value of the second symbol. Again, the other symbols, such as first and third adjacent symbols, have zero crossings at t = 0 and have no contribution. If we were to reduce the bandwidth of the filter to less than 500 kHz (R/2) then in the frequency domain, these pulses would widen (remember that the narrower the frequency spectrum, the longer the time response, and vice versa). This is shown below in Fig. 9.7 if the F_{cutoff} of the pulse shaping filter is narrowed to 250 kHz or R/4.

In this case, notice how at time t = 0, the receiver will be sampling contributions from multiple pulses. At each sampling point of t equal to ...−3T, −2T, −T, 0, T, 2T ... the signal is going to have contributions from many nearby symbols, preventing detection of any specific symbol. This phenomenon is known as intersymbol interference (ISI) and shows that to transmit symbols at a rate R, one needs to have at least R Hz (or 1/T Hz) in the passband frequency spectrum. At baseband, the equivalent two dimensional (complex) spectrum is from −R/2 to +R/2 Hz to avoid creating ISI. Therefore to transmit a 1 MSPS signal over the air, at least 1 MHz of RF frequency spectrum will be required. The baseband filters will need a cutoff frequency of at least 500 kHz.

Notice that the frequency spectrum or bandwidth required depends on the symbol rate, *not* the bit rate. We can have a much higher bit rate, depending on the constellation type used. For example, each 256-QAM symbol carries 8 bits of information, while a QPSK symbol carries only 2 bits of information. But if they both have a common symbol rate, both constellations require the same bandwidth.

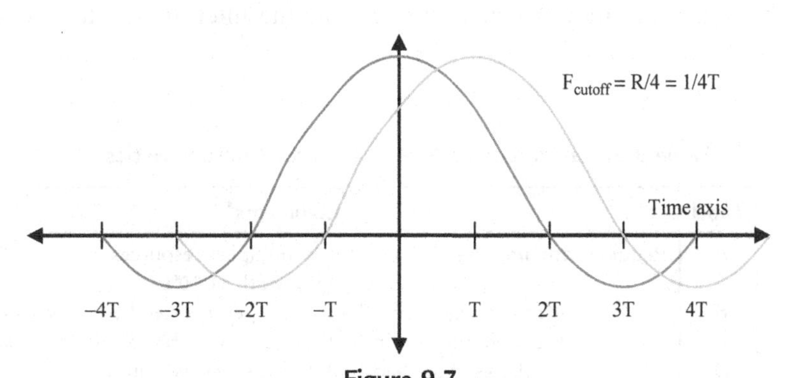

Figure 9.7
Reduced frequency impulse response.

We still have two problems, however. One is that the sinc impulse response decays very slowly, and so will take a long filter (many multipliers) to implement. The second is that although the response of the other symbols does go to zero at the sampling time when $t = N \cdot T$, where N is any integer, we can see visually that if our receiver samples just a little bit to either side, the adjacent symbols will contribute. This makes the receiver symbol detection performance very sensitive to our sampling timing.

Ideally, what we want is an impulse response that still goes to zero at intervals of T, but decays faster, and has lower amplitude lobes or tails. This way, if we sample a bit to one side of the ideal sampling point, the lower amplitude tails will make the unwanted contribution of the neighboring symbols smaller. By making the impulse response decay faster, we can reduce the number of taps and therefore multipliers, required to implement the pulse shaping filter.

9.4 Raised Cosine Filter

There is a type of filter commonly used to meet these requirements. It is called the "raised cosine filter", and it has an adjustable bandwidth, controlled by the "roll-off" factor. The trade-off will be that as bandwidth of the signal will become a bit wider, more frequency spectrum will be required to transmit the signal (See Table 9.3).

The following table summarizes the raised cosine response shown in Fig. 9.8 for different roll-off factors.

In Fig. 9.8 above, the frequency response of the raised cosine filter is shown. To better see the passband shape, it is plotted linearly, rather than logarithmically (dB). It has a cutoff frequency of 500 kHz, the same as our ideal low-pass filter. A raised cosine filter response is wider than the ideal low-pass filter, due to the transition band. This excess frequency bandwidth is controlled by a parameter called the "roll-off" factor. The frequency response is plotted for several different roll-off factors. As the roll-off factor gets closer to zero, the transition becomes steeper, and the filter approaches the ideal low-pass filter.

Table 9.3: Raised cosine Roll-Off Factor Characteristics

Roll-Off Factor	Label	Comments[a]
0.10	A	Requires long impulse response (high multiplier resources), has small frequency excess bandwidth of 10%
0.25	B	A commonly used roll-off factor, excess bandwidth of 25%
0.50	C	A commonly used roll-off factor, excess bandwidth of 50%
1.00	D	Excess bandwidth of 100%, never used in practice

[a]Excess bandwidth refers to percentage of additional bandwidth required compared to ideal low pass filter.

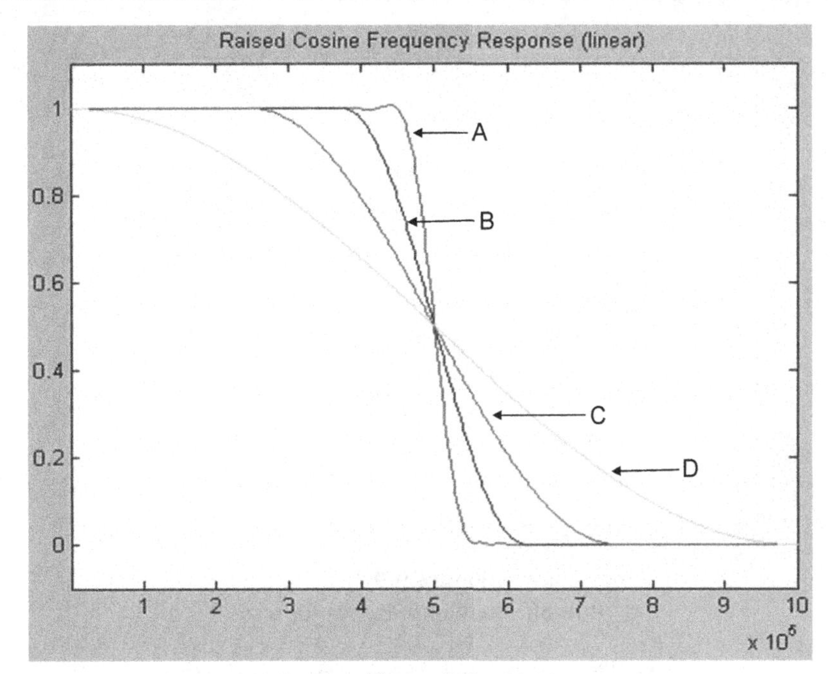

Figure 9.8
Raised cosine roll-off filters frequency response.

The impulse response of the raised cosine filter is shown above. Fig. 9.9 shows the filter impulse response, and Fig. 9.10 zooms in to better show the lobes of the filter impulse. Again, it is plotted for several different roll-off factors. It is similar to the sinc impulse response in that it has zero crossings at time intervals of T (as this is shown in sample domain, rather than time domain, it is not readily apparent from the diagram).

As the excess bandwidth is reduced to approach the ideal low-pass filter frequency response, the lobes in the impulse response become higher, approaching the sinc impulse response. The signal with the smaller amplitude lobes has a larger excess bandwidth or wider spectrum (See Fig. 9.9).

Let us review this idea of pulse-shaping filter again, in light of the Figs. 9.9 and 9.10. We need a pulse shaping filter which will have a zero response at intervals of T in time so that a given symbol's pulse response will not have a contribution to the signal at the sampling times of the neighboring symbols. We would also like to minimize the height of the lobes of the impulse (time) response and have it decay quickly, as to reduce our sensitivity to ISI, if the receiver does not sample precisely at the correct time for each symbol. As the roll-off factor increases, we can see this is exactly what happens in the figures (signal "D"). The impulse response goes to zero very quickly, and the lobes of the filter impulse

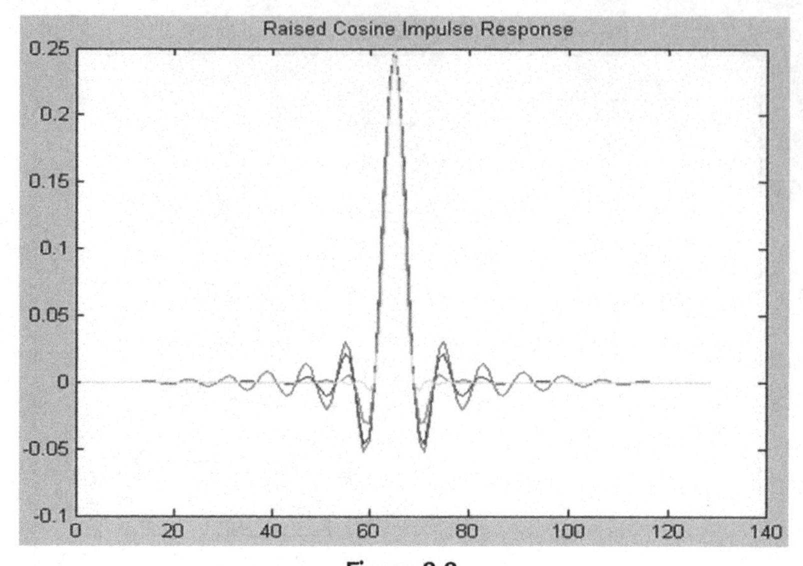

Figure 9.9
Roll-off filters impulse response.

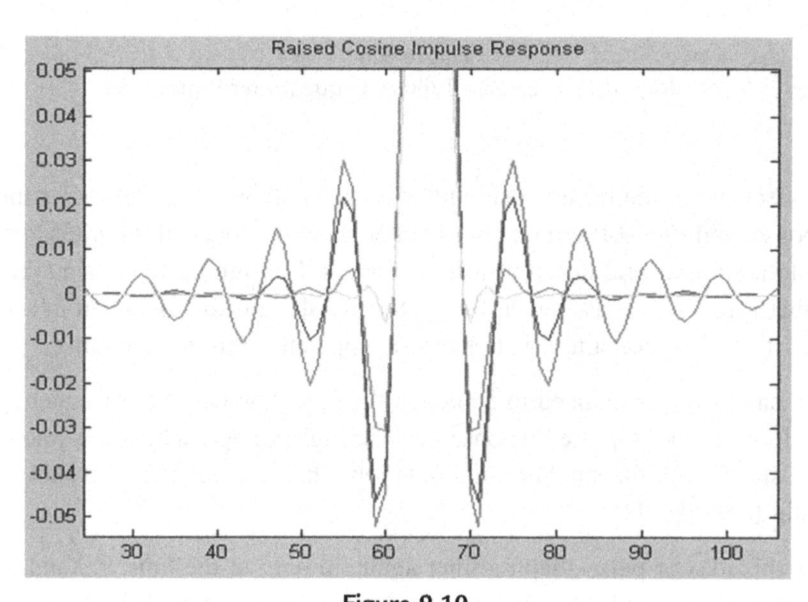

Figure 9.10
Roll-off filters impulse response detail.

response are very small. On the other hand, we have a frequency spectrum, which is excessively wide. A better compromise would be a roll-off factor somewhere between 0.25 and 0.5 (signals "B" and "C"). Here, the impulse response decays relatively quickly with small lobes, requiring a pulse-shaping filter with small number of taps, while still keeping the required bandwidth reasonable.

The roll-off factor controls the compromise between:

- spectral bandwidth requirement,
- length or number of taps of pulse shaping filter, and
- receiver sensitivity to ISI.

Another significant aspect of the pulse-shaping filter is that it is always an interpolating filter. In our figures, this is shown as a four times interpolation filter. If you look carefully at the impulse response in the Fig. 9.10, you can see that the zero crossing are every four samples. This corresponds to $t = N \cdot T$ in the time domain, due to the $4\times$ interpolation.

The pulse-shaping filter must be an interpolating filter, as the I and Q baseband signals must meet the Nyquist criterion. In this example, the symbol rate is 1 MSPS. If we use a high roll-off factor, the baseband spectrum of the I and Q signals can be as high as 1 MHz. So we require a minimum sampling rate of 2 MHz or twice the symbol rate. Therefore, the pulse-shaping filter will need to interpolate by at least a factor of two and is often interpolated quite a bit higher than this, for reasons we will discuss in the digital upconversion chapter.

Once we have our pulse shaped and interpolated I and Q baseband digital signals, we can use digital to analog converters to create the analog I and Q baseband signals. These signals can be used to drive an analog mixer that can create a passband signal. A passband signal is a baseband signal which has been upconverted or mixed with a carrier frequency.

For example, we might use a 0.25 roll-off filter for our 1 MSPS modulator. The baseband I and Q signals will have a bandwidth of 625 kHz. If we use a carrier frequency of 1 GHz, then our transmit signal will require about 1.25 MHz of spectrum centered at 1 GHz.

So far, we have discussed the process which occurs in the transmission path. The receive path is quite similar. The signal is down converted or mixed down to baseband. Again, we will discuss this in more detail in a later chapter. The demodulation process starts with baseband I and Q signals. The receiver is more complex, as it must deal with several additional issues. First of all, there may be nearby signals that can interfere with the demodulation process. These must be filtered out, usually with a combination of analog and digital filters. The final stage of digital filtering is often the same pulse-shaping filter used in the transmitter. This is called a matched filter. The idea is that if the same filter that was used to create the signal is also used to filter the spectrum prior to sampling, we can maximize the amount of signal energy used in the detection (or sampling process). There is a bit of mathematics to prove this, so we will just take it at face value. Due to this idea of using the same filter in the transmitter and receiver, the raised cosine filter is usually modified to a square root raised cosine filter. The frequency response of the raised cosine filter is modified to be the square root of the amplitude across the passband. This also modifies the impulse response as well. This is shown in the figures below, for the same roll-off factors (See Figs. 9.11 and 9.12).

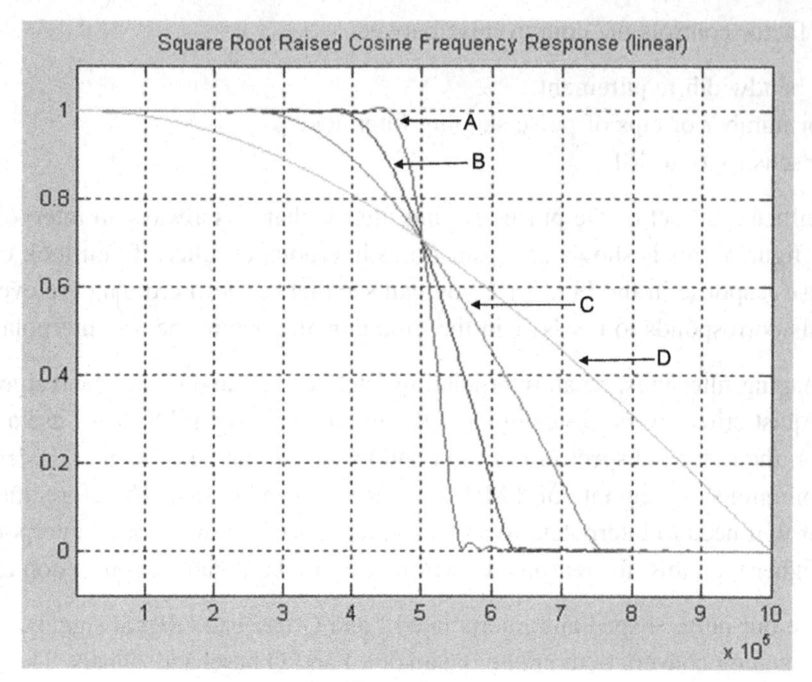

Figure 9.11
Square root raised cosine frequency response.

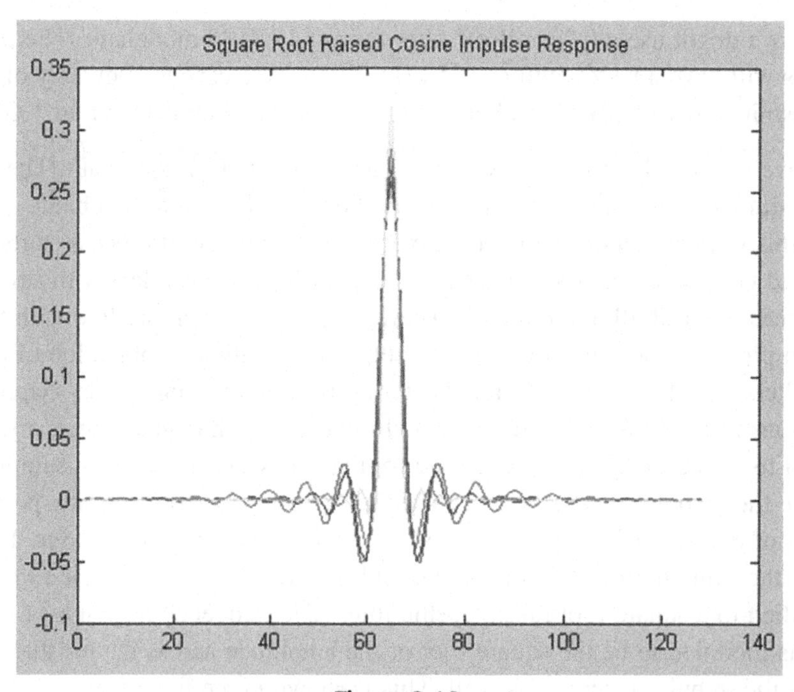

Figure 9.12
Square root raised cosine impulse response.

Since the signal passes through both filters, the net frequency response is the raised cosine filter. After passing through the receive pulse-shaping (also called matched) filter, the signal is sampled. Using the sampled I and Q value, the receiver will choose the constellation point in the I-Q plane closest to the sampled value. The bits corresponding to that symbol are recovered, and if all went well, the receiver has chosen the same symbol point selected by the transmitter. We have very much simplified this whole process, but this is the essence of digital communications.

We can see why it will be easier to have errors when transmitting 64-QAM as compared to QPSK. The receiver has 64 closely spaced symbols to select from in the case of 64-QAM, whereas in QPSK, there are only four widely spaced symbols to select from. This makes 64-QAM systems much more susceptible to ISI, noise or interference. You might think, let us transmit the 64-QAM signal with higher power, as to spread the symbols further apart. This is an effective, but very expensive way, to mitigate the noise and interference which prevents correct detection of the symbol at the receiver. Also, the transmit power is often limited by the regulatory agencies, or the transmitter may be battery powered or have other constraints.

The receiver also has a number of other problems to contend with. We assumed that we always sample at the correct instant in time when one symbol has a nonzero value in the signal. The receiver must somehow determine this correct sampling time, usually by a combination of trial and error during initial part of the reception, and sometimes by having the transmitter send a predetermined (or training) sequence known by both transmitter and receiver. This process is known as acquisition, where the receiver tries to fine-tune the sampling time, the symbol rate, the exact frequency and phase of the carrier and other parameters, which may be needed to demodulate the received signal with a minimum of errors. And once all this information is determined, it must still be tracked, to account for differences in transmit and receive clocks, Doppler shifts due to relative motion between the receiver and transmitter, and changes in the path the signal takes from transmitter to receiver, causing various reflections, distortions, and attenuations.

These problems are what make digital receivers so difficult and interesting to work with. Unfortunately, there is usually a lot of mathematics associated with most receiver algorithms and methods, so we will not go into this in any depth. But later chapters will describe the basic principles of several common types of digital communication systems.

Below in Figs. 9.13 and 9.14 are plots from both a 16-QAM and 64-QAM constellation after being sampled by an actual digital receiver. Each receiver signal has the same average energy. This is from a WiMax wireless system, operating in the presence of noise. But the receiver does manage to do a sufficiently good job at detection so that each of the constellation points is clearly visible. But we can imagine that as the receiver noise level increases, the constellation samples would quickly start to drift together on the 64-QAM

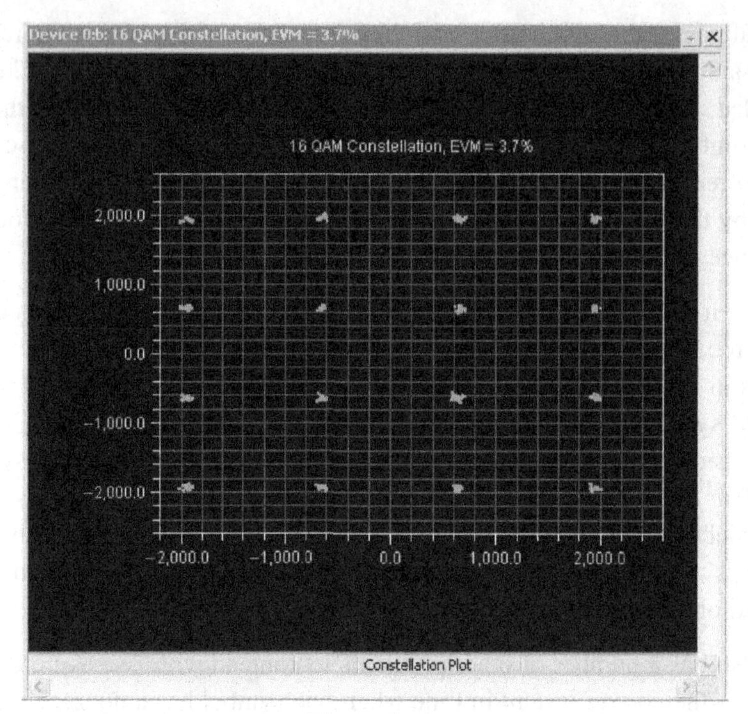

Figure 9.13
16-QAM constellation plot.

constellation, and we would be unable to accurately determine to which constellation point a given symbol should map to. The 16-QAM system is more robust in the presence of additive noise and other impairments, compared to the 64-QAM.

The *mo*dulation and *dem*odulation (modem) ideas presented in this chapter are used in most digital communication systems, including satellite, microwave, cellular (OFDMA, CDMA, and TDMA), wireless LAN (OFDM), DSL, fax, and data dialup modems. Actually, the lowly dialup modem is among the most complicated of all—a V.34 modem can have over 1000 constellation points.

Hopefully, by now the name conventions of the modulation methods is starting to make more sense. In QPSK, all four symbols have the same amplitude. The phase in the complex plane is what distinguishes the different symbols, each of which is located in a different quadrant. For QAM, the amplitude and phase of the symbol is needed to distinguish a particular symbol.

In general, communication systems are full of trade-offs. The most important comes from a famous theorem developed by Claude Shannon, which gives the maximum theoretical data bit rate that can be communicated over a communications channel depending upon

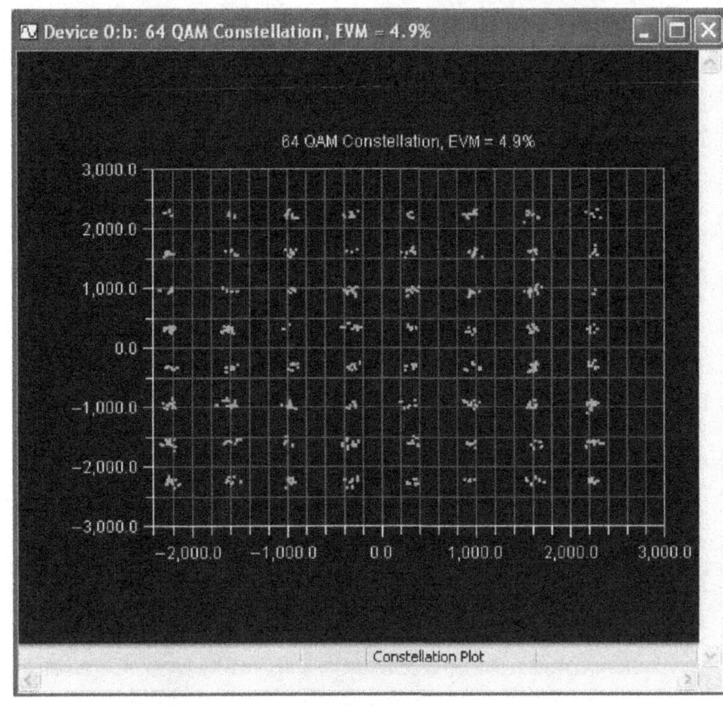

Figure 9.14
64-QAM constellation plot.

bandwidth, transmit power, and receiver noise level. It gives the maximum data rate that can be sent over a channel or link with a given noise level and bandwidth.

This is known as the Shannon limit and is somewhat analogous to the speed of light, which can be approached with ever increasing amounts of cleverness and effort, but can never be exceeded. This will be discussed further in the chapter on error correction codes.

CHAPTER 10

Discrete and Fast Fourier Transforms (DFT, FFT)

Many of us have heard the term "FFT." In this chapter, we are going to discuss the DFT (Discrete Fourier Transform) and its more popular cousin, the FFT (Fast Fourier Transform). We will again try to approach this intuitively, although a bit of math is unavoidable. This chapter will make liberal use of the complex exponential, so if you are not comfortable with this, you may want to review Chapter 4.

The key is to understand the DFT operation. The FFT is simply a highly optimized implementation of the DFT. They both produce identical results. While the implementation of the FFT is interesting, unless you are actually building it, you do not really need to know the details. As the FFT is such a basic building block in DSP, there are many implementations already available for implementation on CPUs, DSPs, FPGAs, and GPUs.

So what is the DFT? To start with, the DFT is simply a transform. It takes a sequence of sampled data (a signal), and computes the frequency content of that sampled data sequence. This will give the representation of the signal in the frequency domain, as opposed to the familiar time domain representation. This is very similar to what we did when computing frequency response of FIR filters. This can be a very powerful tool in signal processing applications, because it allows one to examine any given signal (not just a filter) in the frequency domain, which provides the spectral content of a given signal.

There is also an IDFT (Inverse Discrete Fourier Transform) and IFFT (Inverse Fast Fourier Transform). Again, the IFFT is simply an optimized form of the IDFT. They both compute the time domain representation of the signal from the frequency domain information. Using these transforms, it is possible to go back and forth from the time domain signal and the frequency domain spectral representation.

Conceptually, the DFT tries to take any complex signal, and break it up into a sum of different frequency cosine and sine waves of different frequencies. This is a bit of a novel idea if this is your first exposure to this.

To appreciate what is happening, we are going to examine a few simple examples. This will involve multiplying and summing up complex numbers, which, while it is not difficult, can be tedious. We will minimize the tedium by using a short transform length,

Digital Signal Processing 101. http://dx.doi.org/10.1016/B978-0-12-811453-7.00010-X

but it cannot really be avoided to understand the DFT, and later if you wish to delve in to FFT (the optimized form of DFT).

Let us begin with the equation we used earlier for frequency response in the chapter on FIR filters.

$$\text{Frequency response} = H(\omega) = \sum_{i=-\infty \text{ to } \infty} C_i e^{-j\omega i}$$

Let us see if we can start simplifying. For example, we will decide to perform the calculation over a finite length of sampled data signal "x," which contains N samples, rather than infinite sequence of coefficients C_i. This gets rid of the infinity, and makes this something we can actually build.

$$H(\omega) = \sum_{i=0 \text{ to } N-1} x_i e^{-j\omega i}$$

Next, notice that ω is a continuous variable, which we evaluated over a 2π interval, usually from $-\pi$ to π. Instead, we will be transforming a sampled time domain signal to a sampled frequency domain spectral plot. So rather than computing ω continuously from $-\pi$ to π, we will instead compute ω at M equally spaced points over an interval of 2π. Now it turns out that to avoid aliasing in the frequency domain, we must make $M \geq N$. The reverse transform is the IDFT (or IFFT), which reconstructs the sampled time domain signal of length N, does not require more than N points in the frequency domain. The end result is that we will set $N = M$, and the frequency domain representation will have the same number of points as the time domain representation. We will use the convention "x_i" to represent the time domain version of the signal, and "X_k" to represent the frequency domain representation of the signal. Both indexes i and k will run from 0 to $N - 1$.

10.1 Discrete Fourier Transform and Inverse Discrete Fourier Transform Equations

$$\text{DFT (time} \rightarrow \text{frequency)}\ X_k = H(2\pi k/N) = \sum_{i=0 \text{ to } N-1} x_i e^{-j2\pi ki/N}, \quad \text{for } k = \{0, \ldots, N-1\}$$

$$\text{IDFT (frequency} \rightarrow \text{time)}\ x_i = 1/N \cdot \sum_{k=0 \text{ to } N-1} X_k e^{+j2\pi ki/N}, \quad \text{for } i = \{0, \ldots, N-1\}$$

These equations do appear very similar. The differences are the negative sign on the exponent on the DFT equation, and the factor of 1/N on the IDFT equation. The DFT equation requires that every single sample in the frequency domain has a contribution from each and every one of the time domain samples. And the IDFT equation requires that every single sample in the time domain has a contribution from each and every one of the frequency domain samples. To compute a single sample of either transform requires N complex multiplies and additions.

To compute the entire transform will require computing N samples, for a total of N^2 multiplies and additions. This can become a computational problem when N becomes large. As we will see later, this is the reason the FFT and IFFT were developed.

The values of X_k represent the amount of signal energy at each frequency point. Imagine taking a spectrum of 1 MHz. Then we divide it into N bins. If $N = 20$, then we will have 20 frequency bins, each 50 kHz wide. The DFT output, X_k, will represent the signal energy in each of these bins. For example, X_0 will represent the signal energy at 0 kHz, or the DC component of the signal. X_1 will represent the frequency content of the signal at 50 kHz. X_2 will represent the frequency content of the signal at 100 kHz. X_{19} will represent the frequency content of the signal at 950 kHz.

Now a few comments on these transforms. First, they are reversible. We can take a signal represented by N samples, and perform the DFT on it. We will get N outputs representing the frequency response or spectrum on the signal. If we take this frequency response and perform the IDFT on it, we will get back our original signal of N samples back. Secondly, when the DFT output gives the frequency content of the input signal, it is assuming that the input signal is periodic in N. To put it another way, the frequency response is actually the frequency response of an infinite long periodic signal, where the N long sequence of x_i samples repeat over and over. Lastly, the input signal x_i is usually assumed to be a complex (two dimensional) signal. The frequency response samples X_k are also complex. Often we are more interested in only the magnitude of the frequency response X_k, which can be more easily displayed. But to get back the original complex input x_i using the IDFT, we would need the complex sequence X_k.

At this point, we will do a few examples, selecting $N = 8$.

For our $N = 8$ point DFT, the output gives us the distribution of input signal energy into eight frequency bins, corresponding to the frequencies in the Table 10.1 below. By computing the DFT coefficients X_k, we are performing a correlation, or trying to match,

Table 10.1: 8-Point Discrete Fourier Transform (DFT) Coefficients

k	X_k	Compute by Correlating to Complex Exponential Signal	Δphase Between Each Sample of Complex Exponential Signal
0	X_0	e^0 for $i = 0,1, \ldots, 7$	0
1	X_1	$e^{-j2\pi i/8}$ for $i = 0,1, \ldots, 7$	$-\pi/4$ or $-45°$
2	X_2	$e^{-j4\pi i/8}$ for $i = 0,1, \ldots, 7$	$-2\pi/4$ or $-90°$
3	X_3	$e^{-j6\pi i/8}$ for $i = 0,1, \ldots, 7$	$-3\pi/4$ or $-135°$
4	X_4	$e^{-j8\pi i/8}$ for $i = 0,1, \ldots, 7$	$-4\pi/4$ or $-180°$
5	X_5	$e^{-j10\pi i/8}$ for $i = 0,1, \ldots, 7$	$-5\pi/4$ or $-225°$
6	X_6	$e^{-j12\pi i/8}$ for $i = 0,1, \ldots, 7$	$-6\pi/4$ or $-270°$
7	X_7	$e^{-j14\pi i/8}$ for $i = 0,1, \ldots, 7$	$-7\pi/4$ or $-315°$

our input signal to each of these frequencies. The magnitude DFT output coefficients X_k represent the degree of match of the time domain signal x_i to each frequency component.

10.2 First Discrete Fourier Transform Example

Let us start with a simple time domain signal consisting of {1,1,1,1,1,1,1,1}. Remember, the DFT assumes this signal keeps repeating, so the frequency output will actually be that of an indefinite string of 1s. As this signal is unchanging, then by intuition, we will expect that zero frequency component (DC of signal) is going to be the only nonzero component of the DFT output X_k.

$$\text{Starting with } X_k = \sum_{i=0 \text{ to } N-1} x_i e^{-j2\pi ki/N} \text{ and setting } N = 8 \text{ and all } x_i = 1$$

$$X_k = \sum_{i=0 \text{ to } 7} 1 \cdot e^{-j2\pi ki/8} \text{ and setting } k = 0 \ \left(\text{recall that } e^0 = 1\right)$$

$$X_0 = \sum_{i=0 \text{ to } 7} 1 \cdot 1 = 8$$

Next, evaluate for $k = 1$

$$X_1 = \sum_{i=0 \text{ to } 7} 1 \cdot e^{-j2\pi i/8} = 1 + e^{-j2\pi/8} + e^{-j4\pi/8} + e^{-j6\pi/8} + e^{-j8\pi/8} + e^{-j10\pi/8} + e^{-j12\pi/8} + e^{-j14\pi/8}$$

$$X_1 = 1 + (0.7071 - j0.7071) - j + (-0.7071 - j0.7071) - 1 + (-0.7071 + j0.7071) + j + (0.7071 + j0.7071)$$

$$X_1 = 0$$

The eight terms of the summation for X_1 cancel out. This makes sense if you think about it. This is a sum of eight equally spaced points about the origin on the unit circle of complex plane. The summation of these points must equal the center, in this case zero.

Next, evaluate for $k = 2$

$$X_2 = \sum_{i=0 \text{ to } 7} 1 \cdot e^{-j2\pi i/8} = 1 + e^{-j\pi/2} + e^{-j\pi} + e^{-j3\pi/2} + e^{-j2\pi} + e^{-j5\pi/2} + e^{-j3\pi} + e^{-j7\pi/2}$$

$$X_2 = 1 - j - 1 + j + 1 - j - 1 + j = 0$$

We will find out similarly that X_3, X_4, X_5, X_6, X_7 also are zero. Each of these will represent eight points equally spaced about the unit circle. X_1 has its points spaced at $-45°$ increments, X_2 has its points spaced at $-90°$ increments, X_3 has its points spaced at $-135°$ increments, and so forth (the points may wrap around the unit circle in the frequency domain multiple times). So as we expected, the only nonzero term is X_0, which is the DC term. There is no other frequency content of the signal.

Now, let us use the IDFT to get the original sequence back.

$$x_i = 1/N \cdot \sum_{k=0 \text{ to } N-1} X_k e^{+j2\pi ki/N} \quad \text{for } N = 8 \text{ and } X_0 = 8, \text{ all other } X_k = 0$$

$$x_i = 1/8 \cdot \sum_{k=0 \text{ to } N-1} X_k e^{+j2\pi ki/8}$$

Since $X_0 = 8$ and the rest are zero, we only need to evaluate the summation for $k = 0$.

$$x_i = 1/8 \cdot 8 \cdot e^{+j2\pi 0i/8} = 1$$

This is true for all values of i (the 0 in the exponent means the value of "i" is irrelevant). So we get an infinite sequence of 1s.

In general, however, we would evaluate for i from 0 to $N - 1$. Due to the periodicity of the transform, there is no point in evaluating when $i = N$ or greater. If we evaluate for $i = N$, we will find we get the same value as $i = 0$, and for $i = N + 1$, we will get the same value as $i = 1$.

10.3 Second Discrete Fourier Transform Example

Lets us consider another simple example, with a time domain signal $\{1, j, -1, -j, 1, j, -1, -j\}$. This is actually the complex exponential $e^{+j2\pi i/4}$. This consists of a single frequency, and corresponds to one of the frequency "bins" that the DFT will measure. So we can expect a nonzero DFT output in this frequency bin, but zero elsewhere. Let us see how this works out.

Starting with $X_k = \sum_{i=0 \text{ to } N-1} x_i \, e^{-j2\pi ki/N}$ and setting $N = 8$ and all

$$x_i = \{1, j, -1, -j, 1, j, -1, -j\}$$

$$X_0 = \sum_{i=0 \text{ to } 7} x_i \cdot 1, \quad \text{as } k = 0 \left(e^0 = 1\right)$$

$X_0 = 1 + j - 1 - j + 1 + j - 1 - j = 0$, so the signal has no DC content, as expected. Notice that to calculate X_0 which is the DC content of x_i, the DFT reduces to just summing (essentially averaging) the input samples.

Next, evaluate for $k = 1$

$$X_1 = \sum_{i=0 \text{ to } 7} x_i \cdot e^{-j2\pi i/8} = 1 \cdot 1 + j \cdot e^{-j2\pi/8} - 1 \cdot e^{-j4\pi/8} - j \cdot e^{-j6\pi/8}$$

$$+ 1 \cdot e^{-j8\pi/8} + j \cdot e^{-j10\pi/8} - 1 \cdot e^{-j12\pi/8} - j \cdot e^{-j14\pi/8}$$

$$X_1 = 1 + [j\cdot(0.7071 - j0.7071)] + j - [j\cdot(-0.7071 - j0.7071)]$$
$$- 1 + [j\cdot(-0.7071 + j0.7071)] - j - [j\cdot(0.7071 + j0.7071)]$$
$$X_1 = 0$$

If you take time to work this out, you will see that all eight terms of the summation cancel out. This also happens for X_3, X_4, X_5, X_6, and X_7. Let us look at X_2 now. We will also express x_i using complex exponential format of $e^{+j2\pi i/4}$.

$$X_k = \sum_{i=0 \text{ to } 7} x_i e^{-j2\pi ki/8}$$

$$X_2 = \sum_{i=0 \text{ to } 7} x_i \cdot e^{-j4\pi i/8} = \sum_{i=0 \text{ to } 7} e^{+j2\pi i/4} \cdot e^{-j4\pi i/8} = \sum_{i=0 \text{ to } 7} e^{+j2\pi i/4} \cdot e^{-j2\pi i/4}$$

Remember, that when exponentials are multiplied, the exponents are added ($x^2 \cdot x^3 = x^5$). Here the exponents are the identical, except of opposite sign. So they add to zero.

$$X_2 = \sum_{i=0 \text{ to } 7} e^{+j2\pi i/4} \cdot e^{+j4\pi i/8} \sum_{i=0 \text{ to } 7} e^0 = \sum_{i=0 \text{ to } 7} 1 = 8$$

The sole frequency component of the input signal is X_2. This is because our input is a complex exponential frequency at the exact frequency that X_2 represents.

10.4 Third Discrete Fourier Transform Example

Next, we can try modifying x_i such that we introduce a phase shift, or delay (like substituting a sine wave for a cosine wave). Suppose we introduce a delay, so x_i starts at j instead of 1, but is still the same frequency. The input x_i is still rotating around the complex plane at the same rate, but starts at j (angle of $\pi/2$) rather than 1 (angle of 0). Now the sequence $x_i = \{j,-1,-j,1,j,-1,-j,1\}$ or $e^{+j(2\pi(i+1)/4)}$.

The DFT output will result in X_0, X_1, X_3, X_4, X_5, X_6, and $X_7 = 0$, as before. Changing the phase cannot cause any new frequency to appear in the other bins.

Next, evaluate for $k = 2$.

$$X_k = \sum_{i=0 \text{ to } 7} x_i \, e^{-j2\pi ki/8}$$

$$X_2 = \sum_{i=0 \text{ to } 7} x_i \cdot e^{-j4\pi i/8} = \sum_{i=0 \text{ to } 7} e^{+j((2\pi(i+1)/4)+1)} \cdot e^{-j4\pi i/8}$$

We need to sum the two values of the two exponents.

$$+j(2\pi(i+1)/4) + -j4\pi i/8 = +j2\pi i/4 + j2\pi/4 - j2\pi i/4 = j\pi/2$$

Substituting back this exponent value

$$X_2 = \sum_{i=0 \text{ to } 7} e^{+j((2\pi i/4)+1)} \cdot e^{-j4\pi i/8} = \sum_{i=0 \text{ to } 7} e^{+j\pi/2} = \sum_{i=0 \text{ to } 7} j = j8$$

So we get exactly the same magnitude at the frequency component X_2. The difference is the phase of X_2. So the DFT does not just pick out the frequency components of a signal, but is sensitive to the phase of those components. The phase as well as amplitude of the frequency components X_k can be represented because the DFT output is complex.

The process of the DFT is to correlate the N sample input data stream x_i against N equally spaced complex frequencies. If the input data stream is one of these N complex frequencies, then we will get a perfect match, and get zero in the other $N - 1$ frequencies which do not match. But what happens if we have an input data stream with a frequency in between one of the N frequencies?

To review, we have looked at three simple examples. The first was a constant level signal, so the DFT output was just the zero frequency or DC component. The second example was a complex frequency that matched exactly to one of the frequency bins, X_k, of the DFT. The third was the same complex frequency, but with a phase offset. The fourth will be a complex frequency not matched to one of the N frequencies used by the DFT. Next, we will look at an example where the frequency is somewhere in between the DFT bins.

10.5 Fourth Discrete Fourier Transform Example

We will look at an input signal of frequency $e^{+j2.1\pi i/8}$. This is pretty close to $e^{+j2\pi i/8}$, so we would expect a pretty strong output at X_1. Let us see what the $N = 8$ DFT result is—hopefully the arithmetic is all correct. Slogging through this arithmetic is purely optional; the details are shown to provide a complete example.

Generic DFT equation for $N = 8$: $X_k = \sum_{i=0 \text{ to } 7} x_i e^{-j2\pi ki/8}$

$$X_0 = \sum_{i=0 \text{ to } 7} e^{+j2.1\pi i/8} \cdot 1 = \sum_{i=0 \text{ to } 7} e^{+j2.1\pi i/8}$$

$$= [1 + j0] + [0.6788 + j0.7343]$$

$$+ [-0.0785 + j0.9969] + [-0.7853 + j0.6191] + [-0.9877 - j0.1564]$$

$$+ [-0.5556 - j0.8315] + [0.2334 - j0.9724] + [0.8725 - j0.4886] = 0.3777 - j0.0986$$

$$X_1 = \sum_{i=0 \text{ to } 7} e^{+j2.1\pi i/8} \cdot e^{-j2\pi i/8} = \sum_{i=0 \text{ to } 7} e^{+j0.1\pi i/8}$$

$$= [1 + j0] + [0.9992 + j0.0393]$$

$$+ [0.9969 + j0.0785] + [0.9931 + j0.1175] + [0.9877 + j0.1564] + [0.9808 + j0.1951]$$

$$+ [0.9724 + j0.2334] + [0.9625 + j0.2714] = 7.8925 + j1.0917$$

$$X_2 = \sum_{i=0 \text{ to } 7} e^{+j2.1\pi i/8} \cdot e^{-j4\pi i/8} = \sum_{i=0 \text{ to } 7} e^{-j1.9\pi i/8}$$

$$= [1 + j0] + [0.7343 - j0.6788]$$

$$+ [0.0785 - j0.9969] + [-0.6191 - j0.7853] + [-0.9877 - j0.1564] + [-0.8315 + j0.5556]$$

$$+ [-0.2334 + j0.9724] + [0.4886 + j0.8725] = -0.3703 - j0.2170$$

$$X_3 = \sum_{i=0 \text{ to } 7} e^{+j2.1\pi i/8} \cdot e^{-j6\pi i/8} = \sum_{i=0 \text{ to } 7} e^{-j3.9\pi i/8}$$

$$= [1 + j0] + [0.0393 - j0.9992]$$

$$+ [-0.9969 - j0.0785] + [-0.1175 + j0.9931] + [0.9877 + j0.1564] + [0.1951 - j0.9808]$$

$$+ [-0.9724 - j0.2334] + [-0.2714 + j0.9625] = -0.1362 - j0.1800$$

$$X_4 = \sum_{i=0 \text{ to } 7} e^{+j2.1\pi i/8} \cdot e^{-j8\pi i/8} = \sum_{i=0 \text{ to } 7} e^{-j5.9\pi i/8}$$

$$= [1 + j0] + [-0.6788 - j0.7343]$$

$$+ [-0.0785 + j0.9969] + [0.7853 - j0.6191] + [-0.9877 - j0.1564] + [0.5556 + j0.8315]$$

$$+ [0.2334 - j0.9724] + [-0.8725 + j0.4886] = -0.0431 - j0.1652$$

$$X_5 = \sum_{i=0 \text{ to } 7} e^{+j2.1\pi i/8} \cdot e^{-j10\pi i/8} = \sum_{i=0 \text{ to } 7} e^{-j7.9\pi i/8}$$

$$= [1 + j0] + [-0.9992 - j0.0393]$$

$$+ [0.9969 + j0.0785] + [-0.9931 - j0.1175] + [0.9877 + j0.1564] + [-0.9808 - j0.1951]$$

$$+ [0.9724 + j0.2334] + [-0.9625 - j0.2714] = 0.0214 - j0.1550$$

$$X_6 = \sum_{i=0 \text{ to } 7} e^{+j2.1\pi i/8} \cdot e^{-j12\pi i/8} = \sum_{i=0 \text{ to } 7} e^{-j9.9\pi i/8}$$

$$= [1 + j0] + [-0.7343 + j0.6788]$$

$$+ [0.0785 - j0.9969] + [0.6191 + j0.7853] + [-0.9877 - j0.1564] + [0.8315 - j0.5556]$$

$$+ [-0.2334 + j0.9724] + [-0.4886 - j0.8725] = 0.0849 + -j0.1449$$

$$X_7 = \sum_{i=0 \text{ to } 7} e^{+j2.1\pi i/8} \cdot e^{-j14\pi i/8} = \sum_{i=0 \text{ to } 7} e^{-j11.9\pi i/8}$$

$$= [1 + j0] + [-0.0393 + j0.9992]$$

$$+ [-0.9969 - j0.0785] + [0.1175 - j0.9931] + [0.9877 + j0.1564]$$

$$+ [-0.1951 + j0.9808] + [-0.9724 - j0.2334] + [0.2714 - j0.9625] = 0.1730 - j0.1310$$

That was a bit tedious. But there is some insight to be gained from the results of these simple examples.

Table 10.2 shows how the DFT is able to represent the signal energy in each frequency bin. The first example has all its energy at DC. The second and third examples are complex exponentials at frequency $\omega = \pi/2$ radians/sample, which corresponds to DFT output X_2. The magnitude of the DFT outputs is the same for both examples, since the only difference of the inputs is the phase. The fourth example is the most interesting. In this case, the input frequency is close to $\pi/4$ radians/sample, which corresponds to DFT output X_1. So X_1 does capture most of the energy of the signal. But small amounts of energy spill into other frequency bins, particularly the adjacent bins.

We can increase the frequency sorting ability of the DFT by increasing the value of N. Then each frequency bin is narrower (since the frequency spectrum is divided in N

Table 10.2: Sample Discrete Fourier Transform (DFT) Results

DFT Output Magnitude	$x_i = \{1,1,1,1,1,1,1,1\}$	$x_i = e^{+j2\pi i/4}$	$x_i = e^{+j(2\pi(i+1)/4)}$	$x_i = e^{+j2.1\pi i/8}$
Output X_0	8	0	0	0.39
Output X_1	0	0	0	7.99
Output X_2	0	8	8	0.43
Output X_3	0	0	0	0.23
Output X_4	0	0	0	0.17
Output X_5	0	0	0	0.16
Output X_6	0	0	0	0.17
Output X_7	0	0	0	0.22

Table 10.3: Fast Fourier Transform Computational Efficiency

N	DFT—N^2 Complex Multiplies & Additions	FFT—$N \cdot \log_2 N$ complex Multiplies & Additions	Computational Effort of FFT Compared to DFT
8	64	24	37.50%
32	1024	160	15.62%
256	65,536	2048	3.12%
1024	1,048,576	10,240	0.98%
4096	16,777,216	49,152	0.29%

sections in the DFT). This will result in a given any frequency component being more selectively represented by a particular frequency bin. For example, the frequency response plots of the filters contained in the FIR chapter are computed with a value of N equal to 1024. This means the spectrum was divided into 1024 sections, and the response computed for each particular frequency. When plotted together, this gives a very good representation of the complete frequency spectrum.

Please note this also requires taking a longer input sample stream x_i, equal to N. This in turn, will require much greater number of operations to compute.

At some point, some smart people searched for a way to compute the DFT in a more efficient way. The result is the FFT, or fast Fourier transform. Rather than requiring N^2 complex multiplies and additions, the FFT requires $N \cdot \log_2 N$ complex multiplies and additions. This may not sound like a big deal, but look at the comparison in the Table 10.3 below.

So by using the FFT algorithm on a 1024 point (or sample) input, we are able to reduce the computational requirements to less than 1%, or by two-orders of magnitude, of what the DFT algorithm would require.

10.6 Fast Fourier Transform

Let us start with the calculation of the simplest DFT: N = 2 DFT.

Generic DFT equation for N = 2: $X_k = \sum_{i=0 \text{ to } 1} x_i e^{-j2\pi ki/2}$

$$X_0 = x_0 \cdot e^{-j2\pi 0/2} + x_1 \cdot e^{-j2\pi 0/2}$$

$$X_1 = x_0 \cdot e^{-j2\pi 0/2} + x_1 \cdot e^{-j2\pi 1/2}$$

Simplifying since $e^0 = 1$, we find

$$X_0 = x_0 + x_1$$
$$X_1 = x_0 + x_1 \cdot e^{-j\pi} = x_0 - x_1$$

Next, we will do the 4-point (N = 4) DFT.

Generic DFT equation for N = 4: $X_k = \sum_{i=0 \text{ to } 3} x_i e^{-j2\pi ki/4}$

$$X_0 = x_0 \cdot e^{-j2\pi 0/4} + x_1 \cdot e^{-j2\pi 0/4} + x_2 \cdot e^{-j2\pi 0/4} + x_3 \cdot e^{-j2\pi 0/4}$$

$$X_1 = x_0 \cdot e^{-j2\pi 0/4} + x_1 \cdot e^{-j2\pi 1/4} + x_2 \cdot e^{-j2\pi 2/4} + x_3 \cdot e^{-j2\pi 3/4}$$

$$X_2 = x_0 \cdot e^{-j2\pi 0/4} + x_1 \cdot e^{-j2\pi 2/4} + x_2 \cdot e^{-j2\pi 4/4} + x_3 \cdot e^{-j2\pi 6/4}$$

$$X_3 = x_0 \cdot e^{-j2\pi 0/4} + x_1 \cdot e^{-j2\pi 3/4} + x_2 \cdot e^{-j2\pi 6/4} + x_3 \cdot e^{-j2\pi 9/4}$$

The term $e^{-j2\pi k/4}$ repeats itself with a period of k = 4, as the complex exponential makes a complete circle and begins another. This periodicity means that $e^{-j2\pi k/4}$ is equal when evaluated for k = 0, 4, 8, 12... It is again equal for k = 1, 5, 9, 13... Because of this, we can simplify the last two terms of expressions for X_2 and X_3 (shown in bold below). We can also remove the exponential when it is to power of zero.

$$X_0 = x_0 + x_1 + x_2 + x_3$$

$$X_1 = x_0 + x_1 \cdot e^{-j2\pi 1/4} + x_2 \cdot e^{-j2\pi 2/4} + x_3 \cdot e^{-j2\pi 3/4}$$

$$X_2 = x_0 + x_1 \cdot e^{-j2\pi 2/4} + x_2 \cdot \mathbf{e^{-j2\pi 0/4}} + x_3 \cdot \mathbf{e^{-j2\pi 2/4}}$$

$$X_3 = x_0 + x_1 \cdot e^{-j2\pi 3/4} + x_2 \cdot \mathbf{e^{-j2\pi 2/4}} + x_3 \cdot \mathbf{e^{-j2\pi 1/4}}$$

Now we are going to rearrange the terms of the 4-point (N = 4) DFT. The even and odd terms will be grouped together.

$$X_0 = [x_0 + x_2] + [x_1 + x_3]$$

$$X_1 = \left[x_0 + x_2 \cdot e^{-j2\pi 2/4}\right] + \left[x_1 \cdot e^{-j2\pi 1/4} + x_3 \cdot e^{-j2\pi 3/4}\right]$$

$$X_2 = [x_0 + x_2] + \left[x_1 \cdot e^{-j2\pi 2/4} + x_3 \cdot e^{-j2\pi 2/4}\right]$$

$$X_3 = \left[x_0 + x_2 \cdot e^{-j2\pi 2/4}\right] + \left[x_1 \cdot e^{-j2\pi 3/4} + x_3 \cdot e^{-j2\pi 1/4}\right]$$

Next we will factor x_1 and x_3 to get this particular form.

$$X_0 = [x_0 + x_2] + [x_1 + x_3]$$

$$X_1 = \left[x_0 + x_2 \cdot e^{-j2\pi 2/4}\right] + \left[x_1 \cdot e^{-j2\pi 1/4} + x_3 \cdot e^{-j2\pi 3/4}\right]$$

$$= \left[x_0 + x_2 \cdot e^{-j2\pi 2/4}\right] + \left[x_1 + x_3 \cdot e^{-j2\pi 2/4}\right] \cdot e^{-j2\pi/4}$$

$$X_2 = [x_0 + x_2] + \left[x_1 \cdot e^{-j2\pi 2/4} + x_3 \cdot e^{-j2\pi 2/4}\right]$$
$$= [x_0 + x_2] + [x_1 + x_3] \cdot e^{-j2\pi 2/4}$$
$$X_3 = \left[x_0 + x_2 \cdot e^{-j2\pi 2/4}\right] + \left[x_1 \cdot e^{-j2\pi 3/4} + x_3 \cdot e^{-j2\pi 1/4}\right]$$
$$= \left[x_0 + x_2 \cdot e^{-j2\pi 2/4}\right] + \left[x_1 + x_3 \cdot e^{-j2\pi 2/4}\right] \cdot e^{-j2\pi 3/4}$$

Here is the result:

$$\mathbf{X_0 = [x_0 + x_2] + [x_1 + x_3]}$$
$$\mathbf{X_1 = [x_0 + x_2 \cdot e^{-j2\pi 2/4}] + [x_1 + x_3 \cdot e^{-j2\pi 2/4}] \cdot e^{-j2\pi/4}}$$
$$\mathbf{X_2 = [x_0 + x_2] + [x_1 + x_3] \cdot e^{-j4\pi/4}}$$
$$\mathbf{X_3 = [x_0 + x_2 \cdot e^{-j2\pi 2/4}] + [x_1 + x_3 \cdot e^{-j2\pi 2/4}] \cdot e^{-j6\pi/4}}$$

Now comes the insightful part. Comparing the four equations above, you can see that the bracketed terms used for X_0 and X_1 are also present in X_2 and X_3. So we do not need to recompute these terms during the calculation of X_2 and X_3. We can simply multiply them by the additional exponential outside the brackets. This reusing of partial products in multiple calculations is the key to understanding the FFT efficiency, so at the risk of being repetitive, this is shown again more explicitly below.

$$\rightarrow \text{define } A = [x_0 + x_2], B = [x_1 + x_3]$$

$$X_0 = [x_0 + x_2] + [x_1 + x_3] = A + B$$

$$\rightarrow \text{define } C = \left[x_0 + x_2 \cdot e^{-j2\pi 2/4}\right], D = \left[x_1 + x_3 \cdot e^{-j2\pi 2/4}\right]$$

$$X_1 = \left[x_0 + x_2 \cdot e^{-j2\pi 2/4}\right] + \left[x_1 + x_3 \cdot e^{-j2\pi 2/4}\right] \cdot e^{-j2\pi/4} = C + D \cdot e^{-j2\pi/4}$$

$$X_2 = [x_0 + x_2] + [x_1 + x_3] \cdot e^{-j4\pi/4} = A + B \cdot e^{-j4\pi/4}$$

$$X_3 = \left[x_0 + x_2 \cdot e^{-j2\pi 2/4}\right] + \left[x_1 + x_3 \cdot e^{-j2\pi 2/4}\right] \cdot e^{-j6\pi/4} = C + D \cdot e^{-j6\pi/4}$$

This process quickly gets out of hand for anything larger than a 4-point ($N = 4$) FFT. So we are going to use a type of representation called a flow graph, shown in Fig. 10.1 below.

The flow graph is an equivalent way of representing the equations, and moreover, represents the actual organization of the computations. You should check for yourself in the simple example above that the flow graph gives the same results as the DFT equations. For example, $X_0 = x_0 + x_1 + x_2 + x_3$, and by examining the flow graph, it is apparent that $X_0 = A + B = [x_0 + x_2] + [x_1 + x_3]$, which is the same result. The order of computations

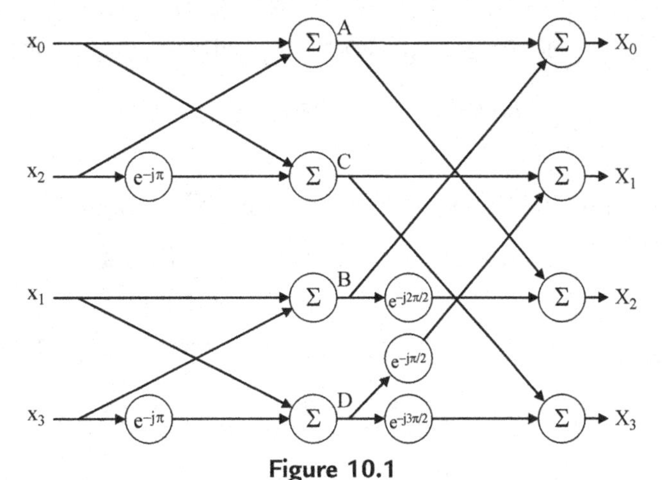

Figure 10.1
Fast Fourier transforms flow graph or butterflies.

would be to compute pairs {A,C} and {B,D} in the first stage. The next stage would be to compute $\{X_0, X_2\}$ and $\{X_1,X_3\}$.

These stages (our example above has two stages) are composed of "butterflies." Each butterfly has two complex inputs and two complex outputs. The butterfly involves one or two complex multiplications and two complex additions. In the first stage, there are two butterflies to compute the two pairs {A,C} and {B,D}. In the second stage, there are two butterflies to compute the two pairs $\{X_0, X_2\}$ and $\{X_1,X_3\}$. The complex exponentials multiplying the data path are known as "twiddle factors." In higher N count FFTs, they are simply sine and cosine values. These are usually stored in a table.

Although you may grumble, next we are going to present an 8-point FFT.

Generic DFT equation for $N = 8$: $X_k = \sum_{i=0 \text{ to } 7} x_i e^{-j2\pi ki/8}$

$$X_0 = x_0 \cdot e^{-j2\pi 0/8} + x_1 \cdot e^{-j2\pi 0/8} + x_2 \cdot e^{-j2\pi 0/8} + x_3 \cdot e^{-j2\pi 0/8}$$
$$+ x_4 \cdot e^{-j2\pi 0/8} + x_5 \cdot e^{-j2\pi 0/8} + x_6 \cdot e^{-j2\pi 0/8} + x_7 \cdot e^{-j2\pi 0/8}$$

$$X_1 = x_0 \cdot e^{-j2\pi 0/8} + x_1 \cdot e^{-j2\pi 1/8} + x_2 \cdot e^{-j2\pi 2/8} + x_3 \cdot e^{-j2\pi 3/8}$$
$$+ x_4 \cdot e^{-j2\pi 4/8} + x_5 \cdot e^{-j2\pi 5/8} + x_6 \cdot e^{-j2\pi 6/8} + x_7 \cdot e^{-j2\pi 7/8}$$

$$X_2 = x_0 \cdot e^{-j2\pi 0/8} + x_1 \cdot e^{-j2\pi 2/8} + x_2 \cdot e^{-j2\pi 4/8} + x_3 \cdot e^{-j2\pi 6/8}$$
$$+ x_4 \cdot e^{-j2\pi 8/8} + x_5 \cdot e^{-j2\pi 10/8} + x_6 \cdot e^{-j2\pi 12/8} + x_7 \cdot e^{-j2\pi 14/8}$$

$$X_3 = x_0 \cdot e^{-j2\pi 0/8} + x_1 \cdot e^{-j2\pi 3/8} + x_2 \cdot e^{-j2\pi 6/8} + x_3 \cdot e^{-j2\pi 9/8}$$
$$+ x_4 \cdot e^{-j2\pi 12/8} + x_5 \cdot e^{-j2\pi 15/8} + x_6 \cdot e^{-j2\pi 18/8} + x_7 \cdot e^{-j2\pi 21/8}$$

$$X_4 = x_0 \cdot e^{-j2\pi 0/8} + x_1 \cdot e^{-j2\pi 4/8} + x_2 \cdot e^{-j2\pi 8/8} + x_3 \cdot e^{-j2\pi 12/8}$$
$$+ x_4 \cdot e^{-j2\pi 16/8} + x_5 \cdot e^{-j2\pi 20/8} + x_6 \cdot e^{-j2\pi 24/8} + x_7 \cdot e^{-j2\pi 28/8}$$

$$X_5 = x_0 \cdot e^{-j2\pi 0/8} + x_1 \cdot e^{-j2\pi 5/8} + x_2 \cdot e^{-j2\pi 10/8} + x_3 \cdot e^{-j2\pi 15/8}$$
$$+ x_4 \cdot e^{-j2\pi 20/8} + x_5 \cdot e^{-j2\pi 25/8} + x_6 \cdot e^{-j2\pi 30/8} + x_7 \cdot e^{-j2\pi 35/8}$$

$$X_6 = x_0 \cdot e^{-j2\pi 0/8} + x_1 \cdot e^{-j2\pi 6/8} + x_2 \cdot e^{-j2\pi 12/8} + x_3 \cdot e^{-j2\pi 18/8}$$
$$+ x_4 \cdot e^{-j2\pi 24/8} + x_5 \cdot e^{-j2\pi 30/8} + x_6 \cdot e^{-j2\pi 36/8} + x_7 \cdot e^{-j2\pi 42/8}$$

$$X_7 = x_0 \cdot e^{-j2\pi 0/8} + x_1 \cdot e^{-j2\pi 7/8} + x_2 \cdot e^{-j2\pi 14/8} + x_3 \cdot e^{-j2\pi 21/8}$$
$$+ x_4 \cdot e^{-j2\pi 28/8} + x_5 \cdot e^{-j2\pi 35/8} + x_6 \cdot e^{-j2\pi 42/8} + x_7 \cdot e^{-j2\pi 49/8}$$

The corresponding flow graph is shown in Fig. 10.2 below. Again, you are encouraged to try a few calculations, and verify that the FFT flow diagram gives the same results as the DFT equations. Now we can see why the FFT is effective in reducing the number of computations. Each time the FFT doubles in size (N increases by factor of 2), we need to add one more stage. For 4-point FFT, we require two stages. For 8-point FFT, we require three stages. For a 16-point FFT, four stages would be required, and so on. The amount of computations required for each stage is proportional to N. The required number of stages is equal to $\log_2 N$. Therefore, the FFT computational load increases by $N \log_2 N$. The DFT computational load increases as N^2.

This is also the reason why FFT sizes are almost always in powers of 2 (2, 4, 8, 16, 32, 64...). This is sometimes called a "radix 2" FFT. So rather than a 1000-point FFT, one will see a 1024-point FFT. In practice, this common restriction to powers of 2 is not a problem.

Another fairly common FFT implementation is the "radix 4" FFT. In this case, the butterfly has four complex inputs and four complex outputs. Although it does not reduce the number of operations, this may sometimes be more efficient to implement in hardware or software. The FFT size would be restricted to powers of 4 in this case (4, 16, 64, 256, 2048...).

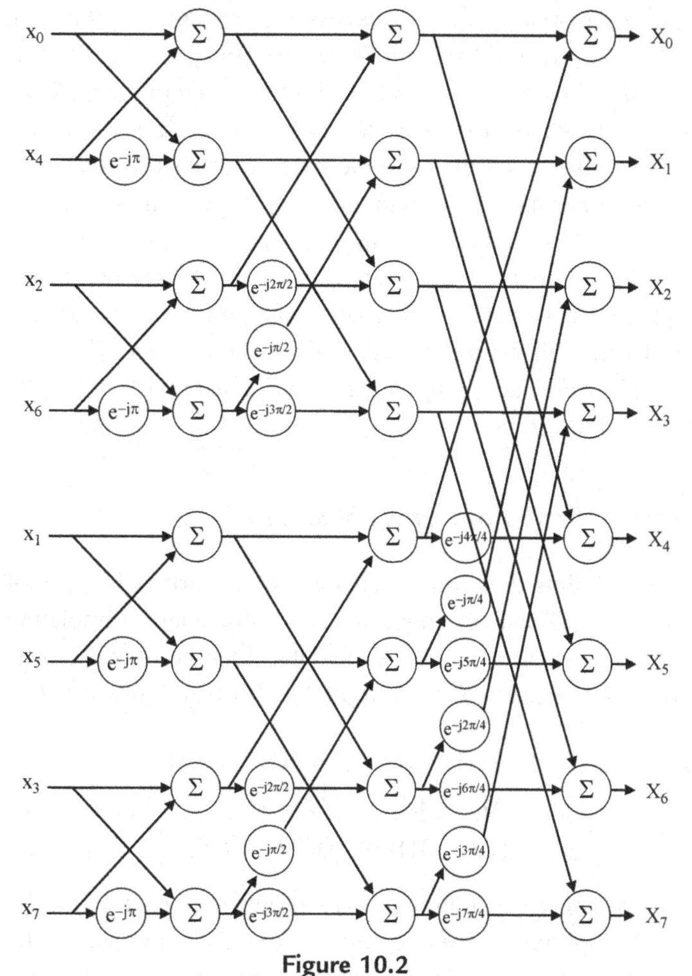

Figure 10.2
8-point fast Fourier transform (FFT) diagram.

10.7 Filtering Using the Fast Fourier Transform and Inverse Fast Fourier Transform

The FFT algorithm is one of the heavily used in many DSP applications. It is used whenever the signal needs to be processed in the spectral, or frequency domain. It is so efficient to implement, that sometimes even FIR filtering functions are performed using an FFT. This can be advantageous when the data is processed in batches. Rather than shifting the input data past a series of multipliers with filter coefficients, a buffer of N input data samples can be transformed into the frequency domain using the FFT. This will create an N sample spectral representation of the input data. The coefficients represent the impulse, or time response of the filter. This filter also has also has a corresponding frequency response, as was discussed in earlier chapters. The spectral representation of the input created by the FFT can be multiplied by the filter frequency response, and the result then converted back to the time domain using the IFFT. This seems like a roundabout way, but this method often requires less work than traditional FIR filtering. In an FIR filter, each data sample must slide past all the coefficients, resulting in M multiplies per data sample in an M tap filter. This process is known as convolution. Whereas in the frequency domain, the entire input spectral response is simply multiplied by the filter frequency response, which is only one multiply per input data sample. This process is performed on groups of every N input data samples in a continuous manner. Even with the additional work of the FFT and IFFT, the net result is still less computational effort than using an FIR filter, particularly when N is chosen to be reasonably large, such as 1024.

10.8 Bit Growth in Fast Fourier Transforms

As is apparent in the FFT flow diagrams, each stage of butterflies has complex exponential multiplies and summations. These are important when considering implementation. First of all, the multipliers are complex, and so is the data. This requires use of four real multipliers, and two adder operations per complex multiply, as shown below.

$$
\begin{aligned}
(A + jB)\cdot(C + jD) &= A\cdot C + jB\cdot C + A\cdot jD + jB\cdot jD \\
&= AC + jBC + jAD - BD \\
&= (AC - BD) + j(BC + AD)
\end{aligned}
$$

Secondly, the summation of the real and complex products in each butterfly can result in a doubling of the signal amplitude if the two signals have the same phase (the multiplies do not increase signal amplitude because the coefficients are all sine and cosine values, which have a range from -1 to 1). This is known as FFT bit growth. It is 1 bit per stage in a Radix 2 FFT, and 2 bits per stage in a Radix 4 FFT. This must be considered in fixed pint arithmetic. It can be most easily compensated for by simply by shifting all the butterfly results right, or dividing by 2, before processing the next stage of butterflies. This ensures there can never be an overflow due to the summations. This is common when implementing fixed point FFTs using a processor-based architecture. However, for small input signals, this truncation of the LSB at each stage can raise the quantization noise floor, leading to a loss in precision, especially in large FFTs.

In a hardware-based architecture, there is more flexibility. In particular, if the multiplier precision for one input operand can be increased, then the precision loss can be avoided. Note that the complex exponential always has a magnitude of one, so the other operand of the multiplier is not required to grow. This is why sometimes asymmetrical-sized multipliers are used in FFT applications.

Another solution to this issue is to use floating point arithmetic, which can allow for very large dynamic ranges. This is less common except in applications where the FFT precision (very low noise floor) is extremely important or when the size of the FFT is very high, due to the high implementation cost of floating point. For example, a 2^{20} point FFT (1 million points) would require double precision floating point to implement the required precision in both the complex data and complex exponential coefficients.

10.9 Bit Reversal Addressing

One last discussion point is "bit-reversal." Notice the order of the x_i inputs to the 8-point FFT. This order is "bit-reversed" from normal sequential order (see Table 10.4 below) to form the symmetry needed in the FFT structure. This bit-reversal addressing of the input order can be easily implemented in hardware by crossing, or reversing the order of address bits. DSP processors can also perform this operation, usually by using a special bit reversing addressing mode.

In practice, the FFT can be set up to have either the input sequence bit-reversed, or the output sequence bit-reversed (but not both). If the input is sequential, then the output will be bit-reversed. If the input is bit-reversed (as in our examples), then the output will be sequential. Either situation provides the needed symmetry. If the input (time domain) sequence is bit-reversed, this called "decimation in time." If on the other hand, it is chosen to have the output (frequency domain) sequence bit-reversed, then this called "decimation in frequency." You may need to be aware of this vernacular when using FFT software or IP modules. Also, most DSP processors have special instructions that support bit reversed addressing for use in FFT implementations.

Table 10.4: Bit Reversal Example

Bit-Reversed Input (Decimal)	Bit-Reversed Input Index (Binary)	Sequential Input Index (Binary)	Sequential Input Index (Decimal)
0	000	000	0
4	100	001	1
2	010	010	2
6	110	011	3
1	001	100	4
5	101	101	5
3	011	110	6
7	111	111	7

CHAPTER 11

Digital Upconversion and Downconversion

Previously, we talked about complex modulation and baseband signals. The whole point of this is to create a signal that can be used to carry the information bits from one location to another, whether over copper wire, a fiber optic cable, or electromagnetically through the air. In nearly all cases, the signal needs to ride on a carrier frequency to be efficiently sent from one location to another. To do this, the signal frequency spectrum needs the ability to be moved up and down the frequency axis, at will. This is a process of up and down (frequency) conversion. All early methods used analog circuits to accomplish this. In the last couple of decades, digital circuits, particularly FPGAs, have developed the computational capacity to perform these functions in many cases and offer important advantages over analog methods. This is known as digital upconversion and downconversion (also known by acronyms of DUC and DDC within the industry).

The process of upconversion is to take a signal that is at baseband (this means the frequency representation of the signal) and move or shift that frequency spectrum up to a carrier frequency. The width of the signal's frequency spectrum does not change; it is just moved to another part of the frequency spectrum. One common area of confusion is what happens to the negative part of the frequency spectrum. In a baseband signal, the negative frequency components overlay the positive components. Positive and negative frequencies are distinguished using complex representation. When the signal is upconverted, the positive and negative components are "unfolded" from on top of each, with the negative components below the carrier and the positive components above the carrier frequency.

A common example is speech. When you speak into a microphone, the sound waves create an electrical baseband signal, with frequency content from near 0 to about 3 kHz. With music, the frequency range can be much greater, up to 20 kHz. When we hear, the vibrations our ears detect are within this frequency range (in fact, our ears cannot hear frequencies beyond this range). This signal can be sampled and converted into a digital baseband signal or remain as an analog signal. To transmit this signal over the air using electromagnetic radio waves, the frequency needs to be increased. For example, the commercial AM radio system in the United States operates between 540 and 1600 kHz. The baseband speech signal is multiplied, or "mixed" with the much higher frequency carrier signal. This process superimposes the baseband signal on top of the carrier. For example, if the carrier frequency is 600 kHz, the upconverted 3 kHz speech signal will now ideally have a frequency or spectral content between $600 - 3$ and $600 + 3$ kHz. This

Digital Signal Processing 101. http://dx.doi.org/10.1016/B978-0-12-811453-7.00011-1

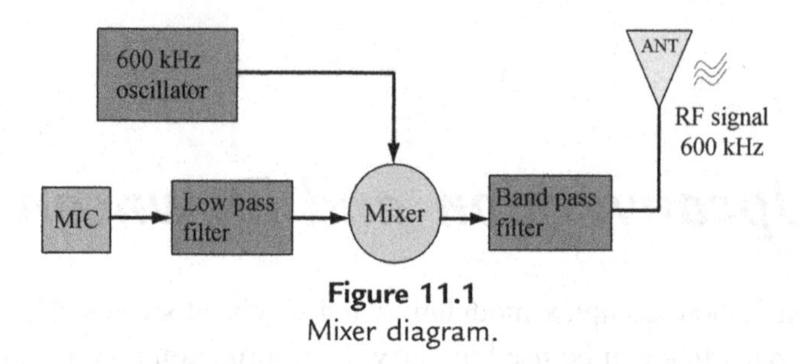

Figure 11.1
Mixer diagram.

process also allows for multiple users to transmit and receive signals simultaneously, as each user can be assigned a different carrier frequency, and occupy different portions of the frequency spectrum. The up and downconversion process can place the baseband signal at any desired "channel" frequency, allowing many different signals to occupy a common frequency band or range without interference. In the AM radio example, the carrier frequencies are spaced 10 kHz apart. This frequency separation is sufficient to prevent interference between stations.

The process of mixing a baseband signal onto a carrier frequency is called upconversion and is performed in the radio transmitter. In the radio receiver, the signal is brought back down to baseband in a process called downconversion. Traditionally, this up and downconversion process was done using analog signals and analog circuits. Analog upconversion with real (not complex) signals is depicted in Fig. 11.1 below.

To see why a mixer works, let us consider a simple baseband signal of 1 kHz tone and a carrier frequency of 600 kHz. Each can be represented as a sinusoid. A real signal, such as a baseband cosine, has both a positive and negative frequency component. At baseband, these overlay each other, so this is not obvious. But once upconverted, the two components can be readily seen both above and below the carrier frequency.

The equation for the upconversion mixer is

$$\cos(\omega_{\text{carrier}} t)\cdot\cos(\omega_{\text{signal}} t) = 1/2\cdot[\cos((\omega_{\text{carrier}} + \omega_{\text{signal}})\cdot t) + \cos((\omega_{\text{carrier}} - \omega_{\text{signal}})\cdot t)]$$

or

$$\cos(2\pi\cdot 600000\cdot t)\cdot\cos(2\pi\cdot 1000\cdot t) = 1/2\cdot[\cos(2\pi\cdot 599000\cdot t) + \cos(2\pi\cdot 601000\cdot t)]$$

The result is two tones, of half amplitude, at 599 and 601 kHz.

11.1 Digital Upconversion

Alternately, this process can also be done digitally. Let us assume that the information content, whether it is voice, music, or data, is in a sampled digital form. In fact, as we covered in the earlier chapter on modulation, this digital signal is often in a complex constellation form, such as QPSK or QAM, for example.

To transmit this information signal, at some point, it must be converted to analog domain. In the past, the conversion from digital to analog occurred while the signal was in baseband form, because the data converters could not handle higher frequencies. As the speeds and capabilities of analog-to-digital converters (ADC) and digital-to-analog converters (DAC) improved, it has became possible to perform the up and downconversion digitally, using a digital carrier frequency. The upconverted signal, which has much higher frequency content, can then be converted to analog form using a high speed DAC.

Upconversion is accomplished by multiplying the complex baseband signal (with I and Q quadrature signals) with a complex exponential of frequency equal to the desired carrier frequency.

The complex carrier sinusoid can be generated using a lookup table, or implemented using any circuit capable of generating two sampled sinusoids offset by 90°degrees (Fig. 11.2). To do this digitally, the sample rates of the baseband and carrier sinusoid signal must be equal. Since the carrier signal will usually be of much higher frequency than the baseband signal, the baseband signal will have to be interpolated, or upsampled, to match the sample frequency of the carrier signal. Then the mixing or upconversion process will result in the frequency spectrum shift depicted in Fig. 11.3 below.

To simplify things, let us assume the output of the modulator is a complex sinusoid of 1 kHz. The equation for this upconversion process is

$$\begin{aligned}
&[\cos(\omega_{\text{carrier}}t) + j\sin(\omega_{\text{carrier}}t)]\cdot[\cos(2\pi\cdot 1000\cdot t) + j\sin(2\pi\cdot 1000\cdot t)] \\
&= 1/2\cdot[\cos((\omega_{\text{carrier}} - 2\pi\cdot 1000)\cdot t) + \cos((\omega_{\text{carrier}} + 2\pi\cdot 1000)\cdot t)] \\
&-1/2\cdot[\cos((\omega_{\text{car6rier}} - 2\pi\cdot 1000)\cdot t) - \cos((\omega_{\text{carrier}} + 2\pi\cdot 1000)\cdot t)] \\
&+j\cdot 1/2\cdot[\sin((\omega_{\text{carrier}} + 2\pi\cdot 1000)\cdot t) + \sin((\omega_{\text{carrier}} - 2\pi\cdot 1000)\cdot t)] \\
&+j\cdot 1/2\cdot[\sin((\omega_{\text{carrier}} + 2\pi\cdot 1000)\cdot t) + \sin((-\omega_{\text{carrier}} + 2\pi\cdot 1000)\cdot t)]
\end{aligned}$$

After simplifying, this becomes

$$\cos((\omega_{\text{carrier}} + 2\pi\cdot 1000)\cdot t) + j\cdot\sin((\omega_{\text{carrier}} + 2\pi\cdot 1000)\cdot t)$$

We can discard the imaginary portion. It is not needed because at carrier frequencies, both positive and negative baseband frequency components can be represented by spectrum above and below the carrier frequency.

The final result at the output of the DAC is

$$\cos((\omega_{\text{carrier}} + 2\pi\cdot 1000)\cdot t)$$

Note that there is only a frequency component above the carrier frequency. This is because the input was a complex sinusoid rotating in the positive (counterclockwise) direction—there was no negative frequency component in this baseband signal.

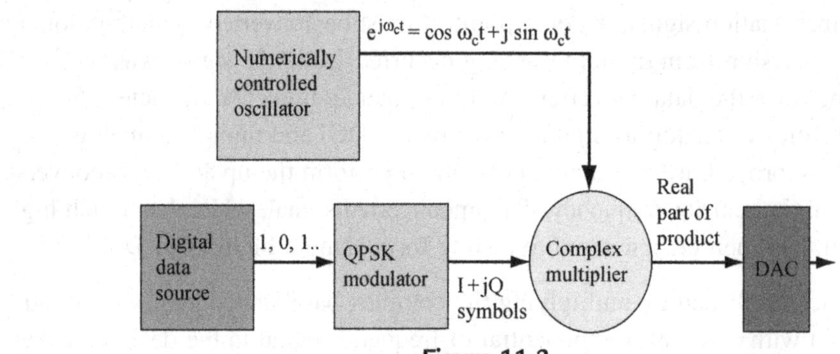

Figure 11.2
Digital upconversion diagram.

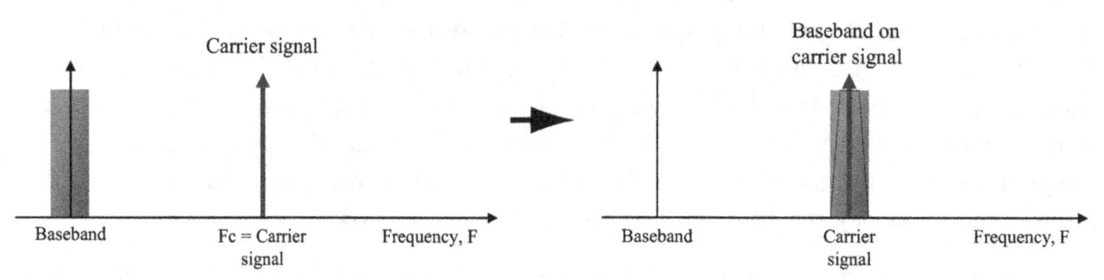

Figure 11.3
Digital upconversion in frequency domain.

Normally, baseband signal will have both positive and negative frequency components. For example, the complex QPSK modulator output can jump both clockwise and counterclockwise depending upon the input data sequence. When upconverted, the positive and negative baseband components will no longer overlay each other, but be unfolded on either side of the carrier frequency. The baseband signal with a frequency spectrum of 0−10 kHz will occupy a total of 20 kHz, with 10 kHz on either side of the carrier frequency. But keep in mind that the baseband signal represents the positive and negative frequencies using quadrature form, so it is in the form of an I and Q signal, each with a frequency spectrum from 0 to 10 kHz.

Frequently, there are several steps in upconverting to the final frequency used for transmission, as shown in Fig. 11.4. There are several advantages in upconverting in steps, which is known as superheterodyne. Generally, the DAC must operate at least 2 ½ times the carrier frequency. For signals in the GHz range, this exceeds the capacity of most DACs (although very high performance DACs can now operate at several GHz conversion rates). Another consideration is filtering. If several upconverting steps are used, filtering can be applied at each step. Often, multiple stages of filtering are required to meet the

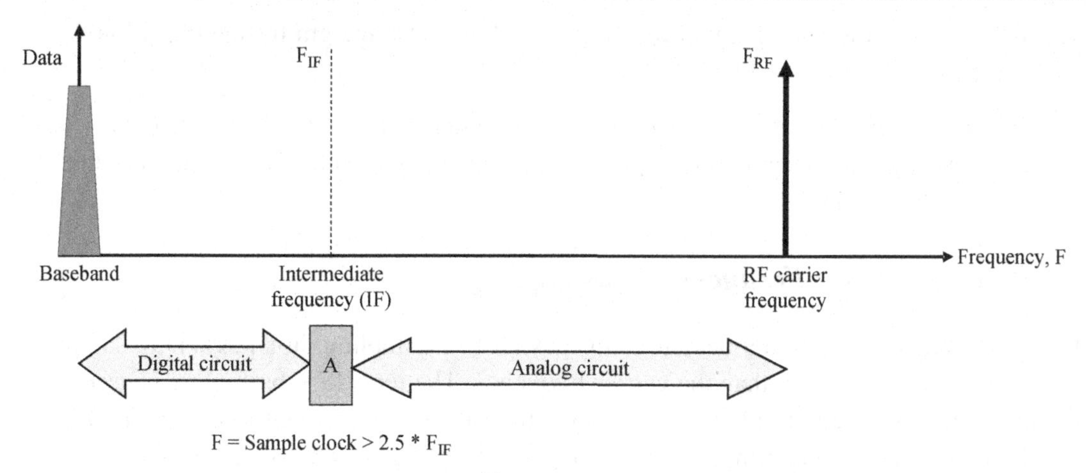

Figure 11.4
Superheterodyne upconversion frequencies.

system requirements in terms of meeting the spectral emission requirements in the transmitted RF signal, to ensure no interference with other users. A typical digital upconverting circuits is shown below (Fig 11.4).

11.2 Digital Downconversion

Digital downconversion is the opposite of upconversion. The circuit diagram looks similar (Fig 11.5).

Digital downconversion involves sampling, so naturally aliasing needs to be considered. The Nyquist sampling rule states that the sampling frequency must be at least twice the highest frequency of signal being sampled. In practice, usually the sampling frequency is

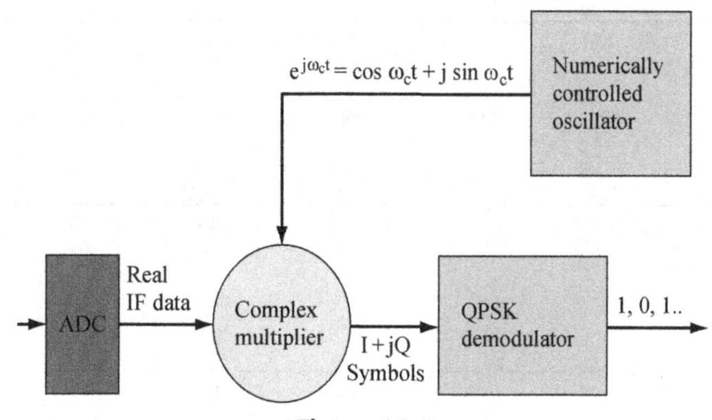

Figure 11.5
Digital downconversion diagram.

at least 2 ½ times the rate of signal frequency, to allow extra margin for the transition band of the digital low-pass filters following.

But this is not quite true. It is possible to sample a signal at frequency lower than its carrier frequency. The Nyquist rule applies to the actual bandwidth of the signal, not the frequency of the carrier.

11.3 Intermediate Frequency Subsampling

Using a technique called intermediate frequency (IF) subsampling, it is possible to sample at a much lower frequency than the carrier frequency. The term IF subsampling is used, as the frequencies typically used in this technique lie somewhere between baseband and RF. The term "IF" refers to intermediate frequency. In this case, we are going to deliberately take advantage of an alias of the signal of interest. This is best illustrated using an example.

We will use a 4G (fourth generation) wireless example. The signal of interest lies at 2500 MHz. Analog circuits are used to downconvert the signal to an IF or intermediate frequency. We will assume that we have an ADC sampling at 200 MHz. Further, let us also assume we have 20 MHz BW IF signal centered at 60 MHz in Fig. 11.6, which we are trying to sample and downconvert for baseband processing. By the Nyquist rule, we

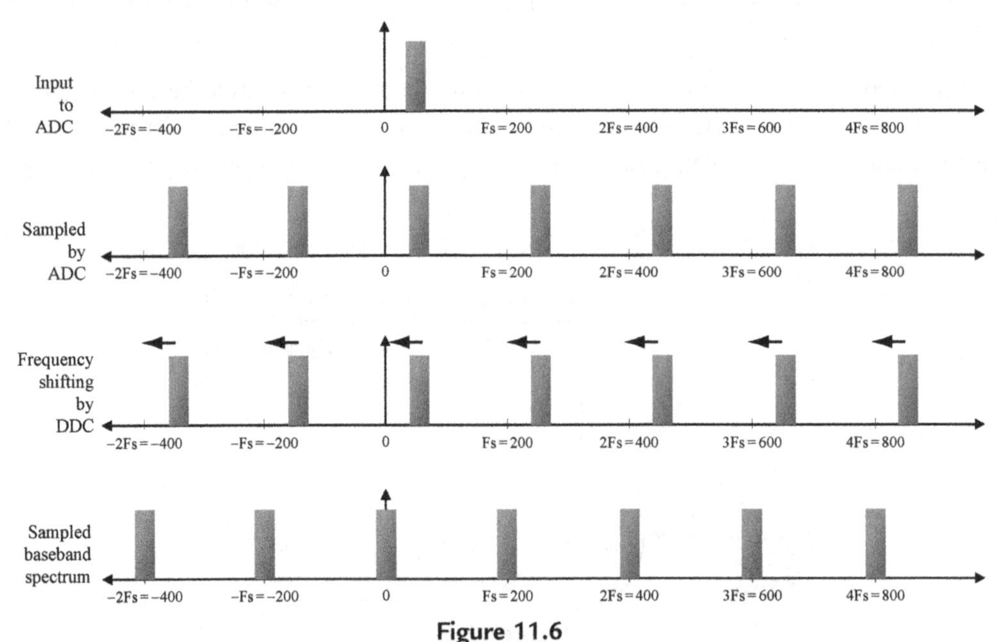

Figure 11.6
Digital downconversion in frequency domain.

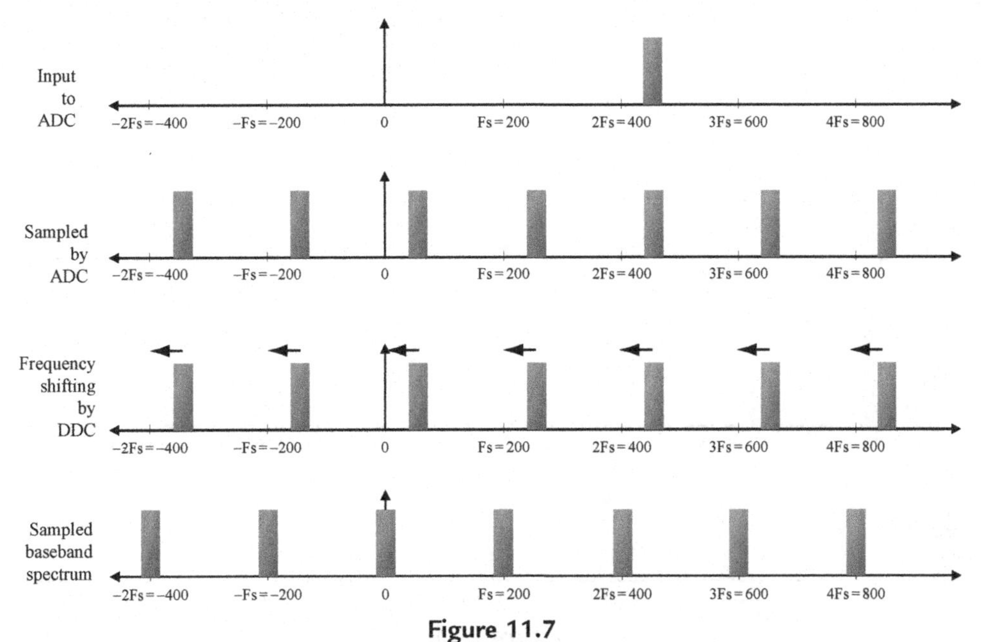

Figure 11.7
Intermediate frequency subsampling in frequency domain.

can sample up to ½ the sample rate, or 100 MHz. To allow easier postsampling filtering, we may want to limit this to 80 MHz instead. Our signal here lies between 50 and 70 MHz, so these conditions are met.

This should be familiar from the chapter on sampling. Now, let us consider what happens if the IF signal is centered at 460 MHz in Fig. 11.7, rather than 60 MHz.

As we discussed in the earlier chapter on sampling and aliasing, there is no way to distinguish between a signal and an alias at a multiple of the sample frequency. When IF subsampling is performed, this aliasing can be used to our advantage.

In our examples, any signal that aliases to a baseband frequency from −80 MHz to +80 MHz can work. Remember, the digital downconversion multiplies by a complex exponential, which can be rotating either counterclockwise or negative, thereby either shifting the spectrum left or right. Below in Fig. 11.8 is another example, showing downconversion from 340 MHz carrier frequency.

The signal at 340 MHz will be aliased as if it was at −60 MHz, and then can be shifted to baseband (the complex NCO runs in opposite rotational direction, to shift alias to right instead of left). The only areas of the spectrum that cannot be properly aliased down are those which alias to the region near the Nyquist frequency, in our case, 100 MHz. So signals close to 300, 500, and 700 MHz cannot be sampled in this way. But the rest of the

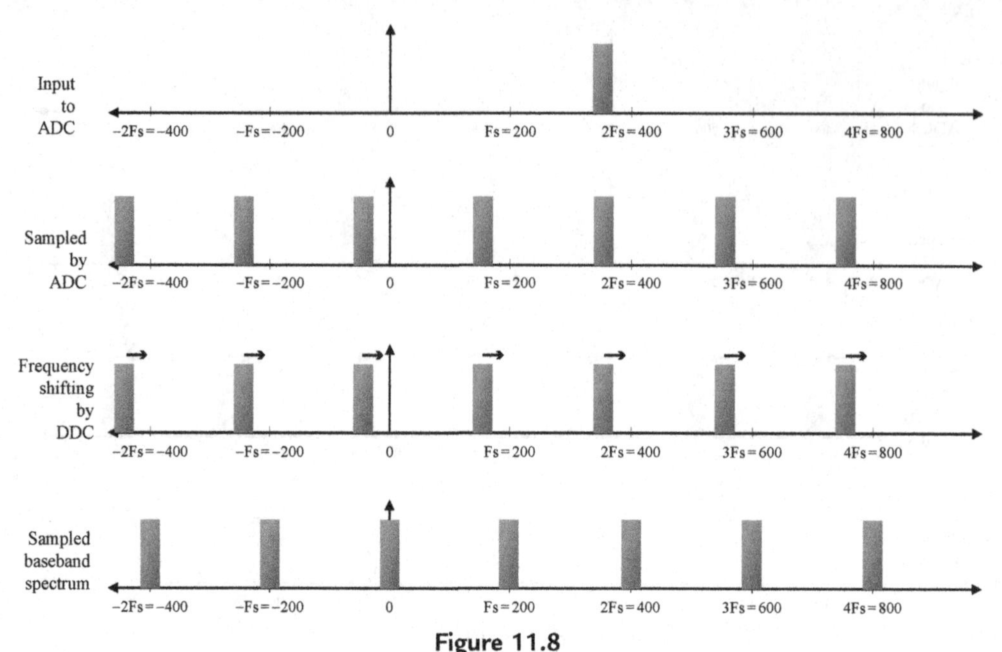

Figure 11.8
Intermediate frequency subsampling of negative alias.

spectrum can be sampled, so long as the sampling frequency meets the requirement of being more than twice the bandwidth of the signal of interest. In practice, the sampling frequency is often much higher than the signal bandwidth, which offers the additional advantage of allowing an increase in signal to noise ratio (SNR) when the baseband signal is decimated to a lower frequency for baseband processing.

In our example, the downconverted signal has a spectrum from -10 to $+10$ MHz. This requires a sampling frequency of at least 20 MHz for both I and Q. If a decimate-by-four finite impulse response (FIR) filter is used to low-pass filter the downconverted signal, and then further complex symbol processing can take place at 25 MHz. By sampling at a much higher rate and then low-pass filtering, a much greater percentage of the sampling quantization noise can be eliminated. Quantization noise, as discussed in an earlier chapter, is due to the effect of the signal being sampled and mapped to specific amplitude levels, which is limited by how many bits of resolution are available in the ADC. This noise is broadband, meaning that is distributed evenly across the frequency spectrum. The effect of the low-pass filter is to attenuate much of this quantization sampling noise, along with unwanted signals, in the stopband region. The net effect is the equivalent of adding 1 bit of precision or 6 dB to the SNR, for every factor of two in decimation and filtering.

In our example, if we used a 12-bit ADC at 200 MHz F_s, and performed the digital downconversion, low-pass filtering and decimate by four, it is the equivalent as if we had sampled at 50 MHz F_s using a 14-bit ADC. The decimation filter has in essence added 2 bits of resolution to the ADC, or 12 dB to the SNR of the sampled signal. This concept can be taken to extremes. There is a class of ADCs, called sigma-delta converters, which run at very high frequencies relative to the signal they are sampling, but only use a 1 bit sampler. An effective 10 or 12 bit ADC can be built using a single-bit sampling front end running at extremely high frequencies followed by decimation filtering stages.

Now let us go back to IF subsampling. In theory, one could sample a signal at any arbitrary frequency, but there must be some practical limitations. So far, we have only discussed the limitation of not sampling signals that alias to near the Nyquist frequency. But there are other limitations, and we will discuss two of these.

First, one must consider the performance of the ADC. There are two principle characteristics that are of concern here. Of course, the first is the maximum sampling rate at which the ADC operates. In our example, we assumed an ADC that could sample at 200 MHz. When IF subsampling, we must also consider the analog bandwidth of the sampling circuit in the ADC. Ideally, the ADC samples the signal for an infinitely small instant of time and converts that measurement into a digital number. In practice, this circuit has a sampling window or period of time in which it samples. The narrower this window, the higher signal frequencies it can sample. This specification is given in the data sheets provided by ADC manufacturers. In our example, we sampled a signal at 460 MHz. So we should check that our ADC has an analog front end bandwidth of 500 MHz or higher.

Another factor that must be considered is clock jitter. Clock jitter is the amount of timing variability of the edge of the ADC sampling clock from cycle to cycle. It can be readily seen on an oscilloscope of sufficient quality. It is more easily seen as clock phase noise on a spectrum analyzer. Jitter shows up as spectral noise tapering off on either side of the clock frequency component. The less jitter, the more closely the clock appears as simply a vertical line in the frequency response. The effect of clock jitter will be proportional to the frequency of the signal being sampled. It will limit the SNR according to the following relationship.

$$SNR_{jitter} = 20\log(1/(2\pi \cdot F_{signal} \cdot t_{jitter}))$$

An example may make this clearer. In our example, the ADC clock was 200 MHz, which has a period of 5 nanoseconds (ns). Let us assume this clock has jitter of 5 picoseconds (ps). This does not sound like too much—only 0.1% of the clock period. First we will compute the SNR limitation due to clock jitter with an input signal at 60 MHz and then with at 460 MHz.

60 MHz IF signal:

$$SNR_{jitter} = 20\log\left(1/\left(2\pi\cdot 60\cdot 10^{6}\cdot 5\cdot 10^{-12}\right)\right) = 54.5\text{ dB}$$

460 MHz IF signal:

$$SNR_{jitter} = 20\log\left(1/\left(2\pi\cdot 460\cdot 10^{6}\cdot 5\cdot 10^{-12}\right)\right) = 36.8\text{ dB}$$

This level of clock jitter will limit the ADC performance for a 60 MHz signal at any level of precision beyond 10 bits. For a 460 MHz signal, the clock jitter will limit the SNR to 36 dB, or about 6 bits. Clearly, this level of clock jitter is excessive in this IF subsampling application.

An analogy can be the same strobe light concept used in the Sampling and Aliasing chapter. The strobe light is assumed to flash at exact intervals. Any variance, or jitter, in the flashing intervals, will cause distortion in the position of the red dot on the wheel. The IF subsampling process would be as if the wheel made multiple rotations between strobe flashs, though it would appear that the wheel was rotating at a slower speed. But as the wheel is actually rotating much faster, any sampling errors due to jitter in the strobe flash will be magnified, making the red dot appear blurred.

A very reasonable question to ask is if we can do something similar with a DAC to utilize aliased versions of the digitally upconverted signal. The answer is yes, but with an important limitation that discourages use of this technique in practice.

A major difference between an ADC and DAC is that while an ADC converts the signal to a digital representation, the DAC converts the digital samples to analog form *and* performs a "sample and hold" function of the analog signal between clocks. It is this sample and hold function which is a critical difference.

The sample and hold function is rectangular filter in the time domain. Each digital sample input is an impulse of a given magnitude. The output is a rectangular shape of the input magnitude. The DAC impulse response, just like in any filter, will define the frequency response. In this case, the rectangular impulse response yields a sinc function (sin(x)/x) response in the frequency domain. This is shown in the below, with the nulls in the frequency response corresponding to multiples of the DAC conversion or clock frequency (F_s) (Fig 11.9).

What this means in practice is that there will be a reduction in signal level at higher frequencies. By the time the DAC frequency response reaches the Nyquist frequency (½ F_s), it will have a droop of nearly 4 dB. The peak of the first lobe is 6 dB below the DC response, and second lobe peak a further 6 dB lower. Due to this attenuation, usefulness of the DAC output at frequencies well above the DAC F_s limited. Moreover, as the frequency response is not flat, it often needs to be compensated for. The closer the

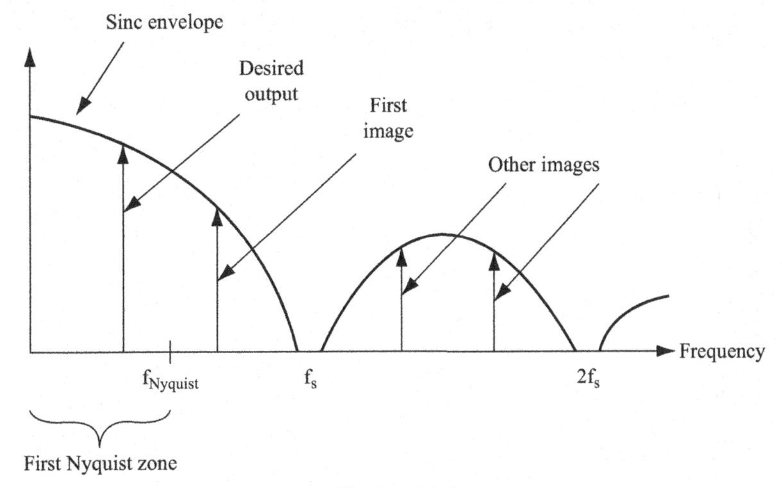

Figure 11.9
Nyquist zones of DAC output.

IF signal lies to multiples of the DAC F_s, the more distorted the frequency response is. Therefore, the IF frequency is usually limited the first Nyquist zone. Furthermore, it is usually limited to about 80% of the Nyquist frequency or to 40% of the DAC F_s. For a DAC being clocked at 250 MHz, the upconverted IF signal should be at 100 MHz or less. For example, if the IF carrier is chosen to be 70 MHz with complex baseband signal extending to 10 MHz, the IF signal will occupy a spectrum from 60 to 80 MHz. After the DAC, the analog output will need to be filtered using an analog filter to remove the higher frequency DAC images and the DAC clock harmonics. Often, a SAW (surface acoustic wave) bandpass filter is used, and the choice of IF frequency is often determined in part by the frequencies where the SAW filters are commercially available.

Many systems, particularly if the IF signals have fairly wide bandwidth (like our 4G wireless example with an IF signal BW of 20 MHz), will need to compensate for the frequency response droop due to the sinc response of the DAC even at these low IF frequencies. To do this, a sinc compensation FIR filter is used, after the NCO and complex multiplier performing the digital upconversion.

Below in Fig. 11.10 is the frequency response of a typical sinc compensation FIR filter, calculated using a popular digital signal processing design tool called Matlab (available

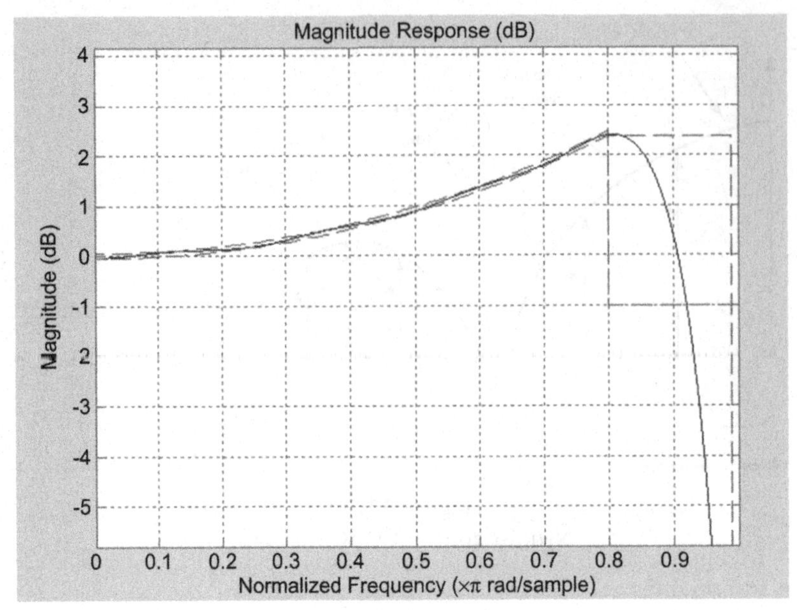

Figure 11.10
Sinc compensation filter frequency response.

Numerator:
-0.0065468956647895373
0.014409998368132634
-0.029151111437169457
0.053017946956878462
-0.091122876704257547
0.15519455343211477
-0.28282698843194176
0.68492126102237871
0.68492126102237871
-0.28282698843194176
0.15519455343211477
-0.091122876704257547
0.053017946956878462
-0.029151111437169457
0.014409998368132634
-0.0065468956647895373

Figure 11.11
Sinc compensation filter coefficients.

from The Mathworks). It will compensate up to 80% of the Nyquist frequency. This digital filter would immediately preced the DAC and be placed after the digital upconversion process. It would have a data throughput rate equal to the DAC conversion rate. The associated coefficients are shown in Fig. 11.11.

CHAPTER 12

Error-Correction Coding

This chapter will give an introduction to error-correction coding and decoding. This is also known as forward error correction or FEC. This is a very complex field and even has its own set of mathematics. A thorough understanding also requires background in probability and statistics theory, which is necessary to quantify the performance of any given coding method. Most texts on this topic will delve into this math to some degree. Appendix E on binary field arithmetic gives a very quick, very basic, summary of the little Boolean arithmetic we will use in this chapter.

As for the approach here, we will skip nearly all mathematics and try to give an intuitive feel for the basics of coding and decoding. This will mostly be done by example, using two types of error-correcting codes. The first is a linear block code, specifically a Hamming code. The second will be a cyclic code, using convolutional coding and Viterbi decoding.

More advanced codes are used in various applications. Reed Solomon is very popular in high-speed Ethernet applications. Turbo codes are used in 3G and 4G wireless systems. Low-density parity codes (LDPCs) are used in 5G wireless, as well as cable (DOCSIS) systems. BCH codes, often used in conjunction with LDPC, are used in storage systems. A fairly recent addition is polar codes, to be used in certain aspects of 5G wireless systems. All of the error-correcting codes provide higher degrees of error correction, at the expense of higher computational cost and often higher latency. Error-corrective coding is a specialty field, and this is just an introduction to some of the basic concepts.

To correct errors, we need to have a basic idea of how and why errors occur. One of the most fundamental mechanisms causing error is noise. Noise is an unavoidable artifact in electronic circuits and in many of the transmission mediums used to transmit information. For example, in wireless and radar systems, there is a certain noise level present in the transmit circuitry, in the receive circuitry, and in the frequency spectrum used for transmission. During transmission, the transmitted signal may be attenuated to less than one billionth of the original power level by the time it arrives at the receive antenna. Therefore, small levels of noise can cause significant errors. There can also be interfering sources as well, but because noise is random, it is much easier to model and used as the basis for defining most coding performance.

It makes sense that given the presence of noise, we can improve our error rate by just transmitting at a higher power level. Since the signal power increases, but the noise does not,

Digital Signal Processing 101. http://dx.doi.org/10.1016/B978-0-12-811453-7.00012-3

the result should be better performance. This also leads to the concept of energy per bit, or E_{bit}. This is computed by dividing the total signal power by the rate of information carrying bits.

One characteristic of all error correction methods is that we need to transmit more bits, or redundant bits, compared to the original bit rate being used. Take a really basic error-correction method. Suppose we just repeat each bit three times, and have the receiver take a majority vote across each set of 3 bits to determine what was originally transmitted. If 1 bit is corrupted by noise, the presence of the other two will still allow the receiver to make the correct decision. But the trade-off is that we have to transmit three times more bits. Now assume we have a transmitter of a fixed power (say 10 Watts). If we are transmitting three times more bits, that means that the energy per bit is only one third what it was before, so each bit is now more susceptible to being corrupted by noise. To measure the effectiveness of a coding scheme, we will define a concept of coding gain. This goal is that with the addition of an error-correcting code, the result is better system performance, and we want to equate the improvement—in error rate due to error-corrective coding to the equivalent improvement—we could achieve if instead we transmitted the signal at a higher power level. We can measure the increase in transmit power needed to achieve the same improvement in errors as a ratio of new transmit power divided by previous transmit power and express using in dB_{power} (please see the last part of the Chapter 3 for further explanation on dB). This measurement is called the "coding gain." The coding gain takes into account that redundant bits must be transmitted for any code that actually reduces the signal transmit power per bit, as the overall transmit power must be divided by the bit rate to determine the actual energy per bit available.

You can probably see that basic idea is to correct errors in an efficient manner. This means being able to correct as many errors as possible using the minimum redundant bits. The efficiency is generally measured by coding gain. Again, the definition of coding gain is how much the transmit power of an uncoded system (in dB) must be increased to match the performance of the coded system. Matching performance means that the average bit error rates (BER) are equal. Coding performance can also change at different BER rates, so sometime this is expressed in graphical form.

Codes are often described by parameters "k" and "n." The number of information bits in a code word is given by k, and the total number of bits in a code word, including the redundant bits, is given by n. The code rate is equal to k/n and tells us how much the data rate must be increased for a given code. Next, we are going to use an example Hamming linear block code where $k = 4$ and $n = 7$.

12.1 Linear Block Encoding

With k information bits, we have 2^k possible inputs sequences. Specifically, with $k = 4$, we have 16 possible code words. We could use a lookup table. But there is a more efficient method. Because this is a linear code, we only need to map the code words generated by each of the k bits. For k = 4, we only need to map four inputs to code words, and use the linear properties to build the remaining possible 12 code words.

Inputs:

$$0001, 0010, 0100, 1000$$

Code mapping rule:

Map four input bits to bits 6–3 of 7 bit code word.

Then create redundant parity bits according to rule below.

$$\text{bit } 2 = \text{bit } 6 + \text{bit } 5 + \text{bit } 3$$
$$\text{bit } 1 = \text{bit } 6 + \text{bit } 4 + \text{bit } 3$$
$$\text{bit } 0 = \text{bit } 5 + \text{bit } 4 + \text{bit } 3$$

Let us take the input 0001 (bit 3 of code word = 1). Therefore, 0001 → 0001111.

With our four inputs, we get the following code words:

$$1000 \rightarrow 1000110$$
$$0100 \rightarrow 0100101$$
$$0010 \rightarrow 0010011$$
$$0001 \rightarrow 0001111$$

Owing to the linear property of the code, this is sufficient to define the mapping of all possible 2^k or 16 input sequences. This process can be easily performed by a matrix.

$$[\text{7 bit codeword}] = [\text{4 bit input sequence}] \cdot \begin{bmatrix} 1000110 \\ 0100101 \\ 0010011 \\ 0001111 \end{bmatrix}$$

Notice that this matrix, called a generator matrix **G**, is simply made up of the four code words, where each row is one of the four single active bit input sequences. A simple example of generating the code word is shown below.

$$[1101] \cdot \begin{bmatrix} 1000110 \\ 0100101 \\ 0010011 \\ 0001111 \end{bmatrix} = [1101100]$$

12.2 Linear Block Decoding

At the receiving end, we will be recovering 7-bit code words. There exists a total of 2^7 or 128 possible 7-bit sequence that we might receive. Only 2^4 or 16 of these are valid code words. If we happen to receive one of the 16 valid code words, then there has been no error in transmission, and we can easily know what 4 bit input sequence was used to generate the code word at the transmit end. But when an error does occur, we will receive one of the remaining 112 code words. We need a method to determine what error occurred, and what the original 4 bit input sequence was.

To do this, we will use another matrix, called a parity check matrix **H**. Let us look again at how the parity bit 2 is formed in this code, using the original parity definition.

$$\text{bit } 2 = \text{bit } 6 + \text{bit } 5 + \text{bit } 3$$

Based on this relationship, bit 2 = 1 only when bit 6 + bit 5 + bit 3 = 1. Given the rules of binary arithmetic, we can say bit 2 + bit 6 + bit 5 + bit 3 = 0 for any valid code word. Extending this, we can say this about any valid code word:

$$\begin{aligned}
&\text{bit } 6 + \text{bit } 5 + \text{bit } 3 + \text{bit } 2 = 0 \quad \text{from parity bit 2 rule} \\
&\text{bit } 6 + \text{bit } 4 + \text{bit } 3 + \text{bit } 1 = 0 \quad \text{from parity bit 1 rule} \\
&\text{bit } 5 + \text{bit } 4 + \text{bit } 3 + \text{bit } 0 = 0 \quad \text{from parity bit 0 rule}
\end{aligned}$$

This comes from the earlier definition of the parity bit relationships we used to form the generator matrix. We can represent the three equations above in the form of parity generation matrix **H**.

$$\mathbf{H} = \begin{bmatrix} 1101100 \\ 1011010 \\ 0111001 \end{bmatrix}$$

As a quick arithmetic check, $\mathbf{G} \cdot \mathbf{H}^{\mathrm{T}}$ must equal zero.

We can use the parity check matrix **H** to decode the received code word. We can compute something called a syndrome **S** as indicated below.

$$[\text{received code word}] \cdot \mathbf{H}^{\mathrm{T}} = \mathbf{S}(\text{the syndrome})$$

When the syndrome is zero, all three parity equations are satisfied, there is no error, a valid code word was received, and we can simply strip away the redundant parity bits from the received code word, thus recovering the input bits. When the syndrome is nonzero, it will indicate that an error has occurred. Because this is a linear code, we can consider any received code word to be the sum of a valid code word and an error vector or word. Errors occur whenever the error vector is nonzero. The leads to the realization that the syndrome depends only on the error vector and not on the valid code word originally transmitted.

In our example, the syndrome is a 3-bit word, capable of representing eight states. When **S** = [0,0,0], no error has occurred. When the syndrome is any other value, an error has occurred, and the syndrome will indicate in which bit position the error is located. But which syndrome value maps to which seven possible error positions in the received code word?

Let us go back to the parity relationship that exists when the **S** = 0.

$$\text{bit } 6 + \text{bit } 5 + \text{bit } 3 + \text{bit } 2 = 0$$
$$\text{bit } 6 + \text{bit } 4 + \text{bit } 3 + \text{bit } 1 = 0$$
$$\text{bit } 5 + \text{bit } 4 + \text{bit } 3 + \text{bit } 0 = 0$$

We have our earlier example of valid code word [1101100]. The parity check matrix is as follows:

$$\mathbf{H} = \begin{bmatrix} 1101100 \\ 1011010 \\ 0111001 \end{bmatrix} \quad \text{and} \quad \mathbf{H}^{\mathrm{T}} = \begin{bmatrix} 110 \\ 101 \\ 011 \\ 111 \\ 100 \\ 010 \\ 001 \end{bmatrix}$$

$$\mathbf{S} = \quad [1101100] \cdot \begin{bmatrix} 110 \\ 101 \\ 011 \\ 111 \\ 100 \\ 010 \\ 001 \end{bmatrix} \quad [0,0,0]$$

By inspection of the parity equations, we can see that syndromes with only one nonzero bit must correspond to an error in the parity bits of the received code word.

S = [1,0,0] indicates the first parity equation is not satisfied, but the second and third are true. By inspection of parity bit definitions, this must be caused by bit 2 of the received code word, as this bit alone is used to create bit 2 of the syndrome. Similarly, an error in bit 1 in the received code word creates the syndrome [0,1,0], and error in bit 0 in the received code word creates the syndrome [0,0,1].

An error in bit 3 of the received code word will cause a syndrome of [1,1,1], as this bit appears in all three parity equations. Therefore, we can map the syndrome value to specific bit positions where an error has occurred.

S = [0,0,0] → no error in received code word
S = [0,0,1] → error in bit 0 of received code word
S = [0,1,0] → error in bit 1 of received code word
S = [0,1,1] → error in bit 4 of received code word
S = [1,0,0] → error in bit 2 of received code word
S = [1,0,1] → error in bit 5 of received code word
S = [1,1,0] → error in bit 6 of received code word
S = [1,1,1] → error in bit 3 of received code word

The procedure to correct errors is to multiply the received code word by $\mathbf{H}^T$, which gives **S**. The value of **S** indicated the position of error in the received code word to be corrected. This can be illustrated with an example.

Suppose we receive a code word [1111100]. We calculate the syndrome.

$$\mathbf{S} = \quad [1111100] \cdot \begin{bmatrix} 110 \\ 101 \\ 011 \\ 111 \\ 100 \\ 010 \\ 001 \end{bmatrix} \quad = \quad [0,1,1]$$

This syndrome indicates an error in bit 4. The corrected code word is [1101100], and the original 4 bit input sequence 1,1,0,1, which is not the 1,1,1,1 of the received code word.

12.3 Minimum Coding Distance

A good question is what happens when there are two errors simultaneously. Hamming codes can only detect and correct one error per received code word. The amount of detection and correction a code can perform is related to something called the minimum distance. For Hamming codes, the minimum distance is three. This means that all the transmitted code words have at least 3 bits different from all the other possible code words. Recall that in our case, we have 16 valid code words out of 128 possible sequences. These are given below as follows:

0000000	0001111	0010011	0011100
0100101	0101010	0110110	0111001
1000110	1001001	1010101	1011010
1100011	1101100	1110000	1111111

Each of these code words has 3 or more bit differences from the other 15 code words. In other words, each code word has a minimum distance of three from neighboring code words. That is why with a single error, it is still possible to correctly find the closest code word, with respect to bit differences. With two errors, the code word will be closer to the wrong code word, again with respect to bit differences. This is analogous to trying to map a symbol in two-dimensional I-Q space to the nearest constellation point in a quadrature amplitude modulation demodulator. There are other coding methods with greater minimum distances, able to correct multiple errors. We will look at one of these next.

12.4 Convolutional Encoding

A second major class of channel codes is known as convolutional coding. The convolutional code can operate on a continuous string of data, whereas block codes operated on words. Convolutional codes also have memory—the behavior of the code depends on previous data.

Convolutional coding is implemented using shift registers with feedback paths. There is a ratio of "k" input bits to "n" output bits, as well as a constraint length "K." The code rate is k/n. The constraint length K corresponds to the length of the shift register and also determines the length of time or memory that the current behavior depends on past inputs.

Next, we will go through a very simple convolutional coding and Viterbi decoding example. We will use an example with k = 1, n = 2, K = 3 (Fig. 12.1). The encoder is described by generator equations, using polynomial expressions to describe the linear shift register relationships. The first register connection to the XOR gate is indicated by the "1" in the equations, the second by "X," the third by X^2 and so forth. Most convolution codes have a constraint length less than 10.

The rate here is ½ as for every input bit M, there are two output bits N.

The output is usually interleaved in a bit stream $N1_j$, $N2_j$, $N1_{j+1}$, $N2_{j+1}$, $N1_{j+2}$, $N2_{j+2}$...

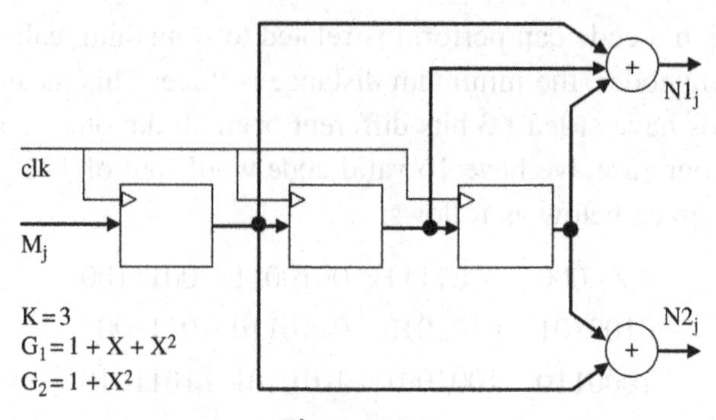

Figure 12.1
Convolutional encoder.

The following table shows encoder operation with the input sequence 1,0,1,1,0,0,1,0,1,1,0,1 for each clock cycle. Register is initialized to zero.

Register Value	N1 Value	N2 Value	Time or Clock Value
1 0 0	1	1	T_1
0 1 0	1	0	T_2
1 0 1	0	0	T_3
1 1 0	0	1	T_4
0 1 1	0	1	T_5
0 0 1	1	1	T_6
1 0 0	1	1	T_7
0 1 0	1	0	T_8
1 0 1	0	0	T_9
1 1 0	0	1	T_{10}
0 1 1	0	1	T_{11}
1 0 1	0	0	T_{12}

The resulting output sequence: {1,1} {1,0} {0,0} {0,1} {0,1} {1,1} {1,1} {1,0} {0,0} {0,1} {0,1} {0,0}.

Notice that the output $N1_j$ and $N2_j$ are both a function of the input bit M_j and the two previous input bits M_{j-1} and M_{j-2}. The previous K−1 bits, M_{j-1} and M_{j-2}, form the state of the encoder state diagram. As shown, the encoder will be at a given state. The input bit M_j will cause a transition to another state at each clock edge, or T_j. Each state transition will result in an output bit pair $N1_j$ and $N2_j$. Only certain state transitions are possible. Transitions due to a 0 bit input are shown in dashed lines, and transitions due to a 1 bit are shown in solid lines. The output bits shown at each transition are labeled on each transition arrow in Fig. 12.2.

Now, let us trace the path of the input sequence through the trellis using Fig. 12.3, and the resulting output sequence. This will help us gain the insight that the trellis is

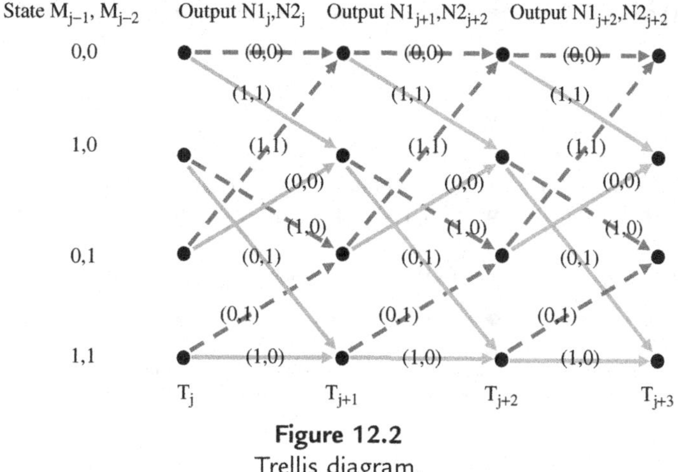

Figure 12.2
Trellis diagram.

representative of the encoder circuit, as use of the trellis will be a key in Viterbi decoding. The highlighted lines show the path of input sequence. The trellis diagram covers from states 0 through 12. It is broken into three sections (A, B, C) in order to fit on the page.

By tracing the highlighted path through the trellis, you can see that the output sequence is the same as our results when computing using the shift register circuit. For constraint length K, we will have $(K-1)^2$ states in our trellis diagram. Therefore, with $K = 3$ in our design example, we have four possible states. For a more typical $K = 6$ or $K = 7$ constraint length, there would be 32 or 64 states respectively; however, this is too tedious to try to diagram.

12.5 Viterbi Decoding

The decoding algorithm takes advantage of the fact that only certain paths through the trellis are possible. For example, starting from state 00, the output on the next transition must be either 0,0 or 1,1 resulting in the next state being either 0,0 or 1,0 respectively. The output of 1,0 or 0,1 is not possible, and if this sequence occurs, then an error must be present in the received bit sequence.

We will first look at Viterbi decoding using the sequence given in the last section as our example. Keep in mind that when decoding, the input data is unknown (whether a "dashed" or "solid" transition), but the output (received) data is known. The job of the decoder is to correctly recover the input data using possibly corrupted received data. The decoder does assume we start from a known state (M_{j-1}, $M_{j-2} = 0, 0$).

We are going to do this by computing the difference between the received data pair, and the transition output for each possible state transition. We will keep track of this cost for

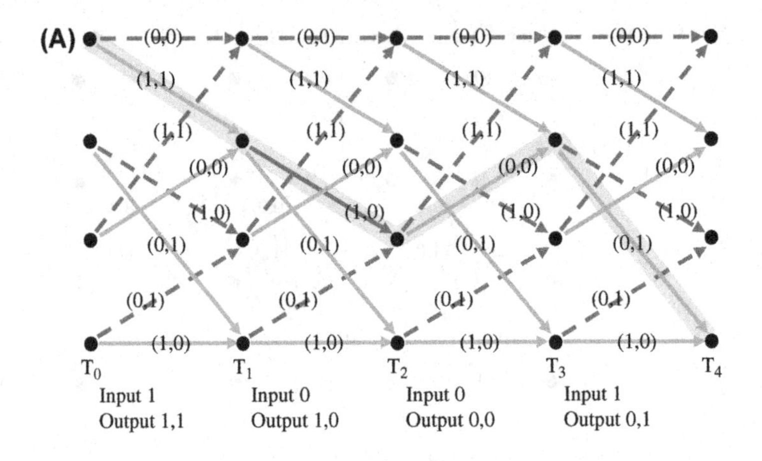

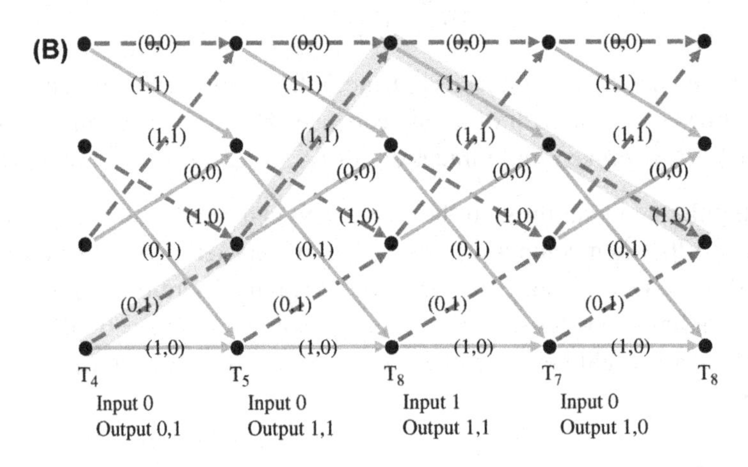

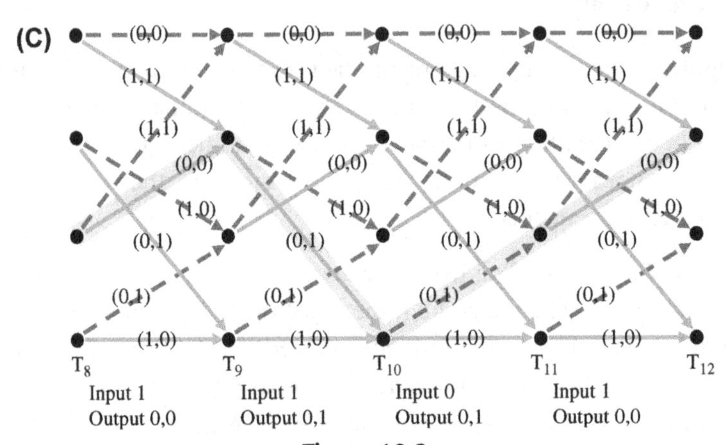

Figure 12.3
Sequence of data through trellis.

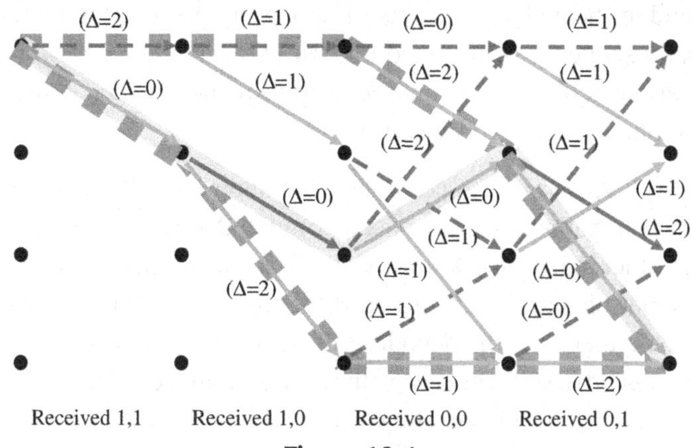

Figure 12.4
Maximum likelyhood decoding.

each transition. Once we get further into the trellis, we can check this cumulative cost entering each of the possible states at each transition, and eliminate the path with the higher cost. In the end, this will yield that lowest-cost valid path (or valid path closest to our received sequence). As always, the best way to get a handle on this is through an example. First, we will look at the cost differences with a correct received sequence (no errors) and then one with errors present. In Fig. 12.4, the figures on each transition arrow are the absolute difference (Δ) between the two received bits and the encoder output generated by that transition (as shown on the encoding trellis diagram above).

Notice that the cumulative Δ is equal to zero as we follow the path the encoder took (solid highlighted) to generate the received sequence. But also notice two other paths that can arrive at the same point (dashed highlighted) which have a nonzero cumulative Δ. The cumulative cost of each these paths is 5. The idea is to find the path with the least cumulative Δ or difference from the received sequence. If we were decoding, we can then recover the input sequence by following the zero cost (solid highlighted) path. The dashed arrows indicate a "zero" input bit, and the solid arrows indicate a "one" input bit. The sequence recovered is therefore 1, 0, 1, 1. Notice that this sequence matches the first 4 bits input to the encoder. This is called "maximum likelihood decoding."

This is all well and good, but as the trellis extends further into time, there are too many possible paths merging and splitting to calculate the cumulative costs. Imagine having 32 possible states, rather than the 4 in our example, and a trellis extending to hundreds of transitions. What the Viterbi algorithm will do is to remove, or prune, less likely paths. Whenever two paths enter the same state, the path having the lowest cost is chosen. The selection of lowest cost, or surviving path, is performed for all states at each transition (excluding some states at beginning and end, when the trellis is either diverging to converging to a known state). The decoded advances through the trellis, eliminating the higher cost paths, and at the end, will backtrack along the path of cumulative least cost to extract the most likely sequence of received bits. In this way, the code word, or valid sequence most closely matching the actual received sequence (which can be considered a valid code sequence plus errors) will be chosen.

Before proceeding further, we need to consider what happens at the end of the sequence. To backtrack after all the paths costs have been computed and the less likely paths pruned, we need to start backtracking from a known state. To achieve this, K−1 zeros are appended to in the input bit sequence entering the encoder. This guarantees that the last state in the encoder and therefore the decoder trellis, must be the zero state (same state as we start from, as the encoder shift register values are initialized to zero). So we will add K −1 or two zeros to our input sequence, shown in bold below, and use in our Viterbi decoding example. The longer the bit sequence, the less impact the addition of K−1 zeros will have on the overall code rate, as these extra bits do not carry information and have to be considered as part of the "redundant bits" (present only to facilitate error correction). So our code rate is slightly less than k/n.

Input bit sequence to encoder:

$$1, 0, 1, 1, 0, 0, 1, 0, 1, 1, 0, 1, \mathbf{0}, \mathbf{0}$$

Output bit sequence from encoder:

$$\{1,1\}\{1,0\}\{0,0\}\{0,1\}\{0,1\}\{1,1\}\{1,1\}\{1,0\}\{0,0\}\{0,1\}\{0,1\}\{0,0\}\{\mathbf{1},\mathbf{0}\}\{\mathbf{1},\mathbf{1}\}$$

The circles identify the four higher cost paths that can be pruned at T_3, based on the higher cumulative path cost ($\Sigma\Delta$ = sum of delta or difference from received bit sequence) (Fig 12.5). The result after pruning is shown in Figs. 12.6 and 12.7 as follows:

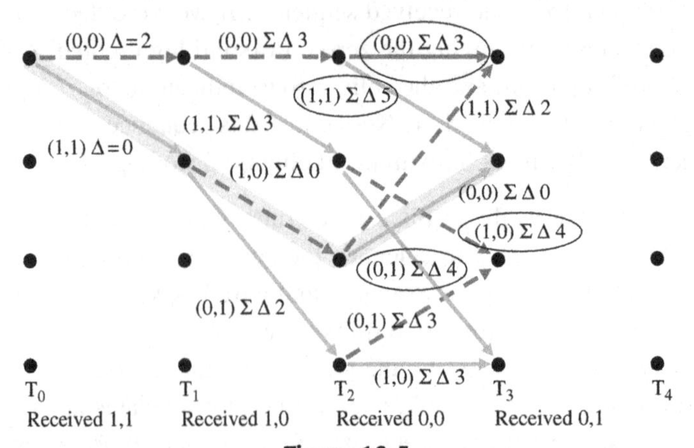

Figure 12.5
Costs of all paths through trellis to T_3.

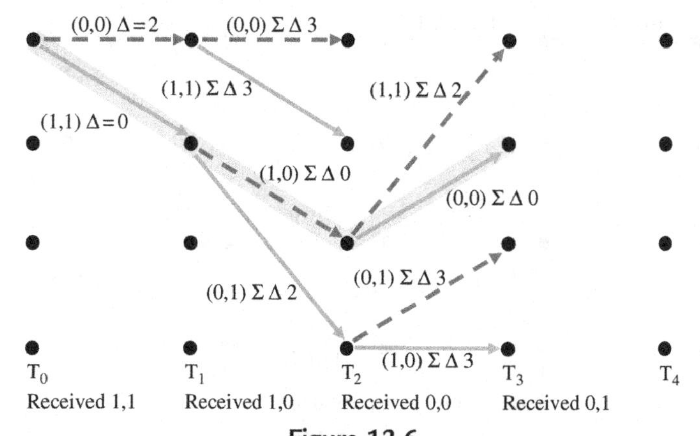

Figure 12.6
After prunning higher cost paths to T_3.

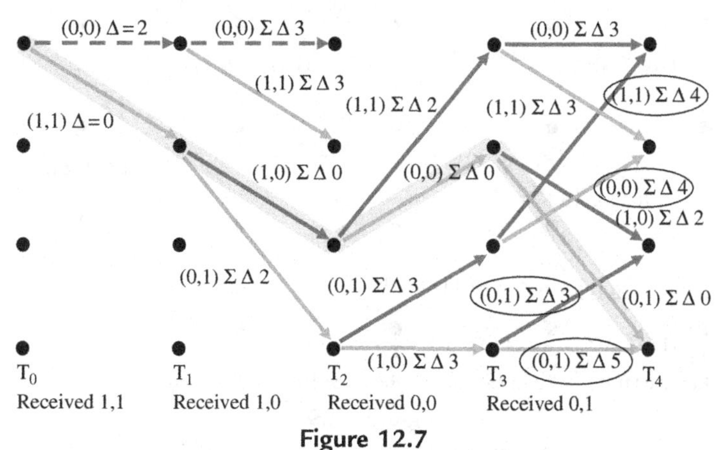

Figure 12.7
Costs of all paths through trellis to T_4.

Next, the higher cost paths (circled) are pruned at T_4, again eliminating half of the paths again. The path taken by the transmitter encoder is again highlighted, at zero cumulative cost as there are no receive errors (Fig 12.8).

Owing to the constant pruning of the high-cost paths, the cumulative costs are not higher than 3 so far (Fig 12.9).

Next, we have the K−1 added bits to bring the trellis path to state 0 (Fig 12.10).

We know, due to the K−1 zeros added to the end of the encoded sequence, that we must start at state 0 at the end of the sequence, at T_{14}.

We can simply follow the only surviving path from state 0,0 at T_{14} backwards, as highlighted (Fig. 12.11). We can determine the bit sequence by the color of the arrows

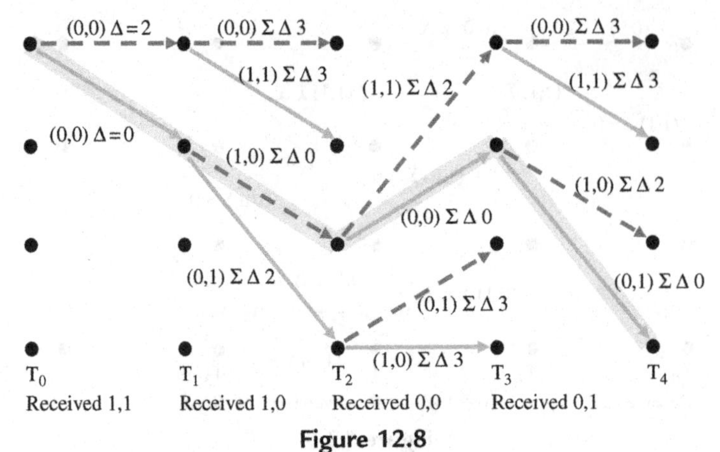

Figure 12.8
After prunning higher cost paths to T_4.

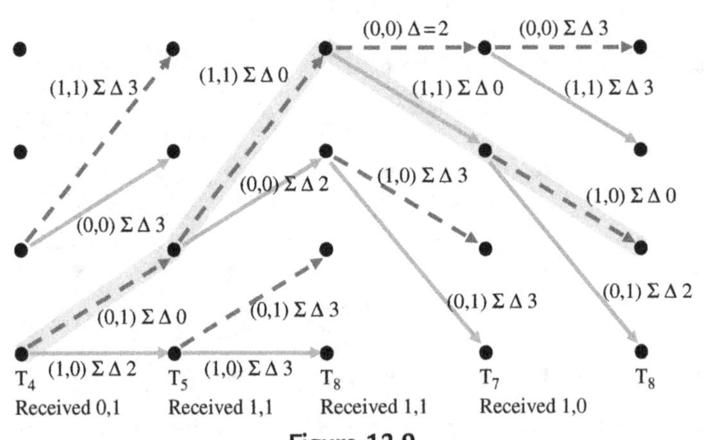

Figure 12.9
After prunning higher cost paths to T_8.

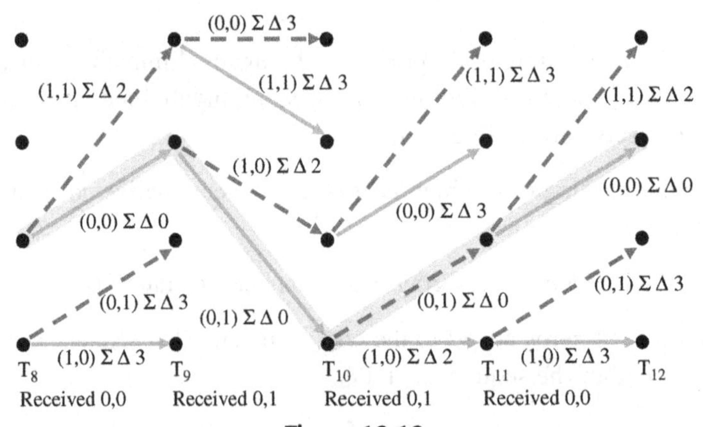

Figure 12.10
After prunning higher cost paths to T_{12}.

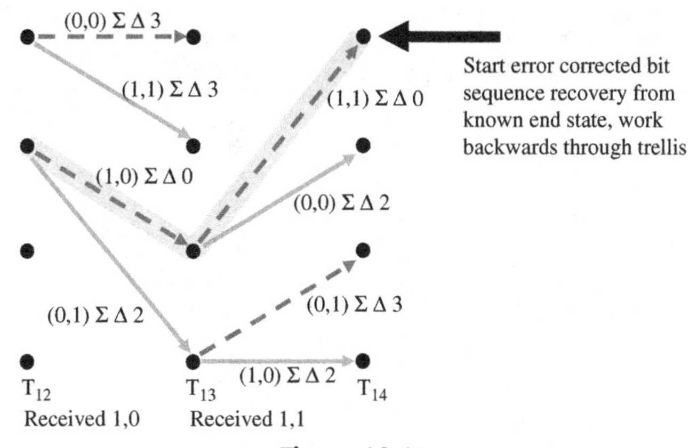

Figure 12.11
After prunning higher cost paths to T_{14}.

(dashed for 0, solid for 1) (Fig.12.12). In a digital system, a 0 or 1 flag would be set for each of the four states at each T_j whenever a path is pruned, identifying the original bit M_j as 0 or 1 associated with the surviving transition.

Now let us reexamine what happens in the event of receive bit errors (indicated in bold in Fig. 12.13). We will assume that there is a bit error at transition T_8 and T_{10}. The correct path (generated by encoder) through the trellis is again highlighted. Note how the cumulative costs change and the surviving paths change. However, in the end, the correct path is the sole surviving path that reaches the end point with the lowest cost.

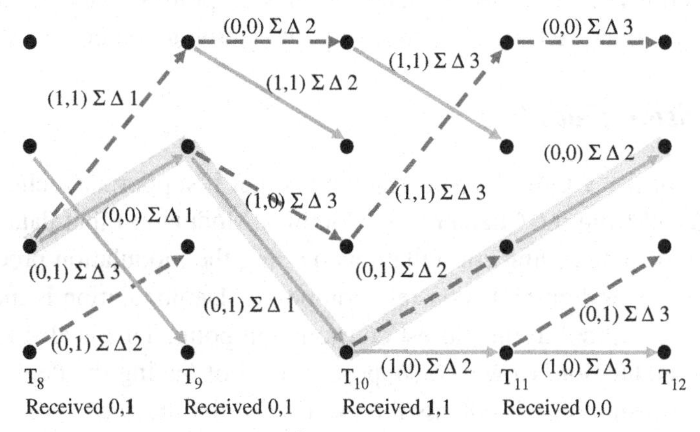

Figure 12.12
Correcting errors in the trellis.

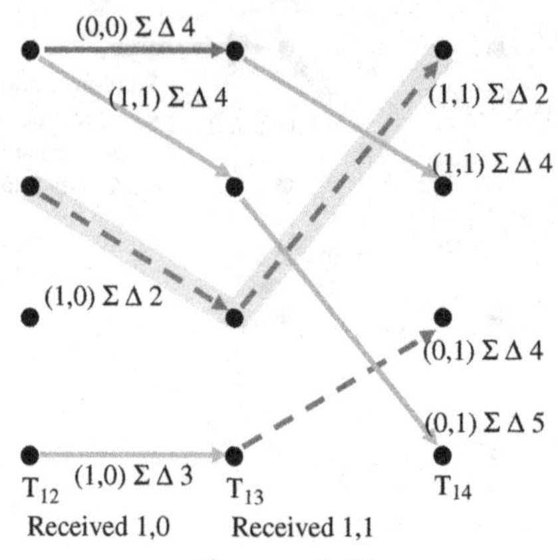

Figure 12.13
Final path cost at end of Trellis.

When we had no errors, we found that all of the competing paths had different costs, and we would always prune away the higher cost path. However, in the presence of errors, we will occasionally have paths of equal cost. In these cases, it does not really matter which path we choose. For example, we can just make an arbitrary rule and always prune away the bottom path when paths of equal cumulative costs merge. And if we use soft decision decoding, which will be explained shortly, the chances of equal cost paths are very low.

The path ending at state 0, at T_{14}, has a cumulative cost of 2, as we corrected two errors along this path. Notice that some of the pruned paths are different than the no-error Viterbi decoding example, but the highlighted path that is selected is the same in both cases. The Viterbi algorithm is finding the most likely valid encoder output sequence in the presence of receive errors and is able to do so efficiently because it prunes away less likely (higher cost) paths continuously at each state where two paths merge together.

12.6 Soft Decision Decoding

Another advantage of the Viterbi decoding method is that it supports a technique called "soft decision." Recall from the Chapter 9 on Complex Modulation that data is often modulated and demodulated using constellations. During the modulation process, the data is mapped to one of several possible constellation points. Demodulation is the process of mapping the received symbol to the closest constellation point. This is also known as "hard decision" demodulation. However, suppose instead of having the demodulator output at the closet constellation point, it outputs the location of the received symbol relative to the constellation points. This is also known as "soft decision" demodulation.

For example, in the constellation in Fig. 12.14, imagine that we receive the symbol labeled S_1, S_2, S_3,and S_4. We will demodulate these symbols as follows:

	S_1	S_2	S_3	S_4
Hard Decision Demodulation	0[a], 0	0, 0	1, 1	0, 1
Soft Decision Demodulation	½, 0	¼, 0	¾, ¾	0, 1

[a]As equal distance between two points, arbitrarily assigned to one of the points.

Basically, we are telling the Viterbi decode how sure we are of the correct demodulation. Instead of giving a yes or no at each symbol, we can also say "maybe" or "pretty sure." The received signal value in the trellis calculation is now any value between 0 and 1 inclusive. This is then factored into the path costs, and the decision on which merged path to prune away.

Simulation and testing have shown over 2 dB improvement in coding gain when using soft decision (16 levels, or 4 bits) compared to hard decision (2 levels, or single bit) representation for the decisions coming from the demodulator. The additional Viterbi decoding complexity for soft decoding is usually negligible.

There are many other error-corrective codes besides the two simple ones we have presented here. Some common codes used in industry are Reed Solomon, BCH, and Turbo decoding. LDPC is another emerging coding technique, which promises even higher performance at a cost of much increased computational requirements. In addition, different codes are sometime concatenated to further improve performance.

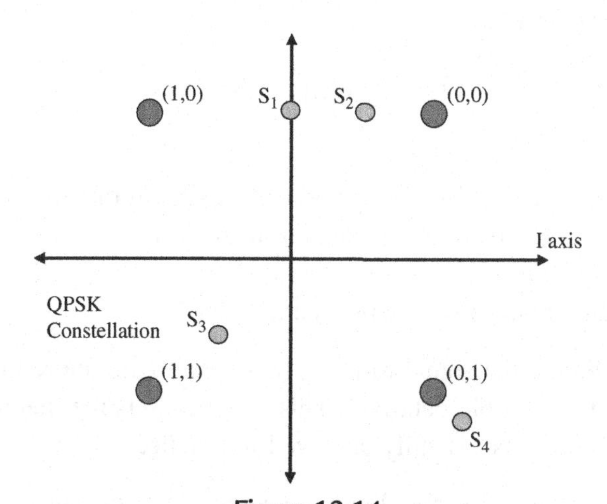

Figure 12.14
Quadrature phase-shift keying (QPSK) constellation diagram.

12.7 Cyclic Redundancy Check

A CRC or cyclic redundancy check word is often appended at the end of a long string or packet of data, prior to the error-correcting encoder. The CRC word is formed by taking all the data words and exclusive OR gate (EXOR) each word together. For example, the data sequence might be 1024 bits, organized as 64 words, each 16 bits. This would require 16-1 EXOR word operations in a DSP or in hardware to form the 16 bit CRC word. This function is analogous to that of a parity check bit for a single digital word being stored and accessed from DRAM memory.

At the conclusion of the error-decoding process, the recovered input stream can again be partitioned in words and EXORed together, and the result compared to the recovered CRC word. If they match, then we can be assured with very high probability that the error correction was successful in correcting any error in transmission. If not, then there were too many errors to be corrected, and the whole data sequence or frame can be discarded. In some cases, there is a retransmission facility built into the higher level communication protocol for these occurrences, and in others, such as a voice packet in mobile phone system, the data is "real time," so alternate mechanisms of dealing with corrupt or missing data are used.

12.8 Shannon Capacity and Limit Theorems

No discussion on coding should be concluded without at least a mention of the Shannon capacity theorem and Shannon limit. The Shannon capacity theorem defines the maximum amount of information, or data capacity, which can be sent over any channel or medium (wireless, coax, twister pair, fiber etc.).

$$C = B \log_2(1 + S/N)$$

where

C is the channel capacity in bits per second (or maximum rate of data)
B is the bandwidth in Hz available for data transmission
S is the received signal power
N is the total channel noise power across bandwidth B

What this says is that higher the signal-to-noise (SNR) ratio and more the channel bandwidth, the higher the possible data rate. This equation sets the theoretical upper limit on data rate, which of course is not fully achieved in practice.

It does not make any limitation on how low the achievable error rate will be. That is dependent on the coding method used.

As a consequence of this, the minimum SNR required for data transmission can be calculated. This is known as the Shannon limit, and it occurs as the available bandwidth goes to infinity.

$$E_b/N_0 = -1.6 \text{ dB}$$

where

E_b is the energy per bit
N_0 is the noise power density in Watts/Hz ($N = B\ N_0$)

If the E_b/N_0 falls below this level, no modulation method or error-correction method will allow data transmission.

These relationships define maximum theoretical limits, against which the performance of practical modulation and coding techniques can be compared against. As newer coding methods are developed, we are able to get closer and closer to the Shannon limit, usually at the expense of higher complexity and computational rates.

CHAPTER 13

Matrix Inversion

Matrix processing is a heavily used technique in communications, radar, medical imaging, and many other applications. This is particularly prevalent in systems with many antennas and performing multiple input, multiple output (MIMO) processing. In 5G wireless, for example, QR decomposition (QRD) is used in both MIMO processing and amplifier digital predistortion adaptation. In radar, QRD is used in space–time adaptive processing (STAP) and can be used to extract signals well below the noise floor. The matrix sizes used are normally modest, but the throughput and processing requirements can be very high. This chapter has more math than most. Note, however, that these functions are usually already available for high performance CPUs, GPUs, and FPGAs.

Matrix inversion techniques, such as QR or Cholesky decomposition, typically involve very small and large numbers in the computation process, which requires a dynamic range that is impractical to support using fixed point numerical representation. Both are computationally intensive and grow in computational requirements by the third power of the matrix dimensions. A matrix of [1000 × 1000] has 1 million elements, and this is a nontrivial effort to find the inverse, assuming it exists.

13.1 Matrix Basics

A matrix is made up of vectors. These can be defined as column or row vectors, depending upon the context. For example, an [M × N] matrix can be defined as N column vectors, where each vector is of length M. This section covers basics of matrix multiplication.

The inverse of matrix is defined as

$$\mathbf{A} \cdot \mathbf{A}^{-1} = \mathbf{I},$$

where $\mathbf{A}^{-1}$ is the inverse of the matrix $\mathbf{A}$.

This simply means that a matrix multiplied by its inverse equals the identity matrix. The identity matrix is a matrix with diagonal elements all equal to 1, and every element off the diagonal equal to zero (Fig. 13.1).

Matrices have a property known as the determinate, which is a value that can be computed from the elements of the matrix. An invertible matrix is called nonsingular and has a

Digital Signal Processing 101. http://dx.doi.org/10.1016/B978-0-12-811453-7.00013-5

$$\begin{pmatrix} 1 & 0 & 0 & 0 \\ 0 & 1 & 0 & 0 \\ 0 & 0 & 1 & 0 \\ 0 & 0 & 0 & 1 \end{pmatrix}$$

Figure 13.1
Identity matrix (dimension = 3).

$$\begin{pmatrix} V_1 \\ V_2 \\ V_3 \\ V_4 \end{pmatrix} * \begin{pmatrix} W_1 \\ W_2 \\ W_3 \\ W_4 \end{pmatrix} = V_1 * W_1 + V_2 * W_2 + V_3 * W_3 + V_4 * W_4$$

Figure 13.2
Vector dot product.

nonzero determinate. When the determinate is zero, the matrix is singular and the inverse matrix does not exist. An "ill-conditioned" matrix has a determinate near zero. In practice, this will mean that computing the inverse will have some very small and/or very large numbers involved and will require a high degree of numerical accuracy.

Matrix multiplication involves taking the dot product between two vectors to compute the each element of the resulting matrix. The dot product means taking two vectors, multiplying each pair of elements, and summing the products. The dot product of two vectors results in a scalar result (Fig. 13.2).

In order to multiply two matrices $\mathbf{A} \cdot \mathbf{B} = \mathbf{C}$, the number of columns in **A** must equal the number of rows in **B**. This is necessary, so when computing the dot product, the vector lengths will be equal. If **A** is an $[N \times M]$ matrix, then **B** must be an $[M \times P]$ matrix to perform the matrix multiplication and C will be of size $[N \times P]$. In the example in Fig. 13.3, the matrix dimension of **A** is $[2 \times 3]$, **B** is $[3 \times 4]$, and **C** is $[2 \times 4]$. The vector dot product is identified to compute output element C_{23}.

For an $[N \times N]$ matrix to be invertible, all of the column vectors need to be independent of each other (Fig. 13.4).

Using of a $[3 \times 3]$ matrix, we can imagine a three-dimensional space. Each column vector represents a vector starting from the origin and terminating at the point defined by the vector in the three-dimensional space. If each vector points in a different direction, then the vectors are independent of each other. If two vectors point in exactly the same

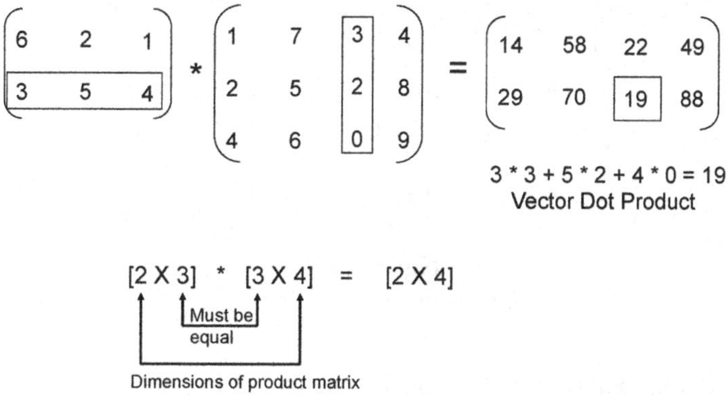

$$\begin{pmatrix} 6 & 2 & 1 \\ 3 & 5 & 4 \end{pmatrix} * \begin{pmatrix} 1 & 7 & 3 & 4 \\ 2 & 5 & 2 & 8 \\ 4 & 6 & 0 & 9 \end{pmatrix} = \begin{pmatrix} 14 & 58 & 22 & 49 \\ 29 & 70 & 19 & 88 \end{pmatrix}$$

3 * 3 + 5 * 2 + 4 * 0 = 19
Vector Dot Product

[2 X 3] * [3 X 4] = [2 X 4]
Must be equal
Dimensions of product matrix

Figure 13.3
Matrix multiply.

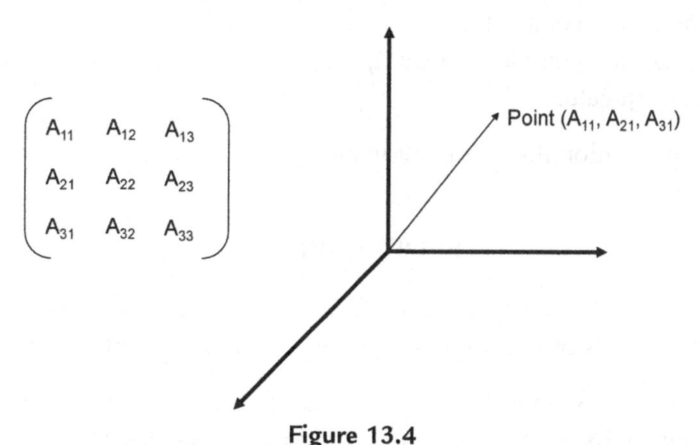

Figure 13.4
Matrix composed of vectors in N-dimensional space.

direction (they may be of different lengths), then the vectors are dependent. If there is a 90-degree angle between two vectors, then they are not only independent but also orthogonal to each other. If all the vectors are 90 degrees with respect to each other (for example, the three axis vectors shown in Fig. 13.4), then the matrix is an orthogonal matrix.

Ill-conditioned matrices have two or more vectors that are close to colinear or nearly in the same direction. Because the vectors are very close to being dependent, this will be a matrix that will require a high degree of accuracy in determining the inverse.

13.2 Cholesky Decomposition

The Cholesky decomposition is used in the special case when **A** is a square, conjugate symmetric matrix. This makes the problem a lot simpler. Recall that a conjugate

$$\begin{pmatrix} A_{11} & A^*_{21} & A^*_{31} & A^*_{41} \\ A_{21} & A_{22} & A^*_{32} & A^*_{42} \\ A_{31} & A_{32} & A_{33} & A^*_{43} \\ A_{41} & A_{42} & A_{43} & A_{44} \end{pmatrix} * \begin{pmatrix} x_1 \\ x_2 \\ x_3 \\ x_4 \end{pmatrix} = \begin{pmatrix} b_1 \\ b_2 \\ b_3 \\ b_4 \end{pmatrix}$$

Figure 13.5
Problem to solve: find x.

symmetric matrix is one where the element A_{jk} equals the element A_{kj} conjugated. This is shown as $A_{jk} = A_{kj}^*$. If A_{jk} is a real value (not complex), then $A_{jk} = A_{kj}$.

Note: A conjugate is then the complex value with the sign on the imaginary component reversed. For example, the conjugate of $5 + j12 = 5 - j12$. And by definition, the diagonal elements must be real (not complex), since $A_{jj} = A_{jj}^*$, or more simply, only a real number can be equal to its conjugate.

The Cholesky decomposition floating point math operations per [N × N] matrix is generally estimated as:

$$\mathbf{FLOPS = 4N^3/3}$$

However, the actual computational rate and efficiency depend on implementation details and the architecture details of the computing device used (CPU, FPGA, GPU, DSP...).

The problem statement is $\mathbf{A} \cdot \mathbf{x} = \mathbf{b}$, where $\mathbf{A}$ is an [N × N] complex symmetric matrix, $\mathbf{x}$ is an unknown complex [N × 1] vector, and $\mathbf{b}$ is a known complex [N × 1] vector. The solution is $\mathbf{x} = \mathbf{A}^{-1} \cdot \mathbf{b}$, which requires the inversion of matrix $\mathbf{A}$ (Fig. 13.5). As directly computing the inverse of a large matrix is difficult, there is an alternate technique using a transform to make this problem easier and require less computations.

The Cholesky decomposition maps matrix A into the product of $\mathbf{A} = \mathbf{L} \cdot \mathbf{L}^H$ where $\mathbf{L}$ is the lower triangular matrix and $\mathbf{L}^H$ is the transposed, complex conjugate or Hermitian, and therefore of upper triangular form (Fig. 13.6). This is true because of the special case of $\mathbf{A}$ being a square, conjugate symmetric matrix. The solution to find $\mathbf{L}$ requires square root and inverse square root operators. The great majority of the computations in Cholesky is to compute the matrix $\mathbf{L}$, which is found to be expanding the vector dot product equations for each element L and solving recursively. Then the product $\mathbf{L} \cdot \mathbf{L}^H$ is substituted for $\mathbf{A}$, and after which $\mathbf{x}$ is solved for using a substitution method. First the equations will be introduced, then an example of the [4 × 4] case will be shown to better illustrate.

(L' is commonly used to indicate L^T or L^H if the matrix is complex)

$$\begin{pmatrix} A_{11} & A^*_{21} & A^*_{31} & A^*_{41} \\ A_{21} & A_{22} & A^*_{32} & A^*_{42} \\ A_{31} & A_{32} & A_{33} & A^*_{43} \\ A_{41} & A_{42} & A_{43} & A_{44} \end{pmatrix} = \begin{pmatrix} L_{11} & 0 & 0 & 0 \\ L_{21} & L_{22} & 0 & 0 \\ L_{31} & L_{32} & L_{33} & 0 \\ L_{41} & L_{42} & L_{43} & L_{44} \end{pmatrix} * \begin{pmatrix} L_{11} & L^*_{21} & L^*_{31} & L^*_{41} \\ 0 & L_{22} & L^*_{32} & L^*_{42} \\ 0 & 0 & L_{33} & L^*_{43} \\ 0 & 0 & 0 & L_{44} \end{pmatrix}$$

Figure 13.6
Matrix substitution.

Solving for the elements of

$$A_{jj} = \sum_{k=1}^{j} L_{jk} \cdot L'_{kj}$$

where j is the column index of the matrix

$$A_{jj} = \sum_{k=1}^{j} L_{jk} \cdot conj(L_{jk})$$

The first nonzero element, in each column, is a diagonal elements and can be found by

Diagonal Elements of L

$$L_{jj} = \sqrt{A_{jj} - \sum_{k=1}^{j-1} L_{jk} \cdot conj(L_{jk})} \tag{13.1}$$

In particular,

$$L_{11} = \sqrt{A_{11}}$$

Similarly, the subsequent elements in the column are related as follows:

$$A_{ij} = \sum_{k=1}^{j} L_{ik} \cdot L'_{kj}$$

where i and j are the row and column indices of the matrix

$$A_{ij} = \sum_{k=1}^{j} L_{ik} \cdot conj(L_{jk})$$

where L_{jk} is the transpose of L_{kj}.

Off-diagonal elements of L

$$L_{ij} = \frac{A_{ij} - \sum_{k=1}^{j-1} L_{ik} \cdot conj(L_{jk})}{L_{jj}} \tag{13.2}$$

By substituting for L_{jj} the full recursion can be seen

$$L_{ij} = \frac{A_{ij} - \sum_{k=1}^{j-1} L_{ik} \cdot conj(L_{jk})}{\sqrt{A_{jj} - \sum_{k=1}^{j-1} L_{jk} \cdot conj(L_{jk})}}$$

Eqs. (13.1) and (13.2) are the equations that will be used to find L_{jj} and L_{ij}. In solving for one element, a vector dot product proportional to its matrix size must calculated.

Although matrices **A** and **L** may be complex, the diagonal elements must be real by definition. Therefore, the square root is taken of a real number. The denominator of the divide function is also real.

Once that **L** is computed, perform the substitution for matrix **A**:

$$\mathbf{A} \cdot \mathbf{x} = \mathbf{b} \rightarrow \mathbf{L} \cdot \mathbf{L}^{\mathrm{H}} \cdot \mathbf{x} = \mathbf{b}$$

Next, we want to introduce an intermediate result, the vector **y**. The product of matrix $\mathbf{L}^{\mathrm{H}}$ and vector **x** is defined as the vector **y**.

$$\mathbf{y} = \mathbf{L}^{\mathrm{H}} \cdot \mathbf{x} \rightarrow \mathbf{L} \cdot \mathbf{y} = \mathbf{b}$$

The vector **y** can be computed by a recursive substitution, called forward substitution, because it is done from top to bottom. For example, $y_1 = b_1/L_{11}$. Once y_1 is known, it is easy to solve for y_2 and so on.

To restate, **L** and $\mathbf{L}^{\mathrm{H}}$ are known after the decomposition, and $\mathbf{L} \cdot \mathbf{L}^{\mathrm{H}} \cdot \mathbf{x} = \mathbf{b}$. Then we define $\mathbf{L}^{\mathrm{H}} \cdot \mathbf{x} = \mathbf{y}$ and substitute. Then we are solving $\mathbf{L} \cdot \mathbf{y} = \mathbf{b}$. **L** is a lower triangular matrix, **y** and **b** are column vectors, and **b** is known. The values y_j can be found from top to bottom using forward substitution. The equations to find **y** is

Intermediate result y

$$y_j = \frac{b_j - \sum_{k=1}^{j-1} y_k \cdot L_{jk}}{L_{jj}} \tag{13.3}$$

Eq. (13.3) is almost the same as Eq. (13.2). If we treat **b** as an extension of **A** and **y** as an extension of **L**, the process of solving **y** is the same as solving **L**. The only difference is, in the multiply operation, the second operand is not conjugated (this consideration may be important for hardware implementation, allowing computational units to be shared).

After **y** is computed, **x** can be solved by backward substitution in $\mathbf{L}^{\mathrm{H}} \cdot \mathbf{x} = \mathbf{y}$ (Fig. 13.7). $\mathbf{L}^{\mathrm{H}}$ is an upper triangular matrix, therefore, **x** has to be solved in reverse order—from bottom to top. That is why it is called backward substitution. Therefore, solving **x** is separate process from the Cholesky decomposition and forward substitution solver.

$$\begin{pmatrix} L_{11} & 0 & 0 & 0 \\ L_{21} & L_{22} & 0 & 0 \\ L_{31} & L_{32} & L_{33} & 0 \\ L_{41} & L_{42} & L_{43} & L_{44} \end{pmatrix} * \begin{pmatrix} y_1 \\ y_2 \\ y_3 \\ y_4 \end{pmatrix} = \begin{pmatrix} b_1 \\ b_2 \\ b_3 \\ b_4 \end{pmatrix}$$

$$\begin{pmatrix} L_{11} & L^*_{21} & L^*_{31} & L^*_{41} \\ 0 & L_{22} & L^*_{32} & L^*_{42} \\ 0 & 0 & L_{33} & L^*_{43} \\ 0 & 0 & 0 & L_{44} \end{pmatrix} * \begin{pmatrix} x_1 \\ x_2 \\ x_3 \\ x_4 \end{pmatrix} = \begin{pmatrix} y_1 \\ y_2 \\ y_3 \\ y_4 \end{pmatrix}$$

Figure 13.7
Solving for y and then for x.

In addition, **y** has to be completely known before solving **x**. The equation to solve **x** is shown below, where VS = N is the length of vectors x and y.

Desired result x

$$x_j = \frac{y_j - \sum\limits_{k=j+1}^{VS} x_k \cdot L'_{jk}}{L'_{jj}} \tag{13.4}$$

The algorithm steps and data dependencies are more easily illustrated using a small [4 × 4] matrix example.

13.3 4 × 4 Cholesky Example

$$A = \begin{bmatrix} L_{11} & 0 & 0 & 0 \\ L_{21} & L_{22} & 0 & 0 \\ L_{31} & L_{32} & L_{33} & 0 \\ L_{41} & L_{42} & L_{43} & L_{44} \end{bmatrix} \begin{bmatrix} L_{11} & L_{21} & L_{31} & L_{41} \\ 0 & L_{22} & L_{32} & L_{42} \\ 0 & 0 & L_{33} & L_{43} \\ 0 & 0 & 0 & L_{44} \end{bmatrix}$$

$$= \begin{bmatrix} L_{11}^2 & & & ConjugateSymmetric \\ L_{21}L_{11} & L_{21}^2 + L_{22}^2 & & \\ L_{31}L_{11} & L_{31}L_{21} + L_{32}L_{22} & L_{31}^2 + L_{32}^2 + L_{33}^2 & \\ L_{41}L_{11} & L_{41}L_{21} + L_{42}L_{22} & L_{41}L_{31} + L_{42}L_{32} + L_{43}L_{33} & L_{41}^2 + L_{42}^2 + L_{43}^2 + L_{44}^2 \end{bmatrix}$$

Notice the diagonal elements to be solved depend on elements to the left in the lower triangle. If the elements are solved one at a time, the top leftmost element is solved first and the remaining matrix can be solved in a horizontal zigzag or vertical zigzag fashion from the top left element to the bottom right element (Fig. 13.8). For the subsequent elements in a column, it only depends on the elements on its left and the row holding the corresponding diagonal element. The vertical zigzag fashion requires less dependency as all subsequent elements in a column can be processed at the same time.

The order of calculations is shown for clarity of the recursion relationships. This can be derived from the matrix equation above by computing each element of A starting with A_{11} and back solving for each element of L. The order of computation is given below.

$$L_{11} = \sqrt{A_{11}} \quad L_{21} = \frac{A_{21}}{L_{11}} \quad L_{31} = \frac{A_{31}}{L_{11}} \quad L_{41} = \frac{A_{41}}{L_{11}}$$

$$L_{22} = \sqrt{A_{22} - L_{21} \cdot conj(L_{21})} \quad L_{32} = \frac{A_{32} - L_{31} \cdot conj(L_{31})}{L_{22}}$$

$$L_{42} = \frac{A_{42} - L_{41} \cdot conj(L_{41})}{L_{22}}$$

$$L_{33} = \sqrt{A_{33} - L_{31} \cdot conj(L_{31}) - L_{32} \cdot conj(L_{32})}$$

$$L_{43} = \frac{A_{43} - L_{41} \cdot conj(L_{41}) - L_{42} \cdot conj(L_{42})}{L_{22}}$$

$$L_{44} = \sqrt{A_{44} - L_{41} \cdot conj(L_{41}) - L_{42} \cdot conj(L_{42}) - L_{43} \cdot conj(L_{43})}$$

Given below are the forward substitution equations for a [4 × 4] matrix, which can be solved by recursion. This is referred to a "forward" substitution, because the solution order is from top to bottom or in the natural sequence of order.

$$\mathbf{y} = \mathbf{L}^{\mathrm{H}} \cdot \mathbf{x} \rightarrow \mathbf{L} \cdot \mathbf{y} = \mathbf{b}$$

$$y_1 = \frac{b_1}{L_{11}} \quad y_2 = \frac{b_2 - y_1 \cdot L_{21}}{L_{22}}$$

$$y_3 = \frac{b_3 - y_1 \cdot L_{31} - y_2 \cdot L_{32}}{L_{33}}$$

$$y_4 = \frac{b_4 - y_1 \cdot L_{41} - y_2 \cdot L_{42} - y_3 \cdot L_{43}}{L_{44}}$$

The example back substitution equations for the [4 × 4] matrix are given below. Here the recursion order is to solve for the bottom term of x and work upward (or backward).

$$A \cdot x = b \rightarrow L \cdot L' \cdot x = b$$

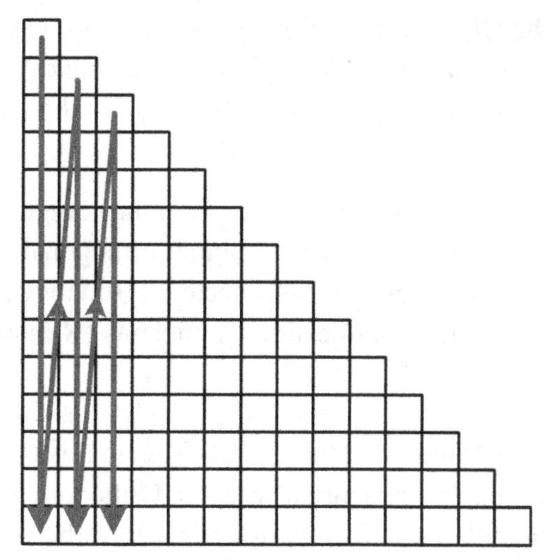

Figure 13.8
Order of solving for elements of matrix to minimize data dependencies.

$$x_4 = \frac{y_4}{L_{44}} \quad x_3 = \frac{y_3 - x_4 \cdot conjL_{43}}{L_{33}}$$

$$x_2 = \frac{y_2 - x_4 \cdot conjL_{42} - x_3 \cdot conjL_{32}}{L_{22}}$$

$$x_1 = \frac{y_1 - x_4 \cdot conjL_{41} - x_3 \cdot conjL_{31} - x_2 \cdot conjL_{21}}{L_{11}}$$

Notice that, from Eqs. (13.2)–(13.4), there is no need to find L_{jj} directly to solve **y** and **x**, all that is needed is the inverse of L_{jj}. Doing so also eliminates the need of the less efficient square root function and divide function.

The numerical accuracy of the results is also improved with less intermediate calculation steps.

13.4 QR Decomposition

Next the QRD algorithm is considered. The problem statement is the same as before $\mathbf{A} \cdot \mathbf{x} = \mathbf{b}$. For this example, **A** is an [N × N] matrix (but is not required to be Hermitian or conjugate symmetric). In addition, **A** does not need to be a square matrix, because the QRD can also work for [M × N] rectangular matrices. The vector **x** is an unknown complex [N × 1] vector, and **b** is a known complex [N × 1] vector. The solution is $\mathbf{x} = \mathbf{A}^{-1} \cdot \mathbf{b}$, which requires the inversion of matrix **A**.

This is avoided by using the QR decomposition, where the matrix **A** is not required to be symmetric, and can even be rectangular, although the rectangular case is not shown here. **A** will be factored into the product of [M × N] matrix Q and [N × N] matrix R, where

$$\mathbf{A} = \mathbf{Q} \cdot \mathbf{R}$$

The Q matrix is orthogonal, and the R matrix is upper right triangular. The method described here for the decomposition is Gram–Schmidt orthogonalization, and it is summarized below. It creates N orthogonal basis vectors (**Q** matrix), which are used to express the vectors in matrix **A**, using projection coefficients (**R** matrix).

The basic idea is this:

The matrix **A** is composed of N independent vectors. The goal is to express each of those vectors as a linear combination of orthonormal vectors. Orthonormal vectors are the same as orthogonal, except in the orthonormal case, in which all of the vectors are unity length.

A key property of an orthonormal matrix is that the conjugate transpose (Hermitian) is also the inverse matrix. So $\mathbf{Q}^H \cdot \mathbf{Q} = \mathbf{I}$. Therefore, the matrix inversion of an orthonormal matrix is trivial. This property is the basis for QR Decomposition.

The problem is solved as follows:

$$\mathbf{A} \cdot \mathbf{x} = \mathbf{b}$$

$$\mathbf{Q} \cdot \mathbf{R} \cdot \mathbf{x} = \mathbf{b}$$

$$\mathbf{Q}^H \cdot \mathbf{Q} \cdot \mathbf{R} \cdot \mathbf{x} = \mathbf{Q}^H \cdot \mathbf{b}$$

$$\mathbf{I} \cdot \mathbf{R} \cdot \mathbf{x} = \mathbf{Q}^H \cdot \mathbf{b}$$

Let's define:

$$\mathbf{y} = \mathbf{Q}^H \cdot \mathbf{b}$$

$$\mathbf{R} \cdot \mathbf{x} = \mathbf{y}$$

As R is an upper triangular [N × N] matrix, **x** can be solved for by back substitution, similar to the Cholesky case.

QRD floating point operations per [M × N] matrix is computed using the standard equation:

$$\mathbf{FLOPS} = \mathbf{8\,MN^2 + 6.5\,N^2 + MN}$$

The exact number of computations will vary per implementation on a given architecture but provides an accepted means to compare computational efficiency of different architectures of this common algorithm.

13.5 Gram–Schmidt Method

The process is to decompose **A** into **Q** · **R**.

A is composed of column vectors $\{\mathbf{a_1}, \mathbf{a_2}, \mathbf{a_3}\ldots \mathbf{a_n}\}$. These vectors are independent, or else the matrix is singular and no inverse exists. Independent means that n vectors can be used to define any point in n-dimensional space. However, unlike the axis vectors (x, y, and z in three-dimensional space), independent vectors do not have to be orthogonal (90-degree angles between vectors).

The next step is to create orthonormal $\{\mathbf{q_1}, \mathbf{q_2}, \mathbf{q_3}\ldots \mathbf{q_n}\}$ vectors from the $\mathbf{a_i}$ vectors. The first one is easy—simply take $\mathbf{a_1}$ and normalize it (make the magnitude equal to 1).

$$\mathbf{q_1} = \mathbf{u_1}/\text{norm}(\mathbf{u_1}) \text{ and } \mathbf{u_1} = \mathbf{a_1},$$

where norm $(\mathbf{u_1}) = \text{sqrt}(\mathbf{u}_{1,1}^2 + \mathbf{u}_{2,1}^2 + \mathbf{u}_{3,1}^2 \ldots + \mathbf{u}_{m,1}^2)$ and $\mathbf{u_1}$ is an m-length column vector. Dividing a vector by its norm (which is a scalar) will normalize that vector or give it a length of 1.

We now need to introduce the concept of projection. The projection of a vector is with respect to a reference vector. The projection is the component of a vector that is colinear or in the same direction as the reference vector. The remaining portion of the vector will be orthogonal to the reference vector. The projection is a scalar value, since the direction of the projection is by definition the reference vector direction (Fig. 13.9).

Dot<$\mathbf{v_1}$, $\mathbf{v_2}$> is the dot product of $\mathbf{v_1}$, $\mathbf{v_2}$ as defined in Fig. 13.2. The dot product produces a scalar result.

$$\text{Scaler proj}_{\mathbf{u}}(\mathbf{a}) = \text{dot} < \mathbf{u}, \mathbf{a} > /\text{dot} < \mathbf{u}, \mathbf{u} >$$

Reference vectors do not need to be orthogonal, just independent. Any vector in n-space can be expressed as a combination of the n reference vectors. The $\mathbf{q_i}$ vectors are both independent and orthogonal. All of the $\mathbf{a_i}$ vectors can be expressed as linear combinations of $\mathbf{q_i}$ vectors, as defined by the values in the **R** matrix. By defining $\mathbf{q_1}$ to be colinear or in same direction as $\mathbf{a_1}$, this means $\mathbf{a_1}$ is defined only by the scale factor of $r_{1,1}$. There is a two-dimensional plane defined by the vectors $\mathbf{a_1}$ and $\mathbf{a_2}$. The vector $\mathbf{q_2}$ lies in this plane

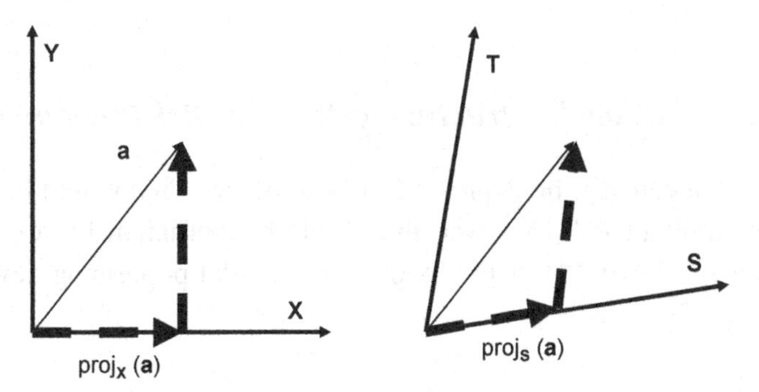

Figure 13.9
Examples of projection.

and is orthogonal to $\mathbf{q_1}$. Therefore, $\mathbf{a_2}$ can be defined as a linear combination of $\mathbf{q_1}$ and $\mathbf{q_2}$, defined by the scale factors of $r_{1,2}$ and $r_{2,2}$.

Using this, the $\mathbf{q_i}$ vectors and the values of the **R** matrix can be computed.

$$\mathbf{q_1} = \mathbf{u_1}/\text{norm}(\mathbf{u_1}) \text{ and } \mathbf{u_1} = \mathbf{a_1}$$

$$\mathbf{q_2} = \mathbf{u_2}/\text{norm}(\mathbf{u_2}) \text{ and } \mathbf{u_2} = \mathbf{a_2} - \text{proj}_{\mathbf{u1}}(\mathbf{a_2})$$

$$\mathbf{q_3} = \mathbf{u_3}/\text{norm}(\mathbf{u_3}) \text{ and } \mathbf{u_3} = \mathbf{a_3} - \text{proj}_{\mathbf{u1}}(\mathbf{a_3}) - \text{proj}_{\mathbf{u2}}(\mathbf{a_3})$$

$$\mathbf{q_4} = \mathbf{u_4}/\text{norm}(\mathbf{u_4}) \text{ and } \mathbf{u_4} = \mathbf{a_4} - \text{proj}_{\mathbf{u1}}(\mathbf{a_4}) - \text{proj}_{\mathbf{u2}}(\mathbf{a_4}) - \text{proj}_{\mathbf{u3}}(\mathbf{a_4})$$

$$\mathbf{q_n} = \mathbf{u_n}/\text{norm}(\mathbf{u_n}) \text{ and } \mathbf{u_n} = \mathbf{a_n} - \text{proj}_{\mathbf{u1}}(\mathbf{a_n}) - \text{proj}_{\mathbf{u2}}(\mathbf{a_n}) - \text{proj}_{\mathbf{u3}}(\mathbf{a_n})\ldots - \text{proj}_{\mathbf{u(n-1)}}(\mathbf{a_n})$$

$\mathbf{Q} = \{\mathbf{q_1}, \mathbf{q_2}, \mathbf{q_3} \ldots \mathbf{q_n}\}$ which is a matrix composed of orthonormal column vectors.

The upper triangular **R** matrix is the scalar coefficients of the projection.

$$\mathbf{R} = \begin{matrix} \text{Dot} < \mathbf{q_1}, \mathbf{a_1} > & \text{Dot} < \mathbf{q_1}, \mathbf{a_2} > & \text{Dot} < \mathbf{q_1}, \mathbf{a_3} > & \ldots & \text{Dot} < \mathbf{q_1}, \mathbf{a_n} > \\ 0 & \text{Dot} < \mathbf{q_2}, \mathbf{a_2} > & \text{Dot} < \mathbf{q_2}, \mathbf{a_3} > & \ldots & \text{Dot} < \mathbf{q_2}, \mathbf{a_n} > \\ 0 & 0 & \text{Dot} < \mathbf{q_3}, \mathbf{a_3} > & \ldots & \text{Dot} < \mathbf{q_3}, \mathbf{a_n} > \\ 0 & 0 & 0 & \ldots & \text{Dot} < \mathbf{q_4}, \mathbf{a_n} > \\ . & . & . & . & . \\ 0 & 0 & 0 & 0 & \text{Dot} < \mathbf{q_n}, \mathbf{a_n} > \end{matrix}$$

Now, $\mathbf{A} = \mathbf{Q} \cdot \mathbf{R}$.

This can be verified by performing the matrix multiplication. For example, multiplying the first row of Q by the first column of R gives

$$a_{1,1} = q_{1,1} \cdot r_{1,1} (\text{since the other values of the } \mathbf{r_1} \text{ column are } 0) = (a_{1,1}/\text{norm}(\mathbf{a_1})) \cdot \text{Dot} < \mathbf{q_1}, \mathbf{a_1} >= \mathbf{a_{1,1}}$$

13.6 QR Decomposition Restructuring for Parallel Implementation

The QR decomposition can also be expressed in code below. This would be typical for a serial implementation on a CPU. However, this would be inefficient for a parallel implementation, such as in a GPU or FPGA or other parallel processing device.

```
for k = 1:n
    r(k,k) = norm(A(1:m, k));
    for j = k+1:n
        r(k, j) = dot(A(1:m, k), A(1:m, j)) / r(k,k);
    end
```

```
        q(1:m, k) = A(1:m, k) / r(k,k);
        for j = k+1:n
            A(1:m, j) = A(1:m, j) - r(k, j) * q(1:m, k);
        end
    end
```

The algorithm can be restructured to allow taking advantage of the parallel implementation in a GPU or FPGA. The `norm` function can be replaced by a dot product and square root operation, which is often more efficient. Second, the first inner loop requires `r(k,k)` to be computed first, which has a longer latency. Third, the second inner loop requires `q(1:m, k)` to be computed first, which also has a long latency. To remove the data dependencies, the order of operations will be changed. All of the **r** terms will be calculated first. This can be before $\mathbf{q}_i$ is known.

```
for k = 1:n
    r(k,k) = norm(A(1:m, k));

r2(k,k) = dot(A(1:m, k), A(1:m, k);

r(k,k) = sqrt(r2(k,k));

    for j = k+1:n
        r(k, j) = dot(A(1:m, k), A(1:m, j)) / r(k,k);
        rn(k, j) = dot(A(1:m, k), A(1:m, j))
        r(k, j) = rn(k,j)/ r(k,k);
    end
    q(1:m, k) = A(1:m, k) / r(k,k);
    for j=k+1:n
        A(1:m, j) = A(1:m, j) - r(k,j) * q(1:m,k);
    end
end
```

Next, the following substitutions can be performed in the second inner loop. Replace `r(k,j)` with `rn(k,j)/r(k,k)` and replace `q(1:m,k)` with `A(1:m,k)/r(k,k)`.

```
for k = 1:n
    r2(k,k) = dot(A(1:m, k), A(1:m,k);
    r(k,k) = sqrt(r2(k,k));
    for j = k+1:n
        rn(k, j) = dot(A(1:m, k), A(1:m, j));
        r(k, j) = rn(k,j)/ r(k,k);
    end
    q(1:m, k) = A(1:m, k) / r(k,k);
    for j = k+1:n
        A(1:m, j) = A(1:m, j) - r(k,j) * q(1:m,k);
        A(1:m, j) = A(1:m, j) - rn(k,j)/ r(k,k) * A(1:m,k) / r(k,k);
    end
end
```

Then operations can be reordered into two functional groups. Every r term depends on the norm of the first column for the current iteration, the calculation of which requires a deep pipeline. The same datapath used for the vector product required for the norm can also be used to calculate the vector operation for each r term. All vector operations for the r term calculation, **<v_i,v_i>** for $\mathbf{r_{ii}}$, and **<v_j,v_i>** for $\mathbf{r_{ij}}$ can be issued on subsequent clock cycles.

Following the vector operations, a separate function can be used for the square roots and inverse square roots. The square root and divide operations are scheduled as late as possible, after the vector operations. All of the **r** terms can therefore be calculated without any data dependencies, once the pipeline is full.

(first computational group)

```
for k = 1:n
    r2(k,k) = dot(A(1:m, k), A(1:m,k);
    for j = k+1:n
        rn(k, j) = dot(A(1:m, k), A(1:m, j));
    end
    for j = k+1:n
         A(1:m, j) = A(1:m, j) - (rn(k,j) / r2(k,k)) * A(1:m,k);
    end
end
```

(second computational group)

```
for k = 1:n
    r(k,k) = sqrt(r2(k,k));
    for j = k+1:n
        r(k, j) = rn(k,j)/ r(k,k);
    end
    q(1:m, k) = A(1:m, k) / r(k,k);
end
```

The upper loop can run with few stalls, as there are no long latency math operations. This is where the bulk of the computation is performed. The **r** terms are then used as one of the inputs to the vector multiplier, the first column for the outer loop is generated, and all of the subsequent columns updated. The only data dependency is between the first **r** term output of the datapath and the start of the $\mathbf{q_i}$ vector calculation, which may introduce a wait state between the two parts of the inner loop. As the pipeline depth of the datapath will be fixed, but the inner loop reduces by one with each pass, wait states may be required at some point in the processing of the matrix, which once started, will gradually increase by each pass.

The lower loop can run as data become available and are relatively latency insensitive. In this way, the computational units can avoid unnecessary stalls, which can improve the matrix processing throughput.

CHAPTER 14

Field-Oriented Motor Control

Electric motors have been ubiquitous for 100 years or more and were controlled using mechanical or analog methods. Today, motors are increasingly digitally controlled, which can provide greater performance, reliability, and efficiency. This chapter will first give an overview of electric motor theory and operation and then introduce a basic digital motor control algorithm known as field-orientated control, or FOC.

14.1 Magnetism Basics

In the 1800s Michael Faraday speculated that magnetic fields exist as lines of force, which we today refer to as "magnetic flux," which is measured in units of "Webers." Essentially, the more flux there is, the stronger the magnet will be. We conventionally think of flux leaving the north pole and reentering the south pole of a magnet, as shown by the arrows in Fig. 14.1. If we measure how much flux cuts through a surface area which is perpendicular to the flux path, it provides a measure of the *flux density* at that particular spot in space. One Weber of flux cutting through one square meter of area constitutes a

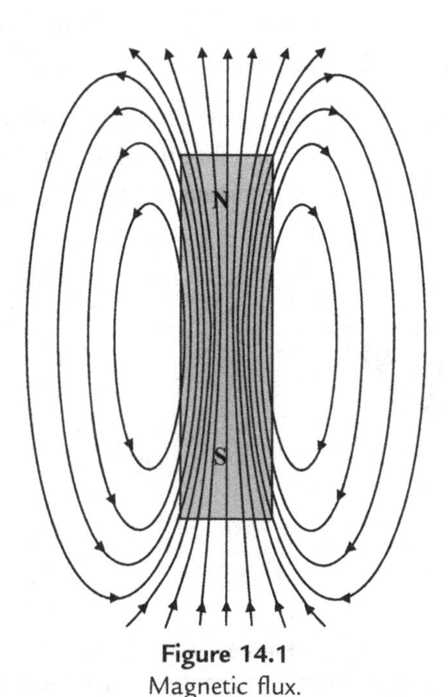

Figure 14.1
Magnetic flux.

Digital Signal Processing 101. http://dx.doi.org/10.1016/B978-0-12-811453-7.00014-7

flux density of one "Tesla," named after Nikola Tesla, who is also the inventor of the AC induction motor (ACIM).

Also in the early 1800s, Hans Christian Oersted discovered that current flowing in a wire creates its own magnetic field, and when this field interacts with a second magnetic field, the result is a force acting on the conductor. This force is proportional to the amount of current flowing in the wire, the strength of the second magnetic field, and the length of wire that is affected by the second magnetic field. The direction of the force can be determined by a technique know as the *right-hand rule*. If your right hand is configured as shown below, where your thumb points in the direction of *positive* current flow and your index finger points in the direction of the second magnetic field's flux (i.e., flowing from the north pole to the south pole), then your middle finger will be pointing in the direction of the force acting on the wire, as shown in Fig. 14.2.

The back EMF (electromotive force) is the voltage generated in a loop of wire caused when the magnetic flux enclosed by that loop is changing. This can result in several ways. The intensity of the magnetic flux level may be controlled by an adjustable source. Or it could be caused if the flux field is moving relative to the loop of wire, or if the loop of wire itself is rotating in the magnetic flux field or both. This effect led to "Faraday's Law," which states that the voltage generated in a loop of wire is equal to the rate at which the magnetic flux threading through that loop of wire is changing.

Using the right-hand rule, this can be applied to a rotating system. In this case, we can see the cross section of a cylinder, which is once again rotating in a uniform magnetic field. A wire is placed about the cylindrical cross section, as shown in Fig. 14.3.

From Faraday's law, the voltage that will be generated through this wire (from one end to the other) is equal to the rate of change of the flux threading through the loop. Even though the flux of the magnetic field is of constant strength, the cross-sectional area of the

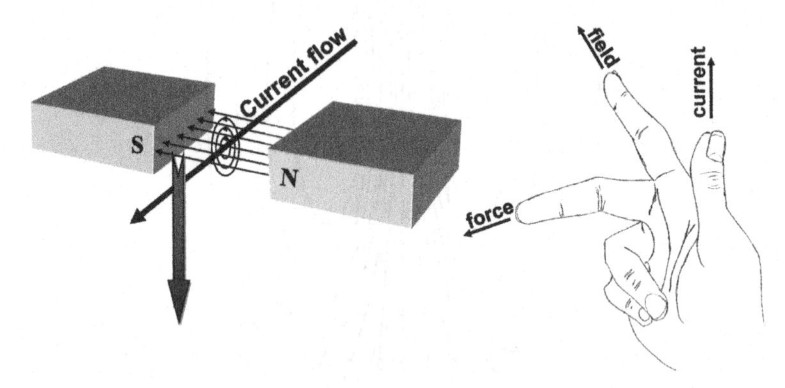

Figure 14.2
Force vector generated by current through magnetic field.

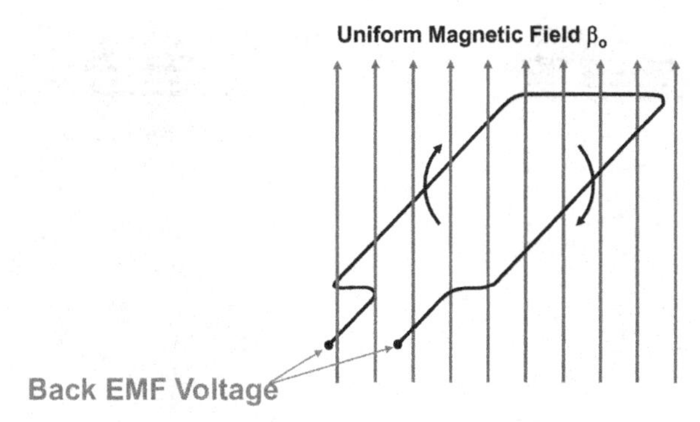

Figure 14.3
Back electromotive force (EMF) voltage.

loop with respect to the magnetic field changes by rotating the wire loop. When the loop of wire is straight up and down (vertically orientated), the loop area collapses to zero, and this is the fastest rate of flux change, and therefore when the peak voltage is generated. When the loop of wire is orientated perpendicular to the flux field, the rate of change of the loop area is zero at that instantaneous moment, producing no voltage. Therefore, as the loop rotated, the voltage induced will be sinusoidal, centered about zero.

This effect can be multiplied, by replacing a single loop of wire with N loops of wire, all connected together in series. The voltages from each loop add together in series, creating a much higher voltage, which has the same effect as multiplying the flux in the loop by a factor of "N." The total rate of change of the flux changes as a function of the rotational angle, the number of turns, the field strength, and the geometry of the machine and generates an AC waveform as the machine rotates.

14.2 AC Motor Basics

Using the same principals, a motor can be driven from an AC voltage (Fig. 14.4). The rotor (rotating part) is composed of a permanent magnet, and the AC voltage is used to magnetize the stator (stationary portion). The stator acts as an electromagnet, excited by the AC waveform. The magnetic field is implemented by coil of wire, and effect of this can be magnified by using soft iron core, which becomes magnetized and conducts the magnetic flux. The rotor will rotate at the same frequency as the AC voltage frequency, as the permanent magnet will follow the magnetic polarity of the stator.

A multiphase AC motor can also be constructed. Three phases are common in high-power industrial motors, with each phase 120 degrees relative to the others. The coils of each phase are interleaved around the stator.

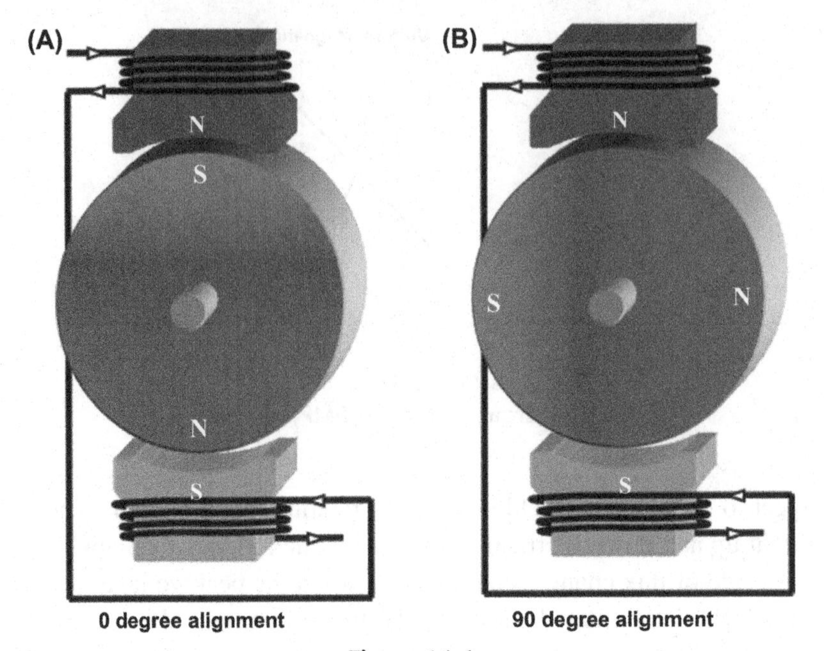

Figure 14.4
Basic AC motor operation.

14.3 DC Motor Basics

A motor can also be driven by a DC voltage. In this a form of commutation must be implemented, to create a rotating magnetic field. In this case, permanent magnets are used in the stator, not in the rotor (Fig. 14.5). The rotor is an electromagnet, which can have multiple coils of wire wound about the iron cores mounted on the rotor shaft. The electromagnets in the rotor are excited when contact is made through the brushes, using

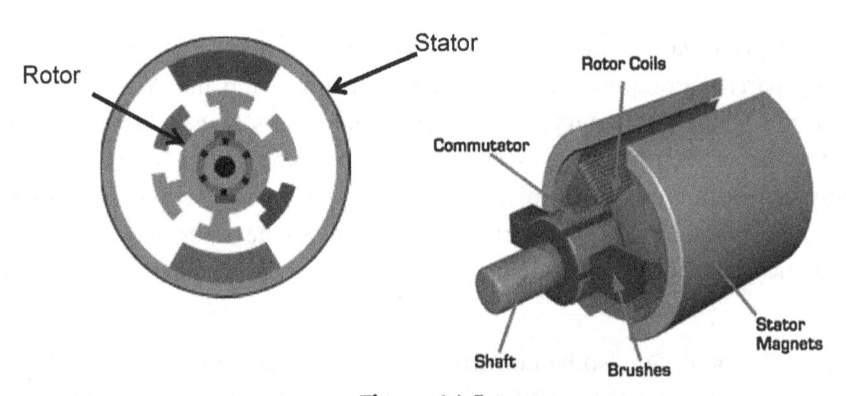

Figure 14.5
Brush DC motor diagram.

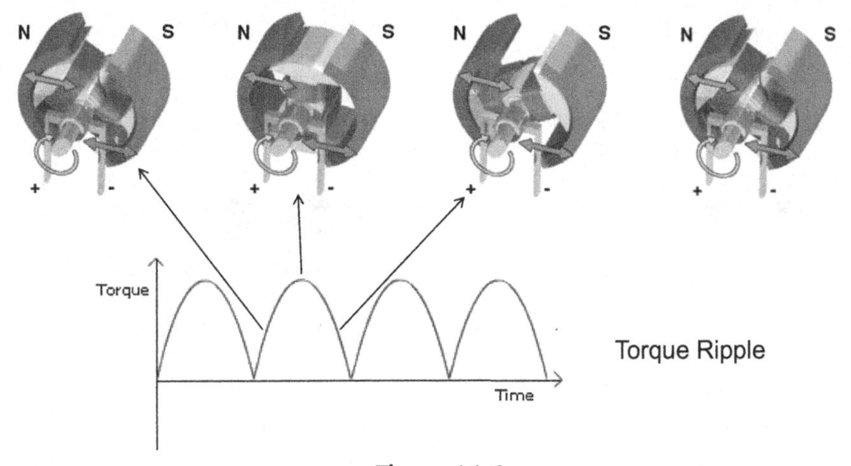

Figure 14.6
Brush DC torque ripple.

DC current. When the shaft has rotated 180 degrees the same DC current will flow in the opposite direction through the wire coil, exciting the same electromagnet in the opposite polarity. Several coils and brush contacts can be interleaved about the shaft, each making contact as the shaft rotates and exciting each rotor coil in turn.

The disadvantage of DC brush motors is that the mechanical commutation provided by the brushes requires periodic brush replacement due to erosion, and the large amount of electromagnetic induction (EMI) generated by the rotor's sparking as they make and break contact. A further disadvantage is torque ripple, which can result in vibration. This is due to changes in torque as the angular relationship changes due to rotation. The peak torque occurs when the rotor electromagnet is orthogonal to the stator magnets and zero torque when the rotor electromagnet is aligned with the stator magnet. At this moment, the brushes will break contact, and then reconnect with current flowing in the opposite direction. In this way, a net torque is maintained in one rotational direction. However, the torque will be sinusoidal in nature, resulting in torque ripple and mechanical vibration (Fig. 14.6).

14.4 Electronic Commutation

Alternately, the DC motor can be electrically commutated. The basic structure used is the "H" bridge (Fig. 14.7), which depending on which transistors are switched on, can drive DC current in either direction through a wire coil. There is no need for brushes, which are a maintenance concern. And the current waveform can be changed as rapidly as necessary, without any mechanical limitation.

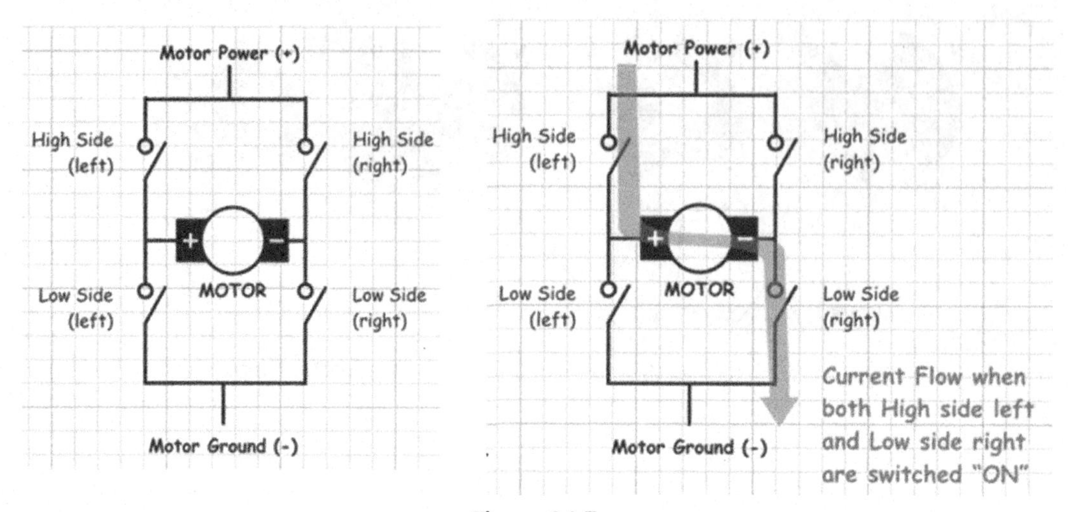

Figure 14.7
H bridge for electronic commutation.

A BLDC (brushless DC) motor is like a brush DC motor turned inside out. To make the motor spin, the current is switched from one set of coils to the next. Normally, three is the minimum number of phases used (Fig. 14.8). Motor speed can be controlled by the rate of the electrical communication, providing for means of regulating the motor's rotational speed. However, this assumes the motor is able to rotate at the rate of the commutation

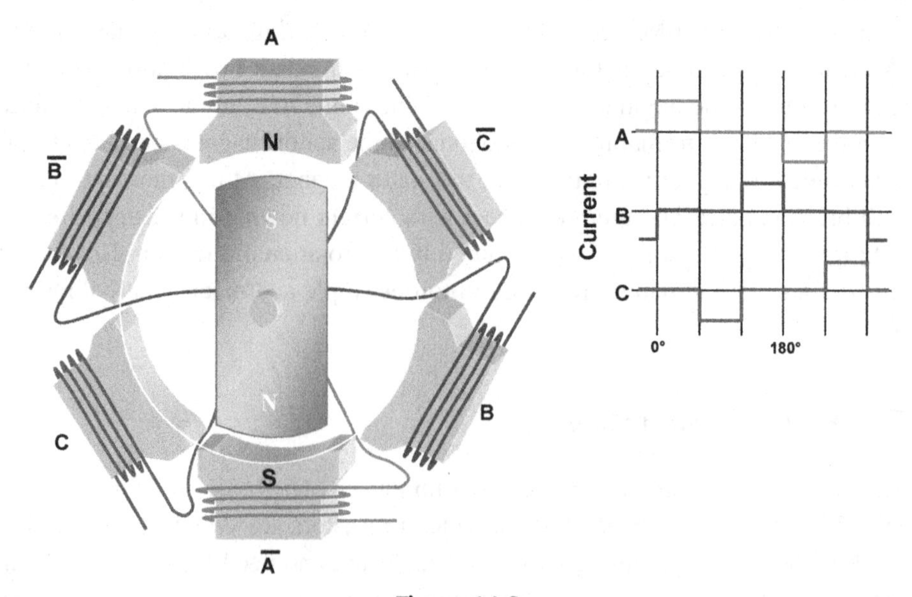

Figure 14.8
Multiphase DC commutation.

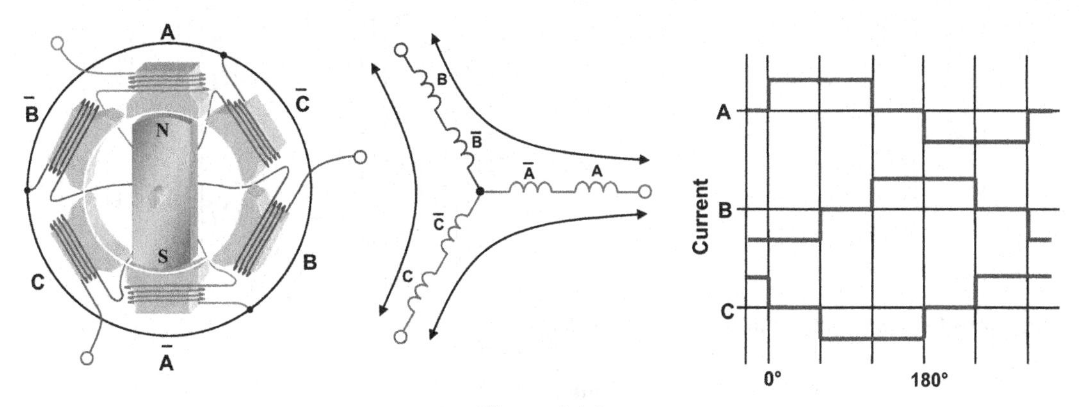

Figure 14.9
Brushless DC motor commutation.

with a given load. Generally, an electrically commutated motor must have some means of feedback providing the rotational position of the rotor at any given instant.

A more efficient commutation is to turn on more than one set of coils at a time, to provide more torque in the motor. Here there are two coils energized at the same time so that when one coil has positive current, the other energized coil has negative current. A clever way to use the same current in both coils is to connect all the coils together inside the motor, called a "Y"-connected winding pattern. The current waveforms for each winding are shown in Fig. 14.9. The commutation sequence provides a complete 360 degrees rotation of the motor, with each phase being 120 degrees relative to the others. Because in each state there are two coils energized, and one coil idle, this provides for greater motor torque that a motor commutated with only one coil active at a given time.

Another benefit over brush commutation is that the current can be switched on and off very rapidly, much of that can be used to create more complex current waveforms. Normally, a linear amplifier or a variable voltage source would be necessary to create a current waveform with different levels. However, this same effect can be created by switching the current very rapidly from a constant or DC power source. This is known as pulse width modulation (PWM). The duty cycle, which is the ratio of "ON" time to the total period of the switching frequency, determines the average amount of current flow. This requires a transistor switching frequency to be much higher than the motor commutation rates. This waveform is then filtered by the inductance in the motor to smooth or average the current output waveform. A sinusoidal waveform is often used and provides the most efficient and smoothest (minimum torque ripples) operation (Fig. 14.10).

Creating a sinusoidal waveform from a DC source to commutate a motor is starting to look a lot like an AC motor, since the stator coils must be driven by some type AC waveforms to cause motor rotation. Therefore, motors of this type are often called BPM or

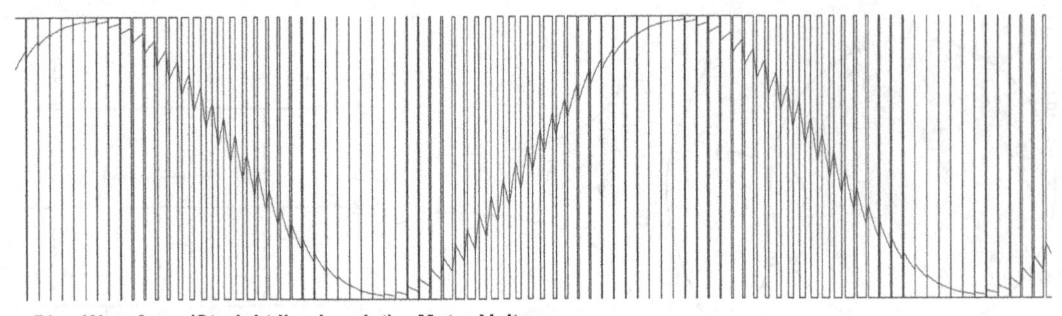

Blue Waveform (Straight line in print) – Motor Voltage
Red Waveform (Curved line in print) – Filtered motor voltage

Figure 14.10
Pulse width modulation motor commutation.

brushless permanent magnet motors. The motor is electrically commutated and has permanent magnets in the rotor. Two types of BPM motors are the BLDC motor, which is driven by "square" wave current modulation, and PMAC (permanent magnet AC) motor, which is driven by sinusoidal current modulation. There are further distinctions with regard to the types of permanent magnets used in rotor.

All of the brushless motors described do require a means of determining the rotational position and speed of the rotor to be able to properly energize the coils in an optimal manner to control the motor behavior (Fig. 14.11). Some applications require precise rotational position. For this method, it is common to mount hall-effect types of sensor on

Figure 14.11
Motor rotational measurements.

the rotor shaft. Using two hall-effect sensors with a 90 degrees shift in their pattern provides the means to determine the direction and amount of rotation and is known as a quadrature encoder. Other applications require precise control of the motor torque. This is commonly done by sensing the motor drive currents, which are proportional to the torque generated by the motor. Other more advanced methods have been developed, which are sensorless. These methods monitor the back EMF of the motor stator windings and use that to estimate the motor speed. A drawback of this method is that back EMF is difficult to measure at low motor speeds.

14.5 AC Induction Motor

The last type of motor that will be covered is one of the most popular. It is the ACIM and is the most common type of motor used in both household and industrial applications. This motor is extremely simple and elegant, as well as can be very efficient, especially under constant load applications. It has no brushes and no magnets. It can be driven by AC waveforms provided directly from the electrical utility (no commutation mechanism required) or by AC waveforms generated using electronic commutation. The use of sensors is optional in the ACIM, particularly when the motor is driven directly from AC utility grid current. The ability to dispense with magnets, commutation circuits or brushes, and sensors makes the ACIM extremely reliable and economical for applications where precise rotational speed is not necessary—one common example is an industrial pump.

The key innovation is the rotor. The rotor contains multiple closed loops of wire. They are actually mounted at a slight angle to the rotor axis, as can be seen in Fig. 14.12. The stator is excited by AC waveforms. This will cause a changing magnetic flux through the closed coil within the rotor. Recall Fig. 14.3, depicting how a changing flux enclosed by a wire

Figure 14.12
AC induction motor.

will cause EMF or voltage to be generated. If the wire is a close loop, this EMF voltage will cause a large flow of current in the wire loop. Current is induced in the rotor circuit from the stator circuit; much the same way that secondary current is induced from the primary coil in a transformer. This in turn will generate a magnetic field with its own flux, just as a permanent magnet in the rotor would. The torque generated will be proportional to the strength of the rotor magnetic field generated in the rotor coils, which will in turn be proportional to the rate of change of the flux through the rotor coils, which in turn is proportional to the rate at which the AC waveforms are changing. Now consider two cases. The first being when the rotor is stationary, which would be the condition on start-up of the motor. This provides the maximum rate of flux change between the rotating stator field and the stationary rotor coils. This will, therefore, be the maximum torque, which is a very desirable condition on start-up of the motor. The second case would be when the rotor is rotating at the same rate as the stator magnetic field. In this case, there is no change of flex through the rotor coils and, therefore, there is no current generated in rotor coils, and as a result no torque will be generated.

The normal operating condition of the ACIM motor is the "slip" condition. At steady state, the motor rotation speed will be such that the difference between the stator magnetic field rotational rate and the rotor rotational rate will produce a torque equal to the mechanical load. Should the load increase, the motor will slow further, having more "slip" that will result in more torque. If all the mechanical load is removed, the rotor will spin at almost the same rate as the stator magnetic field, with just enough slip to generate the torque required to overcome the bearing friction. If the AC stator currents are at a fixed frequency, such as on the grid generated by a centralized power plant, then the rotational speed will depend on the motor load, as just described.

For example, consider a three phase AC motor connected to the public grid with a frequency of 60 Hz. The motor will have stator windings A, B, and C for each of the three phases with each 120 degrees out of phase with the other. If, for example, the windings are replicated four times then the effective rotational speed of the stator commutation will be 60 Hz. Under no load, the rotor will spin at 60 revolutions per second, or 3600 RPM. If a load is placed on the motor to sufficiently cause a 5% slip, then the rotor will spin at 3420 RPM (Fig. 14.13).

The rotor has multiple windings of closed loops of wire, which are energized due the EMF due to the change in stator flux enclosed by the windings.

However, if the both the frequency and amplitude of the stator AC waveforms are generated from a DC source (using three "H" bridge circuits) and controlled, then the motor speed can also be controlled under dynamic load conditions, using an algorithm with closed loop feedback. One example might be an induction motor in an electric car powered by DC voltage batteries.

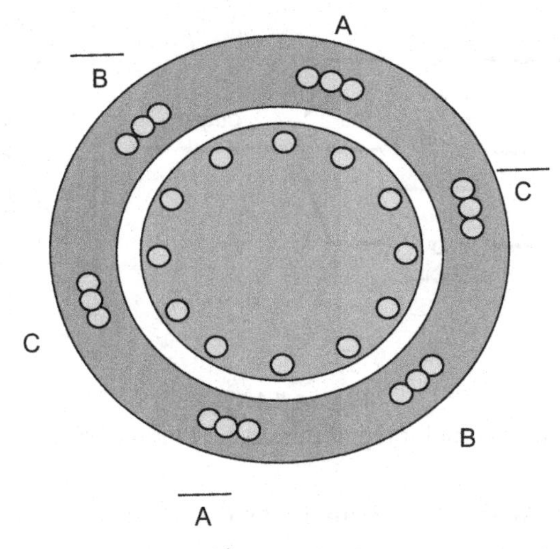

Figure 14.13
AC motor windings.

14.6 Motor Control

Every motor control loop follows the basic diagram in Fig. 14.14. The input to the controller is the desired behavior, which in the case of motors are usually speed and/or position. The controller also receives feedback from the motor indicating the actual motor speed/position. The desired input is subtracted from the actual input state, and this provides an error signal to the controller. The controller will then produce an output to the motor commutator that will minimize, or drive to zero the error signal.

A common controller type is the proportional integral derivator (PID) controller. It is so named because error signal is split into three separate paths of P, I, and D, which each produces a signal to motor circuit, which is summed together (Fig. 14.15). Each of these circuits has a separate gain constant. "P" stands for "proportional" and is the most intuitive. The larger the error signal, the larger will be the signal to the motor, to drive motor error to zero. "I" stands for "Integral," and this term is designed to prevent a

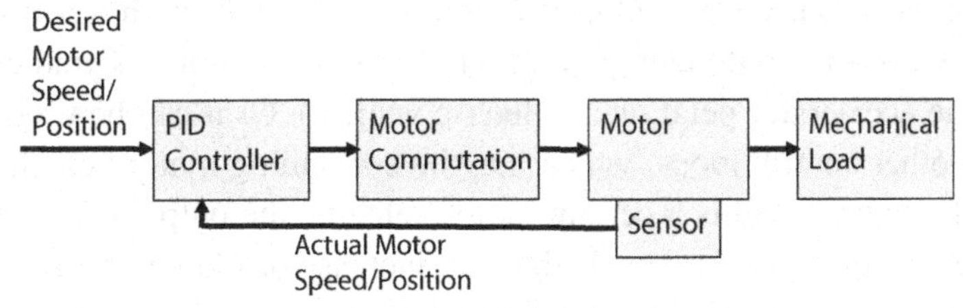

Figure 14.14
Basic motor control. *PID*, proportional integral derivator.

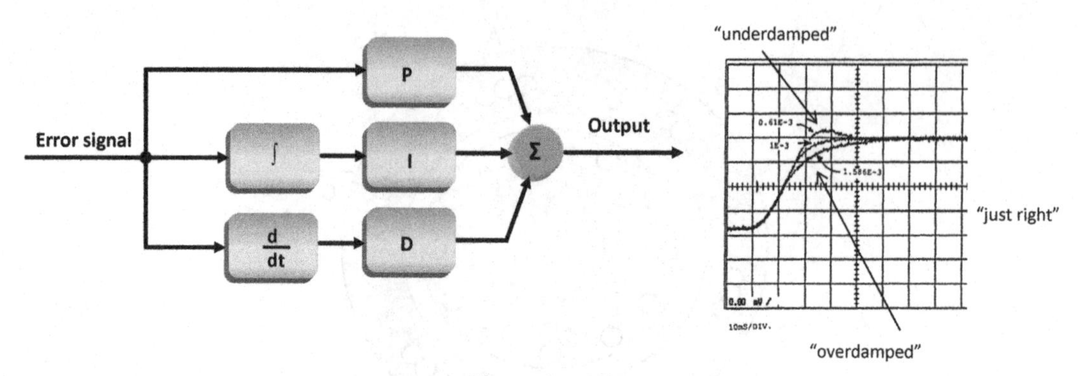

Figure 14.15
Proportional integral derivator (PID) controller.

permanent, residual error. With just a P signal, as the motor approaches the desired position/speed, the correction will approach zero. This can result in a permanent difference. For example, if the motor controller is controlling only motor speed, a "P" control circuit will have the motor approach but never quite reach the desired speed. That is where the "I" circuit comes in. It will integrate the error over time, and if a permanent small lag in motor speed is present, this will build up in the "I" circuit, and drive this error signal to zero. Finally, the "D" stands for "differential." The P and I terms function as described previously. The D path involves taking the derivative of the error signal. This can provide a more rapid response, as when the error signal suddenly begins to increase, it will provoke a vigorous response from the controller. This can cause instability, so often the D path is not used or set to zero. There are detailed mathematical methods to analyze control loops, and analyzing the stability and responsiveness of the control loop. An unstable control loop can oscillate. An underdamped control loop can overshoot, and an overdamped control loop can cause sluggish response. The setting of the P, I, and D gain coefficients is designed to provide proper controller response for a given system. Considerable simulation is normally used to determine the optimum values of P, I, and D.

Control loops can be designed to control a variety of aspects of a motor's operation. Here we are going to consider a common method, called FOC that is used to control the motor's torque, in particular for an AC motor, either induction or with a permanent magnet rotor. This control circuit can be used when the desired input signal corresponds to torque (such as the accelerator petal on an electric vehicle). Or it can be used in conjunction with other control loops, which may be controlling motor velocity or position. For example, if the input signal is a certain motor velocity, the output of the velocity controller will be a torque request signal, since an increase or reduction in torque is used to change motor velocity as needed to match the desired input velocity. Nested control loops are shown in Fig. 14.16.

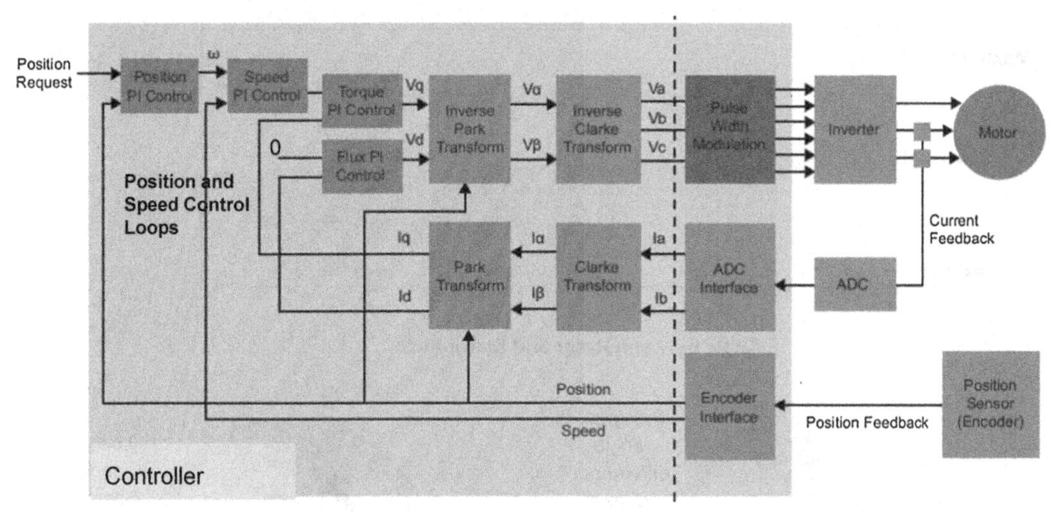

Figure 14.16

Motor control block diagram. *ADC*, analog-to-digital converter.

The control loops can be implemented in hardware or a suitable real-time microprocessor. The PWM circuits, invertor, and analog-to-digital convertor are implemented in hardware. Provision for interfacing and processing motor feedback is also often included.

The current for a three phase electric motor can be generated in the inverter using three "H bridge" circuits, each producing a sinusoidal current that is 120 degrees relative to the others. This will result in a smoothly rotating magnetic flux pattern with frequency equal the frequency of the AC waveform. This can be thought of as a rotating vector, with the direction of the vector indicating the net phase associated with the stator flux at any given instant. A permanent magnet rotor will have a magnetic flux vector that rotates with the rotor mechanical motion. With an ACIM, the rotor will be magnetized by the changing flux of the stator, and the induced EMF will cause current flow and a magnetic flux in the rotor coils. Owing to the rotor slip in an ACIM, the flux is changing in the rotor coils, in response to the relative rotation of the stator flux vector. Note that while the rotor mechanical rotary speed will be less than the stator AC current vector rotation, the induced the magnetic flux vector rotation is synchronous with the stator AC current vector.

The difference between the stator and rotor magnetic flux vectors can be expressed in degrees of phase, similar to describing the phase difference in two sinusoidal waves (Fig. 14.17). At zero phase difference, there will be no torque generated. The forces generated by the interacting magnetic fields are in an axial direction to the motor shaft, which only results in force on the motor bearings, but does not produce any rotational force (designated as "d" direction). A phase difference or lag in the stator

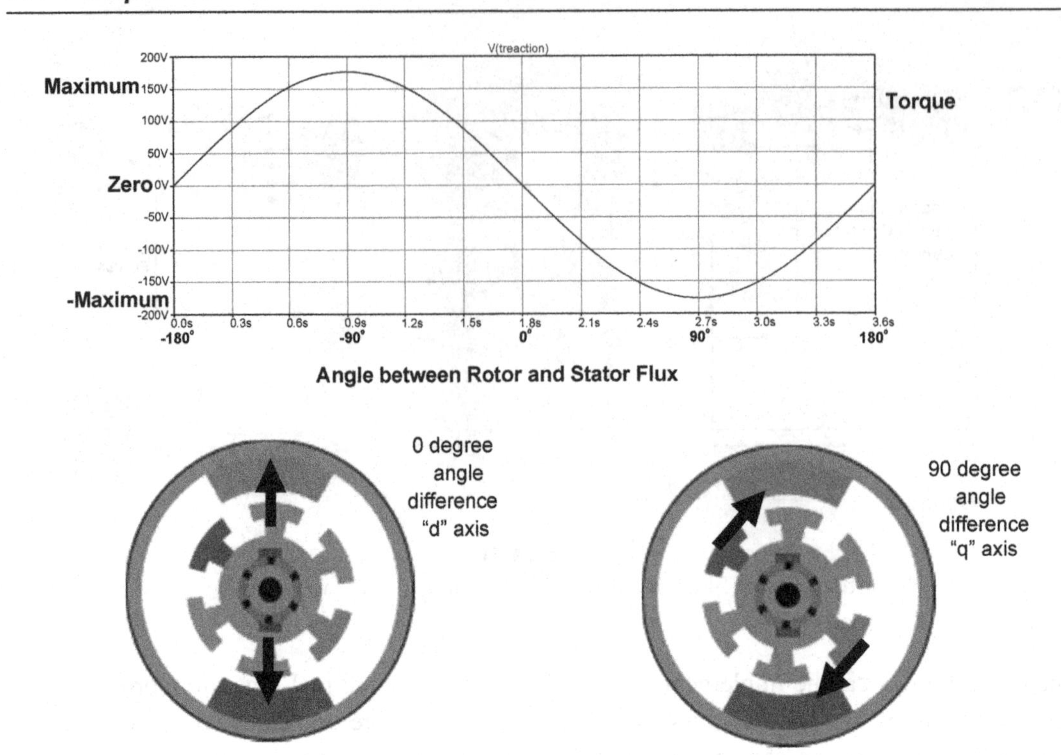

Figure 14.17
Motor torque verses flux angle difference.

and rotor magnetic fluxes will produce a rotational force and generate torque. When the phase lag is 90 degrees, all of this force will be directed toward rotating the motor shaft, generating the maximum torque (designated as "q" direction). This is the desired situation.

First consider the PMAC motor. The goal of the controller is to maintain either the axis of the PM rotor magnets or the induced magnetic field of the rotor at 90 degrees (plus or minus depending on the direction of rotation) with respect to the angle of the stator flux, under varying motor speed and load. When the stator flux is at 90 degrees to the rotor, this is called the quadrature direction or "q." When the stator flux is at 0 degrees to the rotor, this is called the direct direction or "q". As these d and q are orthogonal, the relative angle between the rotor and stator can be described in terms of d and q components. With a permanent magnet, the rotor flux angle is known as long as the rotor position can be determined. Then the AC stator currents can be driven such that the stator flux vector is maintained at 90 degrees with respect to the rotor flux. This is a dynamic procedure, as the motor is spinning and the both the d- and q-axis are spinning. To simplify the controller design, a pair of transforms is used, known as the "Clark" and "Park" transform.

14.7 Park and Clark Transforms

Most AC motors are three phase, with each stator coil wound at 120 degrees relative to each other. Each of these currents creates a magnetic flux vector. These vectors will sum up to produce a net vector, which is rotating. Describing the net vector in terms of three components is inconvenient. The Clark transform solves this, by mapping the stator flex vector from three components A, B, and C of 120 degrees separation into two orthogonal components, α and β (Fig. 14.18).

The Clark transform equations are defined as follows:

$$I_\alpha(\mathrm{t}) = 3/2 \times I_\mathrm{A}(\mathrm{t})$$
$$I_\beta(\mathrm{t}) = \sqrt{3/2} \times I_\mathrm{B}(\mathrm{t}) - \sqrt{3/2} \times I_\mathrm{C}(\mathrm{t})$$

The Clark transform is reversible, and the inverse Clark transform equations are as follows:

$$I_\mathrm{A}(\mathrm{t}) = 2/3 \times I_\alpha(\mathrm{t})$$
$$I_\mathrm{B}(\mathrm{t}) = -1/3 \times I_\alpha(\mathrm{t}) + 1\big/\sqrt{3} \times I_\beta(\mathrm{t})$$
$$I_\mathrm{C}(\mathrm{t}) = -1/3 \times I_\alpha(\mathrm{t}) - 1\big/\sqrt{3} \times I_\beta(t)$$

The control loop processing will be done by using the two orthogonal axis, α and β. The feedback current measurements of the motor will use the Clark transform to generate the I_α and I_β. The output of the control loop processing will use the inverse Clark transform to translate back into the three phase currents I_A, $I_{\mathrm{B},}$, and I_C to be generated by the invertor.

Another transform is used, the Park transform. The currents are rotating at a synchronous rate with the rotor in a PMAC motor. However, the control loop operates at much lower frequencies. The Park transform uses the rotor angular position θ and by calculating

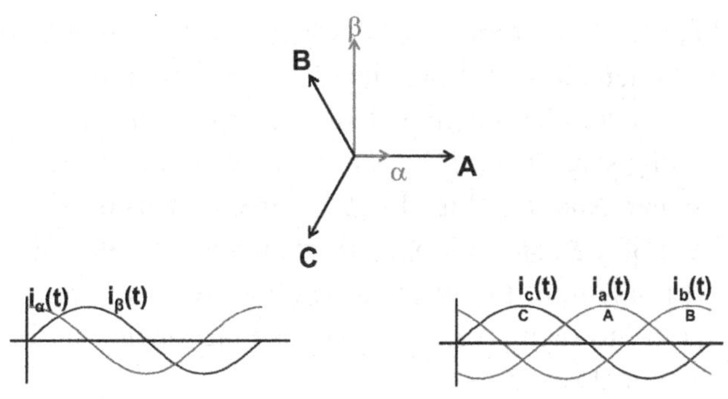

Figure 14.18
Clark transform mapping.

continuously is able to remove the rotation of the currents. Recall that the goal of control loop is to adjust the relative phase or phase difference between the stator magnetic flux and the magnetic flux of the rotor.

The Park transform equations are defined as follows:

$$I_d(t) = I_\alpha(t) \times \cos(\theta_d) + I_\beta(t) \times \sin(\theta_d)$$
$$I_q(t) = -I_\alpha(t) \times \sin(\theta_d) + I_\beta(t) \times \cos(\theta_d)$$

And the inverse Park transform equations are as follows:

$$I_\alpha(t) = I_d(t) \times \cos(\theta_d) - I_q(t) \times \sin(\theta_d)$$
$$I_\beta(t) = I_d(t) \times \sin(\theta_d) + I_q(t) \times \cos(\theta_d)$$

The Park transform is shown pictorially above. To summarize, the three phase rotating currents are described in terms of three axes at 120 degrees (A, B, C). The currents first mapped into two orthogonal components (α, β) using two orthogonal axis at 90 degrees, using the Clark transform. Then, as the rotor angle θ (and therefore rotor flux vector direction) is known, the stator flex vector is then mapped into two other orthogonal current vectors (d, q), using the Park transform. If this was a static situation (the rotor not rotating, the stator currents not rotating), then this would be insignificant. However, the Park transform is computed continually, with update rotor angle θ. This means that d and q orthogonal vectors are rotating along with the rotor. The d-axis by definition is in the same direction as the rotor magnetic flux. The Park transform has mapped the stator currents into two components: a "d" component, which creates a magnetic flux in such a direction as to create zero torque on the rotor, and a "q" component, which will create maximum torque on the rotor. In a system where the motor is running at constant speed at a constant load (torque), the "d" and "q" components will be static or DC values will be static. During motor operation, with the d- and q-axis spinning, stator currents are continually updated as to drive the stator flux pattern to be offset by 90 degrees from the d-axis. So while the I_α and I_β values change in a sinusoidal manner, the I_d and I_q values are DC (or near DC). The I_d and I_q current vectors change only in response to a change in motor operation. For example, when the motor load increases or there is an external command to increase speed, more torque will be required. This requires an increase in magnetic stator flux intensity, by increasing stator current levels (which would make the current vectors longer in Fig. 14.19) longer. Note that Fig. 14.19 is three dimensional, and "C" is coming out of the page. The goal of the control loop is to maintain the stator flux vector at 90 degrees with respect to the rotor flux vector, which is done by making the "d" component to zero, and "q" to such a value as to create the amount of torque to drive the motor at the desired rotational speed.

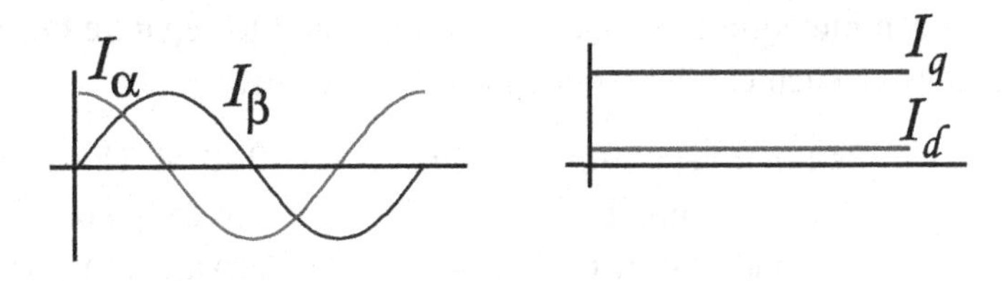

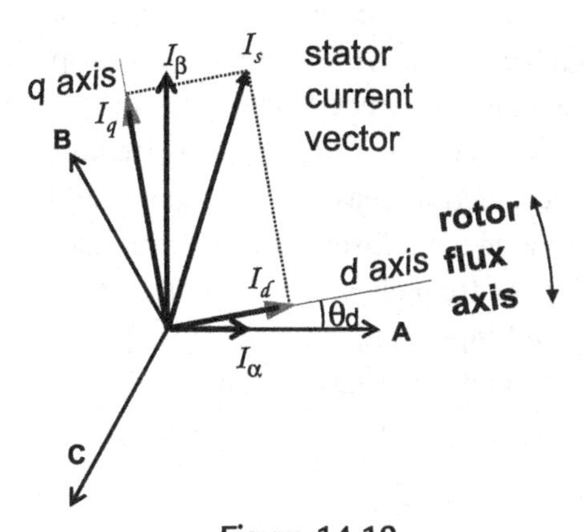

Figure 14.19
Park transform mapping.

To review, the inverse Clark transform is used to take the control loop outputs in the form of orthogonal I_d and I_q current values, and transform them into orthogonal rotating I_α and I_β waveforms. The inverse Park transform is used to then map I_α and I_β signals into three phase I_A, I_B, and I_C s waveforms, which then drive the inverter that creates the actual motor currents using "H" bridge circuits with PWM modulation. Control loops operate from an error signal or the difference between the actual and desired values. The control loops that control the torque (or motor stator current) for "d" and "q" need the actual values. This is done by measuring the actual motor currents, which can be accomplished by using "sense" resistors or more advanced inductive current sensors. In either case, the currents measured are the actual I_A, I_B, and I_C currents. The three phase signals are then mapped to I_α and I_β signals using the Clark transform. Then the I_α and I_β signals are in turn mapped to I_d and I_q signals using the Park transform. The I_d signal should be driven to zero in a permanent magnet synchronous motor (PMSM), and so is subtracted from zero to form the error signal into the P and I control circuits for the direct current vector. The I_q vector is proportional to torque. Therefore, the desired torque level (which is often the output of a velocity control loop) is subtracted from the actual, measured I_q signal to form the error signal into the P and I control circuits for the quadrature current vector. Note that the rotor position is a required input for Park and inverse Park transform computations. The whole point of the Park and Clark transforms is to remove the high frequency, rotational commutation from the control loop, and to separate our the direct and quadrature components so each can be independently controlled.

The ACIM can be controlled in a very similar manner as the PMSM. However, the feedback rotor position θ_d must be modified to introduce the slip frequency into the control circuits (keep in mind that just as velocity is the derivative of position, the rotational speed is the derivative of the rotational angle position θ_d). By adding (or subtracting) an angular offset to each successive rotor position sample, an increase (or decrease) in frequency is affected. For an ACIM, the slip is proportional to the torque. Therefore a slip controller calculator is needed. The controller will attempt to drive the ACIM at the desired speed. However, when the ACIM is running at a lower speed, more torque is needed to achieve the speed increase. This is achieved by having the slip calculator add phase offsets to achieve a faster effective frequency across the θ_d sample train, which will result in a higher frequency stator commutation, which allows the ACIM to slip and still run at the desired input frequency. This will in effect change the phase of the combined I_d and I_q vectors to advance faster than the rotor rotation. The control loops will also increase the vector length of I_d and I_q, to achieve greater torque, until the desired speed is achieved (Fig. 14. 20).

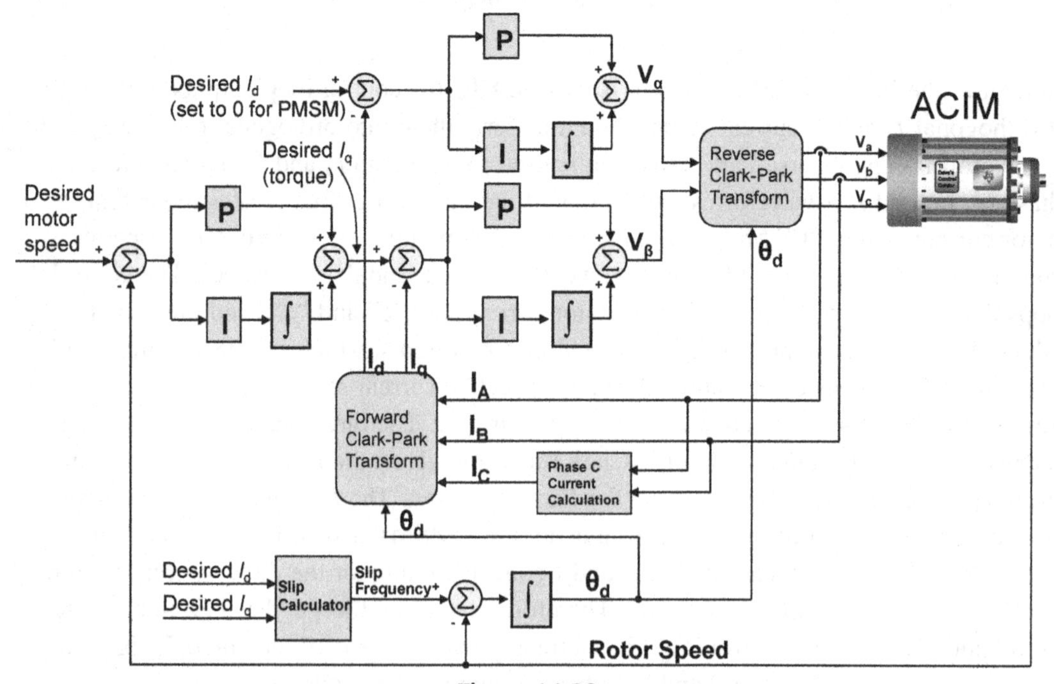

Figure 14.20
AC induction motor and permanent magnet synchronous motor field-oriented control motor loop.

In summary, an ACIM motor can be driven directly from an AC power source, but it will not run synchronously, as greater loads result in greater slip. Since the AC voltage source is generally fixed, the curve of motor torque verses rotational speed is also fixed. This works well for many applications where the load is known, and the motor operational speed is within a fixed range.

The ACIM can also be electronically commutated and controlled from a DC source, in which case the motor speed and operational profile can be controlled under varying load conditions. The motor can also be easily reversed in rotational direction.

CHAPTER 15

Analog and Time Division Multiple Access Wireless Communications

The first mobile phone systems were based on analog technology. Developed in the 1970s, this system in the United States was known as AMPS, or American Mobile Phone System. Similar systems were developed in Europe and Japan. These are known as first generation, or "1G" systems. What differentiated these systems from previous wireless systems (police or public service radio, citizen band radio, military systems) was the concept of frequency reuse. By reusing frequency channels, enough frequency was available so each pair of users could be assigned their own, private communication link. Once the call was finished, this frequency channel could be assigned to other users. An important aspect of frequency reuse was that a large region of phone service or coverage could be divided into sections, or "cells," to allow the same frequency to be used in over and over, providing these locations were not too close to each other. This was the origin of the name "cellular" phone service. This allowed a large pool of users to be serviced with a much smaller set of frequency channels, which allowed for efficient use of the frequency spectrum.

In the cellular diagram in Fig. 15.1, each cell also has its own base station or transmitter/ receiver equipment connected to a central network. The cells are lettered, depicting each different letter cell is assigned a different set of frequency channels. This is a seven-cell reuse pattern. Each of the seven letters uses different frequencies, and no cells of same letter are adjacent to each other. Additionally, separate frequency bands are used for downlink transmission (base station → mobile phones) and uplink (mobile phones → base station). This is known as "frequency division duplex" or FDD. For example, the AMPS

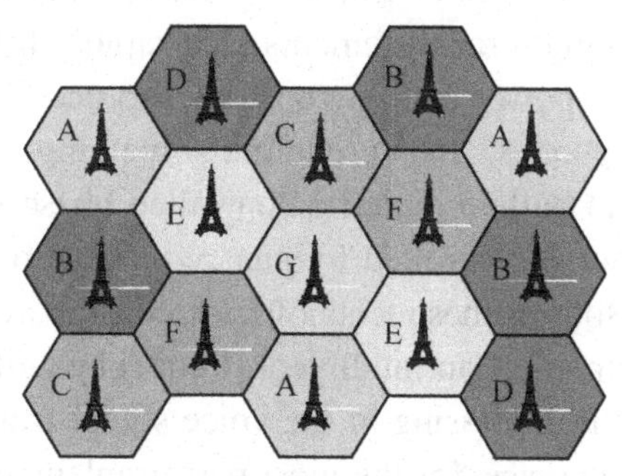

Figure 15.1
Frequency reuse using cells.

Digital Signal Processing 101. http://dx.doi.org/10.1016/B978-0-12-811453-7.00015-9

uses the 824–849 MHz band for downlink and 869–894 MHz band for uplink. Each individual user channel occupies 30 kHz, resulting in 800 channel pairs. If we have frequency reuse every seven cells, then about 114 channel pairs are available in each cell. If we assume that at peak usage times, 10% of cell phone subscribers are making phone calls, then over 1000 users can be serviced by any given cell.

In addition, most base stations use directional panel antennas. There is typically one transmit and two receive antennas per sector. A sector is roughly 120 degrees, with some overlap. There are three sectors per cell site, and the base station antenna tower will have each sectors antenna pointing in three different directions, 120 degrees apart.

Each user will be able to move about within the boundaries of a given sector of a given cell, which is defined by the RF coverage area of base station within that sector. But to move further than this, or roam, requires a centralized network control system. When the mobile phone reaches the edge cell, it can detect the weakening strength of the base station signal. The base station can also detect the lower signal strength of that mobile handset. The mobile phone is instructed by the network via that base station to scan the frequencies of adjacent sectors and of neighboring base stations. When an adjacent sector or neighboring base station's signal is found to be stronger, then the network will perform a handoff. The mobile phone is instructed to switch to an unoccupied frequency channel of the adjacent sector or neighboring base station. Simultaneously, the landline connection of the phone call in progress is switched from the original base station sector to the adjacent sector or to the neighboring base station. In this way, a phone call can be carried on continuously as the mobile phone travels throughout the network coverage area, defined by the contiguous RF coverage of the network or service provider's base stations.

15.1 Early Digital Innovations

Two technology breakthroughs made this type of service possible. For this type of network to function, the mobile phones have to be intelligent enough to receive and reply to commands, perform signal strength measurements, and rapidly tune to new frequencies on command. This required a low-power and low-cost microprocessor, which could be incorporated into the mobile phones. It also required something called a fractional "N" synthesizer, which was used to build a digitally controlled phase-locked loop oscillator circuit. This allowed for software-controlled frequency tuning, to one of many closely spaced frequency channels. Both of these technologies became available in the 1970s and were essential to the development of an intelligent, frequency agile mobile cell phone. However, the baseband and RF processing of the voice signal from microphone/speaker to the antenna in the mobile phone was, for the most part, implemented using traditional analog circuits and techniques.

In each cell, most of the frequency channels are available for use as voice channels, which can be assigned to a given user during a call. But one or more of channels in each cell are used as control channels. A phone that is not in conversation will monitor one of these preassigned control channels. Here, it can monitor RF signal strength, listen for paging messages, and other commands. The modulation used is called frequency shift keying or FSK. It is similar in concept to Morse code, except instead of dashes and dots, two difference frequency tones are transmitted in indicating "ones" and "zeros." All mobile phones would reply to commands using a common uplink channel, known as a random access channel, or RACH. This was a somewhat uncontrolled process, as several mobiles could transmit simultaneously on this channel. To ensure message replies would eventually get through, each phone would wait a random amount of time before retransmitting in the event of a collision. The great majority of channels were designated as voice channels, which carried the voice conversation using frequency modulation (FM), the similar to any FM radio station. In fact, in the early days of analog cell phones, one could eavesdrop on wireless conversations using a simple FM scanner (but could only hear one side of a conversation at a time, as in an FDD system there are separate uplink and downlink frequency bands).

15.2 Frequency Modulation

Information can be carried by a sinusoidal wave using varying the amplitude, frequency, and phase. In quadrature amplitude modulation (QAM), the amplitude and phase are changed. In FM, only the frequency is modified. FM is a modulation method inherently suitable for an analog input or baseband signal. Basically, the instantaneous frequency of the carrier is made to increase or decrease from the carrier frequency by an amount proportional to the modulating or baseband signal. This change in the carrier frequency is known as the frequency deviation. The frequency deviation is proportional to the amplitude of the baseband input. The rate of change (derivative) of the carrier frequency is proportional to the frequency of the baseband input. The AMPS used FM with peak derivative of 12 kHz.

Since there is no amplitude modulation (AM), the FM signal is of constant amplitude. This is the inherent FM superior characteristic over AM, and why FM radio, from its beginnings in the 1930s, was designed for high fidelity compared with AM radio. Any additive noise with an AM signal will cause distortion of the amplitude, which is the baseband signal. In contrast with FM, the frequency carries the baseband signal and is much less affected by additive noise. This additive noise causes phase distortion, which can affect the frequency demodulation, but most of this can be filtered out of the resulting baseband signal. Another important characteristic of FM is that, due to constant amplitude characteristic, it can be very efficiently amplified. This will be further discussed in a subsequent chapter.

15.3 Digital Signal Processor

Analog-based cell phone system brought a landline telephone—like experience to mobile communications. The invention of the digital signal processor, or DSP, paved the way for the next step in the evolution of wireless phone communication.

The DSP is basically a specialized microprocessor. It has at least one dedicated multiplier with an associated accumulator or adder with a feedback path. This can be used to efficiently calculate a sum of products, used in finite impulse response filters. DSPs, unlike most microprocessors, can fetch instruction and data words from memory simultaneously. To do one calculation per clock cycle, at least three data buses are generally required. One data bus is used to fetch the instruction word, and two more buses to fetch the two operands for the multiplier from memory. Sometimes there is a fourth dedicated data bus to be able to simultaneously store data back to memory. This memory is generally all single-cycle access, on-chip, also known as Level 1 memory. In DSP, the data, unlike instructions, are usually read and written in a predictable manner. Therefore, DSPs contain at least two data address generators, which can be preconfigured to calculate addresses in a given pattern, and even in a circular or repeating manner. This allows implementation of virtual shift registers in memory and accessing of filter coefficients in the correct order. There is often a "bit-reversing" mode, which can be used to read or write fast Fourier transform (FFT) data in a decimation in time or decimation in frequency fashion (refer to Chapter 10 on FFT for more detail).

In addition, DSP instructions often have data-shifting capabilities, which allow for the decimal point to be aligned as needed prior to saving data to memory. The data shifting can also be configured as a barrel shifter, and in conjunction with logical operations, used to implement many error-coding operations. Given the popularity of the Viterbi algorithm, there is often a special instruction to implement path metric comparisons and selections. Various accumulator data rounding and saturation modes are often supported. To obtain maximum performance, some of the instructions often had pipeline restrictions, which created exception cases for the programmer.

DSPs were programmed using a manufacturer-specific assembly language, usually by firmware engineers extremely familiar with the details of the DSP hardware architecture. As a consequence, the majority of DSP programmers came not from a software programming background but from an electrical engineering background. Due to the small on-chip memory available, the need to minimize number of clock cycles per calculation, and the intricate, mathematical nature of the algorithms being implemented, DSP programming became as much an art as a skill. Current DSPs come with advanced "C" compilers, making code development much more efficient, although assembly macros are still used for specialized instructions as they often operate in a parallel manner.

With the advent of the first DSPs, it did not take long for many applications to develop. Among the most important were digital mobile phone systems.

These systems were the second generation of mobile phone technology, now known as "2G." These systems are known as "time division multiple access," or TDMA. This fancy term just means that multiple users rotate turns using both the uplink and downlink frequency channels, allowing more simultaneous users.

15.4 Digital Voice Phone Systems

A key feature of a digital mobile phone system is that the voice is digitized. Landline phone systems have long been digital. The actual phones in homes and businesses are analog, and using twisted pair phone lines, connect to the local telephone exchange (this same line also carries the digital subscriber line signal for Internet connection). At the telephone exchange, the voice signal is digitized using ADCs and DACs sampling at 8 kHz. The samples are not mapped linearly, but logarithmically, into an 8-bit digital representation, using a process known as companding. This technique reduces the quantization noise at low-signal levels at the expense of quantization noise at higher signal levels, effectively resulting in a higher dynamic range. The voice signal is now in a digital format, with 8-bit samples at 8 kHz, for a resulting bit rate of 64 kbits/s (kbps). In this form, it can be managed and transmitted by telephone switches and systems world wide. This is known as an uncompressed digital voice signal.

Uncompressed digital voice would require as much frequency bandwidth to transmit as the FM voice signal used in analog mobile wireless phone systems. However, using digital voice compression technology, known as vocoders, the required data rate can be reduced. There is a trade-off in compressed bit rate, voice quality, and complexity of voice compression algorithm used. In North American TDMA systems, the voice was generally compressed from 64 to 8 kbps, using an algorithm known as vector sum excited linear prediction (VSELP). Unfortunately, the voice quality of VSELP was poorer than the previous analog frequency–modulated AMPS. Subsequently, using more powerful DSPs both in base stations and mobile handsets, a more powerful 8-kbps voice compression algorithm known as advanced code excited linear prediction (ACELP) was used, which closed the quality gap. Both vocoders used convolutional encoding and Viterbi decoding error correction, which resulted in a transmitted data rate of about 13 kbps.

15.5 Time Division Multiple Access Modulation and Demodulation

In the United States, the TDMA upgrade system to the AMPS was known as IS-54, later upgraded to IS-136, referring to the interim standards of the Telecommunications Industry Association (TIA). This system used the same 30-kHz channel spacing as AMPS. Each

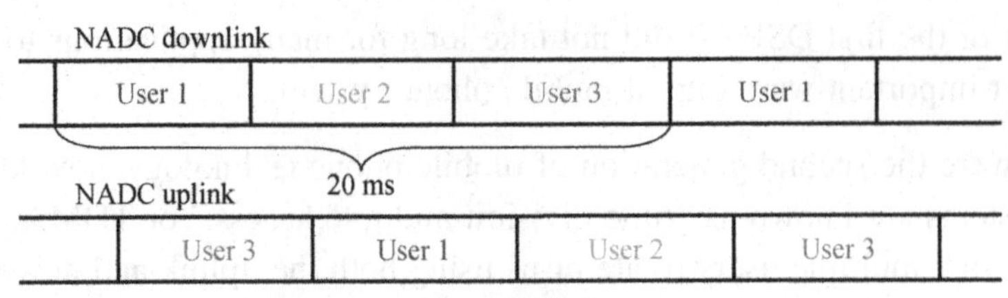

Figure 15.2
Time division channel multiplexing.

frequency channel was organized into frames of 20 milliseconds, or 50 frames per second. Each frame has three time slots, each of which can be assigned to one user, as shown in Fig. 15.2. This increased the capacity of system threefold, compared with AMPS. Since it was compatible and inclusive with AMPS, the digital service could be gradually rolled out by the service provider, allowing for a gradual obsolescence of the AMPS handsets.

Notice that the frame timing is offset between the downlink and uplink. This allows the mobile handset to operate in transmit and receive modes at different, nonoverlapping intervals. Since the mobile handset only needs to transmit for about one-third of the time, power consumption can be reduced and battery life extended.

A number of DSP technologies were used in TDMA systems. In addition to vocoding and error correction, baseband digital modulation and demodulation methods were implemented.

The channel quality between the base station and mobile handsets is often of very poor quality. Unlike satellite and microwave links, it is rare to have a direct line of sight connection between base stations and handsets. The received signals are composed of multiple reflections, often distorted from passing through walls or other obstructions. These signals can sometimes combine out of phase, effectively canceling each other. This phenomenon is known as Rayleigh fading, which can be mitigated by using a second receive antenna (diversity). If the antennas are sufficiently separated in distance, then the phases of the multiple signals will vary differently, and in fact the Rayleigh fading will be uncorrelated, or independent. This means that the likelihood of the signal at one antenna canceling due to Rayleigh fading at the same time as the other receive antenna is very small. By dynamically switching between the antennas, depending on which has the best signal, the impact of this fading can be largely mitigated, compared to using just a single antenna.

A reflection can become delayed and be received on top of other later symbols at the receiver. This is called intersymbol interference (ISI) and can be compensated for by using an adaptive equalizer. Then there is the effect of the handset motion, which causes Doppler frequency shift in the received signals, and rapid changes in the ISI and fading effects, requiring fast adapting digital receivers.

In IS-136, a form of quadrature phase shift keying (QPSK) modulation was used, called differential $\pi/4$ offset QPSK. It is more robust in terms of synchronization and sensitivity to Doppler shift compared with standard QPSK. It also has reduced dynamic range, which is beneficial to RF power amplifier performance. The receivers in both the base stations and handsets were equipped with adaptive equalizers, multiple or diversity receivers, synchronization, and frequency offset compensation algorithms. All of these functions were implemented in DSP software in both the handsets and base stations.

There were alternative forms of TDMA. In Japan, a system called Personal Digital Communications (PDC) was widely used. It had 25-kHz channel spacing and also used three voice slots per 20-ms frame. Later, a half-rate vocoder with sufficient quality, called pitch synchronous innovation-code excited linear prediction, was developed, which allowed six voice slots per frequency channel. This was the most spectrally efficient version of TDMA. In Europe, the Global System for Mobile Communications (GSM) TDMA system was developed, which had eight voice slots, but used a wider 200-kHz channel spacing. While not especially spectrally efficient, it was adopted across all the European countries and came to have widespread commercial adoption around the world. Due to the simplicity and low cost of GSM handsets and infrastructure, it is still popular in the developing world, where often there is little or no landline phone service. The GSM system used Gaussian Minimum Shift Keying (GMSK) modulation, which is a form of phase-shift modulation. Unlike QAM and QPSK modulation, GMSK is a constant amplitude type of modulation. Similar to FM, this made GSM signals very efficient to amplify for transmission in both the base stations and handsets.

The TDMA architectures, both handset and base station, were DSP based. The main difference was that the base station radios supported all of the time slots simultaneously. Mobile handsets, on the other hand, would support one uplink and downlink time slot simultaneously, sufficient for a single call. Later, a derivative system called Enhanced Data GSM Evolution (EDGE) was developed from GSM. Sometimes called a "2.5G" technology, EDGE was developed to be an add-on to GSM networks and supports somewhat higher data rates for mobile internet or email access. It provides these higher date rates by allowing a single handset to occupy all eight time slots of the frequency channel and by using a more efficient modulation method called 8-PSK.

Fig. 15.3 is a basic block diagram of a TDMA base station radio system. There are typically 4–24 radio cards in TDMA base station.

TDMA mobile phone networks work well, are relatively simple and low cost, and provide reliable wireless connectivity rates sufficient for compressed voice. However, new technologies promised higher data rates for internet access, plus more efficient, higher capacity networks. The latter became more important as wireless usage grew, and there was increased demand for frequency spectrum to service the market demand.

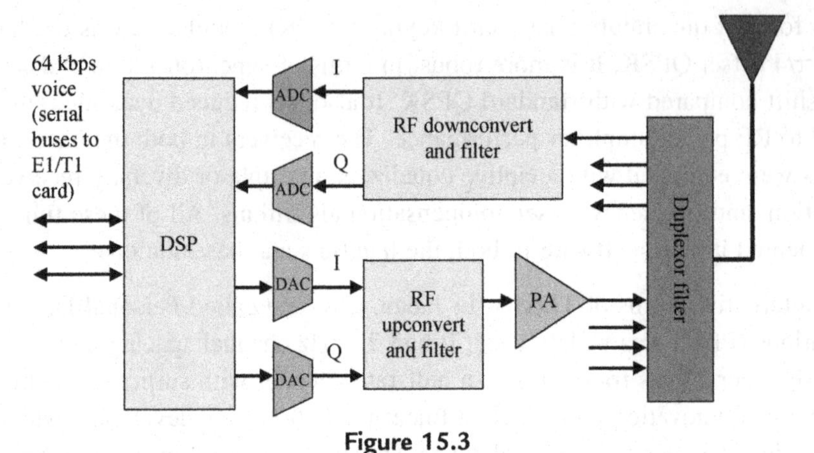

Figure 15.3
Time division multiple access radio block diagram.

However, a rival technology was also being developed, known as CDMA, which stood for code division multiple access. The original goal of second-generation digital systems was a 10-fold increase in capacity compared with AMPS. The IS-136 TDMA standard, with a threefold increase, fell short of that goal. The later versions of PDC used in Japan achieved a 6-fold increase in capacity. Proponents of CDMA claimed that their technology would be able to meet this capacity increase. And that is the subject of the next chapter.

CHAPTER 16

CDMA Wireless Communications

The term CDMA stands for code division multiple access. CDMA modulation and demodulation technology grew out of military spread spectrum techniques. CDMA technology was commercialized by Qualcomm in the early 1990s. The initial Qualcomm CDMA system was known as IS-95. Later, Qualcomm developed and deployed an enhanced version of IS-95, known as CDMA2000 1xRTT, or just CDMA2000.

16.1 Spread Spectrum Technology

The basic idea behind spread spectrum is to start with a narrow band signal, and mix it with a high frequency pseudorandom number (PN) signal, which would have the effect to "spreading" the frequency spectrum occupied by the signal, thereby making it much hard for adversaries to jam or interfere. On reception, the process could be reversed with using the same pseudorandom code, and the original signal recovered (Fig. 16.1).

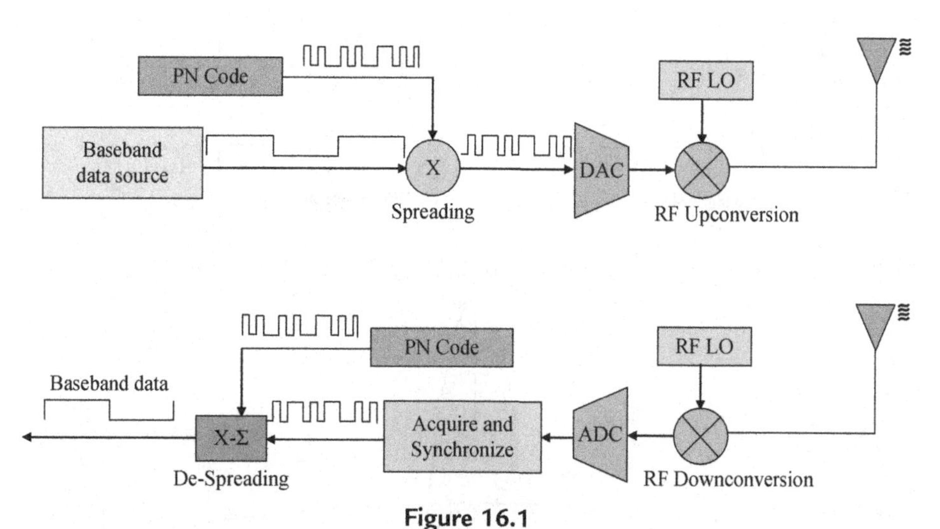

Figure 16.1
Spread spectrum modulation and demodulation. *ADC*, analog-to-digital; *DAC*, digital-to-analog; *PN*, pseudorandom number.

Digital Signal Processing 101. http://dx.doi.org/10.1016/B978-0-12-811453-7.00016-0

16.2 Direct Sequence Spread Spectrum

There are several methods to perform frequency spreading. The method employed in CDMA is known as direct sequence spread spectrum. In direct sequence, the digital data is modulated by a much higher rate sequence of PN data. Each bit of the PN sequence is a "chip", and the higher rate is known as the chip rate. These chips typically modulate the much lower rate digital input data by typically a 180°degree phase shift in the carrier at the chip rate. This is superimposed on the much lower rate phase shifts caused by input data. This will greatly increase the occupied frequency bandwidth of the signal and decrease the concentration of signal energy around the carrier (Fig. 16.2).

CDMA technology takes this concept further. The pseudorandom sequences were replaced by special sequences called Walsh codes. The Walsh codes have a property called orthogonalization. Each codeword is independent or orthogonal of the other. If any of the codes is cross-correlated with another, the result is zero. This property is used to allow multiple users to share the same frequency band, each being assigned a unique codeword, which allows a receiver to pick out the one desired user signal from all the others. The rest of other user signals are removed in the demodulation process.

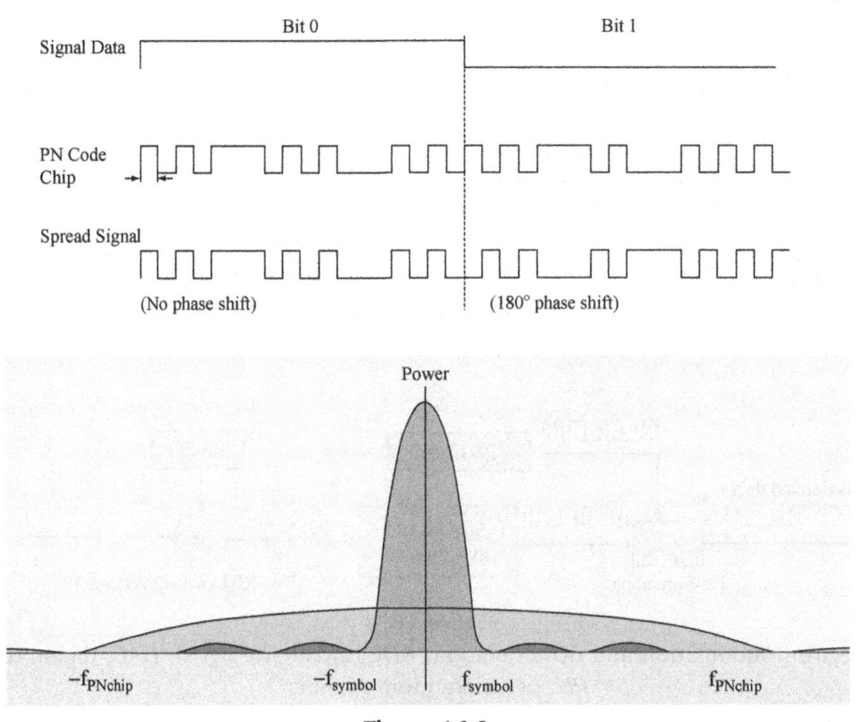

Figure 16.2
Frequency domain spreading using pseudorandom number (PN) codes.

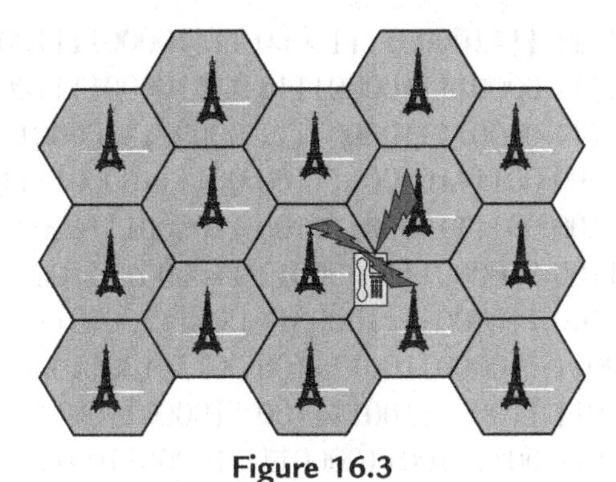

Figure 16.3
Code division multiple access cell diagram.

In CDMA mobile communications, not only do all the users in a given cell or sector share the same frequency channel, but all the cells also use the same frequency channel (Fig 16.3). For CDMA2000, each frequency channel is 1.25 MHz bandwidth. Unlike TDMA or analog systems, there is no frequency channel handoff as the user moves or transitions from sectors or cells. In fact, the mobile phone can be in simultaneous communication with several cells simultaneously. This is one of the remarkable qualities of the CDMA mobile communications system.

16.3 Walsh Codes

There are 64 Walsh codes used in CDMA2000, each 64 bits long, listed below. The Walsh codes are clocked at the chip rate, which is 64 times faster than the data rate. In CDMA2000, the chip rate is 1.2288 Mcps, and the input data rate to the CDMA modulator is 19.2 kbps per Walsh code.

W_0 0000000000000000 0000000000000000 0000000000000000 0000000000000000
W_1 0000000000000000 0000000000000000 1111111111111111 1111111111111111
W_2 0000000000000000 1111111111111111 1111111111111111 0000000000000000
W_3 0000000000000000 1111111111111111 0000000000000000 1111111111111111
W_4 0000000011111111 1111111100000000 0000000011111111 1111111100000000
W_5 0000000011111111 1111111100000000 1111111100000000 0000000011111111
W_6 0000000011111111 0000000011111111 1111111100000000 1111111100000000
W_7 0000000011111111 0000000011111111 0000000011111111 0000000011111111
W_8 0000111111110000 0000111111110000 0000111111110000 0000111111110000
W_9 0000111111110000 0000111111110000 1111000000001111 1111000000001111
W_{10} 0000111111110000 1111000000001111 1111000000001111 0000111111110000
W_{11} 0000111111110000 1111000000001111 0000111111110000 1111000000001111
W_{12} 0000111100001111 1111000011110000 0000111100001111 1111000011110000

W_{13} 0000111100001111 1111000011110000 1111000011110000 0000111100001111
W_{14} 0000111100001111 0000111100001111 1111000011110000 1111000011110000
W_{15} 0000111100001111 0000111100001111 0000111100001111 0000111100001111
W_{16} 0011110000111100 0011110000111100 0011110000111100 0011110000111100
W_{17} 0011110000111100 0011110000111100 1100001111000011 1100001111000011
W_{18} 0011110000111100 1100001111000011 1100001111000011 0011110000111100
W_{19} 0011110000111100 1100001111000011 0011110000111100 1100001111000011
W_{20} 0011110011000011 1100001100111100 0011110011000011 1100001100111100
W_{21} 0011110011000011 1100001100111100 1100001100111100 0011110011000011
W_{22} 0011110011000011 0011110011000011 1100001100111100 1100001100111100
W_{23} 0011110011000011 0011110011000011 0011110011000011 0011110011000011
W_{24} 0011001111001100 0011001111001100 0011001111001100 0011001111001100
W_{25} 0011001111001100 0011001111001100 1100110000110011 1100110000110011
W_{26} 0011001111001100 1100110000110011 1100110000110011 0011001111001100
W_{27} 0011001111001100 1100110000110011 0011001111001100 1100110000110011
W_{28} 0011001100110011 1100110011001100 0011001100110011 1100110011001100
W_{29} 0011001100110011 1100110011001100 1100110011001100 0011001100110011
W_{30} 0011001100,110011 0011001100110011 1100110011001100 1100110011001100
W_{31} 0011001100110011 0011001100110011 0011001100110011 0011001100110011
W_{32} 0110011001100110 0110011001100110 0110011001100110 0110011001100110
W_{33} 0110011001100110 0110011001100110 1001100110011001 1001100110011001
W_{34} 0110011001100110 1001100110011001 1001100110011001 0110011001100110
W_{35} 0110011001100110 1001100110011001 0110011001100110 1001100110011001
W_{36} 0110011001100110 1001100110011001 0110011001100110 1001100110011001
W_{37} 0110011001100110 1001100110011001 1001100110011001 0110011001100110
W_{38} 0110011010011001 0110011010011001 1001100101100110 1001100101100110
W_{39} 0110011010011001 0110011010011001 0110011010011001 0110011010011001
W_{40} 0110100110010110 0110100110010110 0110100110010110 0110100110010110
W_{41} 0110100110010110 0110100110010110 1001011001101001 1001011001101001
W_{42} 0110100110010110 1001011001101001 1001011001101001 0110100110010110
W_{43} 0110100110010110 1001011001101001 0110100110010110 1001011001101001
W_{44} 0110100101101001 1001011010010110 0110100101101001 1001011010010110
W_{45} 0110100101101001 1001011010010110 1001011010010110 0110100101101001
W_{46} 0110100101101001 0110100101101001 1001011010010110 1001011010010110
W_{47} 0110100101101001 0110100101101001 0110100101101001 0110100101101001
W_{48} 0101101001011010 0101101001011010 0101101001011010 0101101001011010
W_{49} 0101101001011010 0101101001011010 1010010110100101 1010010110100101
W_{50} 0101101001011010 1010010110100101 1010010110100101 0101101001011010

W_{51} 0101101001011010 1010010110100101 010110100,011010 1010010110100101
W_{52} 0101101010100101 1010010101011010 0101101010100101 1010010101011010
W_{53} 0101101010100101 1010010101011010 1010010101011010 0101101010100101
W_{54} 0101101010100101 1010010101011010 0101101010100101 1010010101011010
W_{55} 0101101010100101 0101101010100101 0101101010100101 0101101010100101
W_{56} 0101010110101010 0101010110101010 0101010110101010 0101010110101010
W_{57} 0101010110101010 0101010110101010 1010101001010101 1010101001010101
W_{58} 0101010110101010 1010101001010101 1010101001010101 0101010110101010
W_{59} 0101010110101010 1010101001010101 0101010110101010 1010101001010101
W_{60} 0101010101010101 1010101010101010 0101010101010101 1010101010101010
W_{61} 0101010101010101 1010101010101010 1010101010101010 0101010101010101
W_{62} 0101010101010101 0101010101010101 1010101010101010 1010101010101010
W_{63} 0101010101010101 0101010101010101 0101010101010101 0101010101010101

Each user's input data stream of 19.2 kbps is modulated by a different Walsh code.

16.4 Concept of Code Division Multiple Access

A common nontechnical analogy to CDMA is the following. Imagine a round table, where there are multiple one-to-one conversations occurring between various pairs of people who are not adjacent to each other. Ordinarily, this would present a difficult situation, and it would be very difficult for anyone to communicate due to interference from all the other conversations. Now imagine if each pair of people only spoke one language, and for each pair, it was a different language. Now the conversations could proceed much more efficiently. Each pair would hear the other pairs conversations as unintelligible noise, and their own conversation would stand out, as each person could correlate what he heard against familiar words and speech of his own language. Another caveat is that this will only work if there are not too many other conversations, and if everyone cooperates by speaking in a conversational tone at the same volume.

If one pair tries to enhance their conversation by raising their voices, it will degrade everyone else's conversation. And if others in turn respond by raising their voices, things soon degenerate into a shouting match, and all communication is hindered. With this in mind, we will try to outline the essential basics of the CDMA2000 system.

16.5 Walsh Code Demodulation

Imagine each Walsh code W_k sequence is mapped so that a zero is a $+1$, and a one is a -1. There is a single user channel input data bit every 64 chips, or period of the

Walsh code. If the data bit is "0", then the Walsh code W_k is transmitted, or if "1", the inverse, or negative of the W_k is transmitted. Furthermore, the detector is based on a correlator or integrator. The correlator will perform a correlation (cross-multiply and sum) of the received sequence against the same Walsh code W_k. Next, we will go through a few examples. To make easier to represent, we will be using only the first 16 chips of each Walsh code, and picking the Walsh codes that are orthogonal over the first 16 chips. These will be every fourth Walsh code, numbers 0, 4, 8, 12...60. This set of 16 sequences, each 16 chips long, forms a set of 16 orthogonal Walsh codes. The same concept applies to the larger set of 64 Walsh codes, and when the correlation is applied over the complete 64 chips.

Example 1:

Input bit = 0, with Walsh code W_0, gives a transmitted signal of

+1, +1, +1, +1, +1, +1, +1, +1, +1, +1, +1, +1, +1, +1, +1, +1

When we correlate(multiply bit by bit, then sum result) against shortened W_0,

+1, +1, +1, +1, +1, +1, +1, +1, +1, +1, +1, +1, +1, +1, +1, +1

X +1, +1, +1, +1, +1, +1, +1, +1, +1, +1, +1, +1, +1, +1, +1, +1

+1, +1, +1, +1, +1, +1, +1, +1, +1, +1, +1, +1, +1, +1, +1, +1

$\Sigma = +16$, which we decode as a "0" input bit

Example 2:

Input bit = 1, with Walsh code W_0, gives a transmitted signal of

-1, -1, -1, -1, -1, -1, -1, -1, -1, -1, -1, -1, -1, -1, -1, -1

When we correlate against shortened W_0,

+1, +1, +1, +1, +1, +1, +1, +1, +1, +1, +1, +1, +1, +1, +1, +1

X -1, -1, -1, -1, -1, -1, -1, -1, -1, -1, -1, -1, -1, -1, -1, -1

-1, -1, -1, -1, -1, -1, -1, -1, -1, -1, -1, -1, -1, -1, -1, -1

$\Sigma = -16$, which we decode as a "1" input bit

Example 3:

Input bit = 0, with Walsh code W_{32}, gives a transmitted signal of

+1, -1, -1, +1, +1, -1, -1, +1, +1, -1, -1, +1, +1, -1, -1, +1

When we correlate against shortened W_{32},

+1, -1, -1, +1, +1, -1, -1, +1, +1, -1, -1, +1, +1, -1, -1, +1

X +1, -1, -1, +1, +1, -1, -1, +1, +1, -1, -1, +1, +1, -1, -1, +1

+1, +1, +1, +1, +1, +1, +1, +1, +1, +1, +1, +1, +1, +1, +1, +1

$\Sigma = +16$ which we decode as a “0” input bit

Example 4:

Input bit = 1, with Walsh code W_{52}, gives a transmitted signal of

+1, -1, +1, -1, -1, +1, -1, +1, -1, +1, -1, +1, +1, -1, +1, -1

When we correlate against shortened W_{52},

+1, -1, +1, -1, -1, +1, -1, +1, -1, +1, -1, +1, +1, -1, +1, -1

X -1, +1, -1, +1, +1, -1, +1, -1, +1, -1, +1, -1, -1, +1, -1, +1

-1, -1, -1, -1, -1, -1, -1, -1, -1, -1, -1, -1, -1, -1, -1, -1

$\Sigma = -16$ which we decode as a “1” input bit

Next, let us assume we have three different codes in use: W_0, W_{32}, W_{52} and have input bits.

1 for W_0 (example 2), 0 for W_{32} (example 3), and 1 for W_{52} (example 4). Next we sum the transmitted signals together, for a combined signal to be sent to the receiver.

-1, -1, -1, -1, -1, -1, -1, -1, -1, -1, -1, -1, -1, -1, -1, -1

\+ +1, -1, -1, +1, +1, -1, -1, +1, +1, -1, -1, +1, +1, -1, -1, +1

\+ +1, -1, +1, -1, -1, +1, -1, +1, -1, +1, -1, +1, +1, -1, +1, -1

+1, -3, -1, -1, -1, -1, -3, +1, -1, -1, -3, +1, +1, -3, -1, -1

In the receiver, we can recover the original bits correlating to the original Walsh codes.

Correlation with shortened W_0

$$
\begin{array}{r}
+1,+1,+1,+1,+1,+1,+1,+1,+1,+1,+1,+1,+1,+1,+1,+1 \\
\times \quad +1,-3,-1,-1,-1,-1,-3,+1,-1,-1,-3,+1,+1,-3,-1,-1 \\
\hline
+1,-3,-1,-1,-1,-1,-3,+1,-1,-1,-3,+1,+1,-3,-1,-1
\end{array}
$$

$\Sigma = -16$, which we decode as a "1" input bit

When we correlate against shortened W_{32},

$$
\begin{array}{r}
+1,-1,-1,+1,+1,-1,-1,+1,+1,-1,-1,+1,+1,-1,-1,+1 \\
\times \quad +1,-3,-1,-1,-1,-1,-3,+1,-1,-1,-3,+1,+1,-3,-1,-1 \\
\hline
+1,+3,+1,-1,-1,+1,+3,+1,-1,+1,+3,+1,+1,+3,+1,-1
\end{array}
$$

$\Sigma = +16$, which we decode as a "0" input bit

When we correlate against shortened W_{52},

$$
\begin{array}{r}
+1,-1,+1,-1,-1,+1,-1,+1,-1,+1,-1,+1,+1,-1,+1,-1 \\
\times \quad +1,-3,-1,-1,-1,-1,-3,+1,-1,-1,-3,+1,+1,-3,-1,-1 \\
\hline
+1,+3,-1,+1,+1,-1,+3,+1,+1,-1,+3,+1,+1,+3,-1,+1
\end{array}
$$

$\Sigma = +16$, which we decode as a "1" input bit

This simple example shows how a composite of several Walsh codes can be separated into individual contributions by the correlation process, and the input bits used to set the polarity of the individual Walsh codes recovered. During this recovery process, the other Walsh codes are completely excluded due to the nature of orthogonality of the codes.

Next, consider when we correlate the received sequence against a Walsh code that is not present in the composite signal.

When we correlate against shortened W_{20},

$$
\begin{array}{lcccccccccccccccc}
 & +1, & +1, & -1, & -1, & -1, & -1, & +1, & +1, & -1, & -1, & +1, & +1, & +1, & +1, & -1, & -1 \\
X & +1, & -3, & -1, & -1, & -1, & -1, & -3, & +1, & -1, & -1, & -3, & +1, & +1, & -3, & -1, & -1 \\
\hline
 & +1, & -3, & +1, & +1, & +1, & +1, & -3, & +1, & +1, & +1, & -3, & +1, & +1, & -3, & +1, & +1
\end{array}
$$

$\Sigma = 0$. This indicates no correlation, and no W_{20} component in the received signal.

The properties of Walsh codes allow each different code to be perfectly separated from the composite signal of all the Walsh coded user data.

16.6 Network Synchronization

This concept can be extended to build a receiver with a bank of correlators and threshold detectors to recover the data modulated by any given Walsh code. For the correlation process to work, all the Walsh codes must be transmitted with the same start and end timing. This requires close synchronization between all of the base stations and the mobile phones. The base stations are all synchronized using GPS (global positioning system) satellite receivers. In addition to position data, the GPS system also provides timing data with great precision. By comparison, the previous analog and TDMA cellular systems did not require base station synchronization. The mobile phones would reacquire each base station's individual timing at each handoff.

While use of the GPS system provides a means for all the CDMA base stations to be synchronized, this will not work for the mobile phones (GPS signals cannot be received indoors, or when blocked by tall buildings). Instead, a pilot signal is transmitted by each base station. The pilot signal provides a known signal to allow the mobile to determine the basic start and end time of the Walsh code timing, provides a coherent reference for demodulation, and allows for individual base station identification.

Recall in our discussion on Walsh codes, that 180°degree phase shifts in the signal are used to identify $0 \rightarrow 1$ and $1 \rightarrow 0$ transitions. However, wireless signals can only carry relative phase information, not absolute (phase is same as delay, so absolute phase changes whenever distance between transmitter and receiver changes). This limitation can be overcome in two ways. First is to encode all data in a differential form, such as the differential offset QPSK used in IS-136 TDMA. This is known as noncoherent demodulation. It is simpler, but more susceptible to noise. The other method is to detect and track the received carrier phase, and use this to determine actual phase changes. This is known as coherent demodulation. All this sounds complicated, and it is. Suffice to say here that the pilot signal is required to perform coherent demodulation and provides a phase reference for the correlators.

16.7 RAKE Receiver

The receiver architecture used for CDMA is known as a "RAKE" receiver. The RAKE receiver has multiple correlators. These correlators can be programmed for different Walsh codes. Some of the correlators can also be programmed for slight differences in arrival timing (remember multipath in the last chapter?) of the same Walsh code. The CDMA receiver constantly tunes the RAKE receiver multiple correlators, also called "fingers," for optimal reception. This is required in the presence of multiple delayed versions of the received signal or multipath, which would cause ISI in a TDMA system. Multipath is much more prevalent in CDMA, due to the much higher chip rate than TDMA's symbol rate, and the RAKE receiver architecture is well suited to compensate for this effect.

16.8 Pilot Pseudorandom Number Codes

PN sequences are important in CDMA systems. One key characteristic of PN sequences the autocorrelation property. When a PN sequence correlated to itself, the result will be a perfect match if the sequences are perfectly time aligned, as with any other type of sequence. But if PN sequence is correlated with an offset, even just one sample, the result is near zero correlation. This is also a property of a true random noise signal of infinite length.

One of the most important uses of PN sequences is in the pilot signal. The pilot signal is a pair of PN sequences, which operate at the chip rate, and repeats about every 26.66 milliseconds. The pilot signal is sent in quadrature, using QPSK. There is a separate sequence defined for I and Q, given by the generator equations below (Fig. 16.4).

I PN sequence: $G_I = 1 + X^2 + X^6 + X^7 + X^8 + X^{10} + X^{15}$
Q PN sequence: $G_Q = 1 + X^3 + X^4 + X^5 + X^9 + X^{10} + X^{11} + X^{12} + X^{15}$

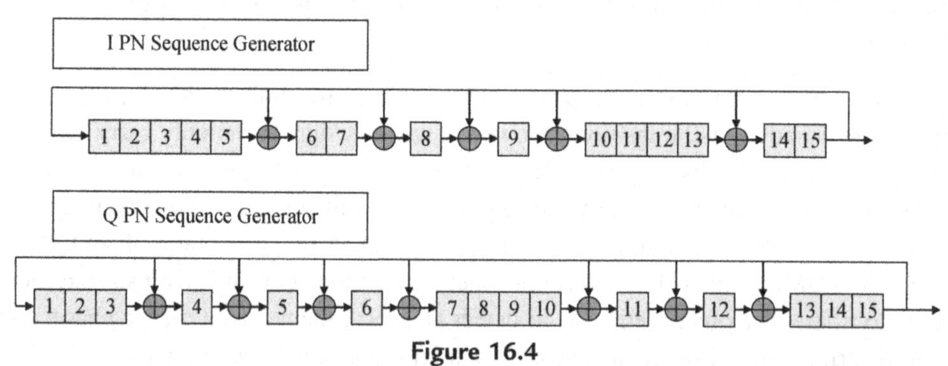

Figure 16.4
I and Q pseudorandom number sequence generation circuits. *PN*, pseudorandom number.

The PN sequences can be generated using a very simple circuit, shown in Figure 13.7. The PN circuit output will repeat every 32,767 bits, and an extra zero is inserted to bring this to the full 32,768 length pilot sequence. The shift register state has 2^{15} or 32,768 possible values. The only invalid state is the all zero state, as this state is self-perpetual (that is why the extra zero is added separately to the output sequence). Any nonzero state can be used at start up, and this state will determine where the repetitive 32,767 sequence begins, or the startup "phase" of the PN sequence. The pilot signal, with the zero is inserted, has 32,768 possible phases.

16.9 Code Division Multiple Access Transmit Architecture

The transmitter circuit for IS-95 and CDMA2000 is the diagram shown below. The I and Q quadrature paths are each mixed or EX-ORed with separate PN sequences. These PN sequences make up the pilot channel. As pilot channel uses W_0, it is the only channel not modified by the Walsh code mixing. The pilot tone is a simple QPSK signal, modulated by the pilot PN sequences. Different cells or base stations distinguish themselves by using different offsets, or phases of the pilot sequence. The offsets are in increments of 512 chips, providing 64 possible offsets. When a CDMA mobile phone scans for the nearby cell sites, it will identify them by the relative signal strength and their PN offset of the pilot channel. As transmission delays due to distance between base station and mobile are usually in the tens of chips, a separation in phase or delay of 512 chips clearly indicates that another base station is the source. These known PN sequence phase delays of 512 chips allow the mobile phone to perform a correlator search across increments of 512 chip delays to detect nearby base station pilots.

The mobile phones uses the strength of the pilot signal to determine if it is nearing the edges of the cell, defined by the area of strong pilot signal coverage. By measuring the strength of other phase offset pilots, the mobile phone can report to the network of which the adjacent cells are in range, information which is used in handoff decisions (Fig. 16.5).

Note that the pilot I and Q sequences are used both in the pilot and all other channels, after Walsh code modulation. This provides for the individual quadrature components for each channel. The synchronization channel uses W_{32}, and provides the GPS time reference used to obtain all the system timing information needed. Included in this is the data needed to setup the long code generator. This is a much longer PN sequence, 2^{42} bits long (repeats about every 41 days!). Each mobile will be assigned a unique long code phase, to provide privacy for the user channel data. There is often only one paging channel, although the system allows up to seven paging channels. Paging channels are used similarly as control channels in TDMA systems. The mobile phones will monitor the paging channels to determine if there is an incoming call. The rest of the Walsh codes, up

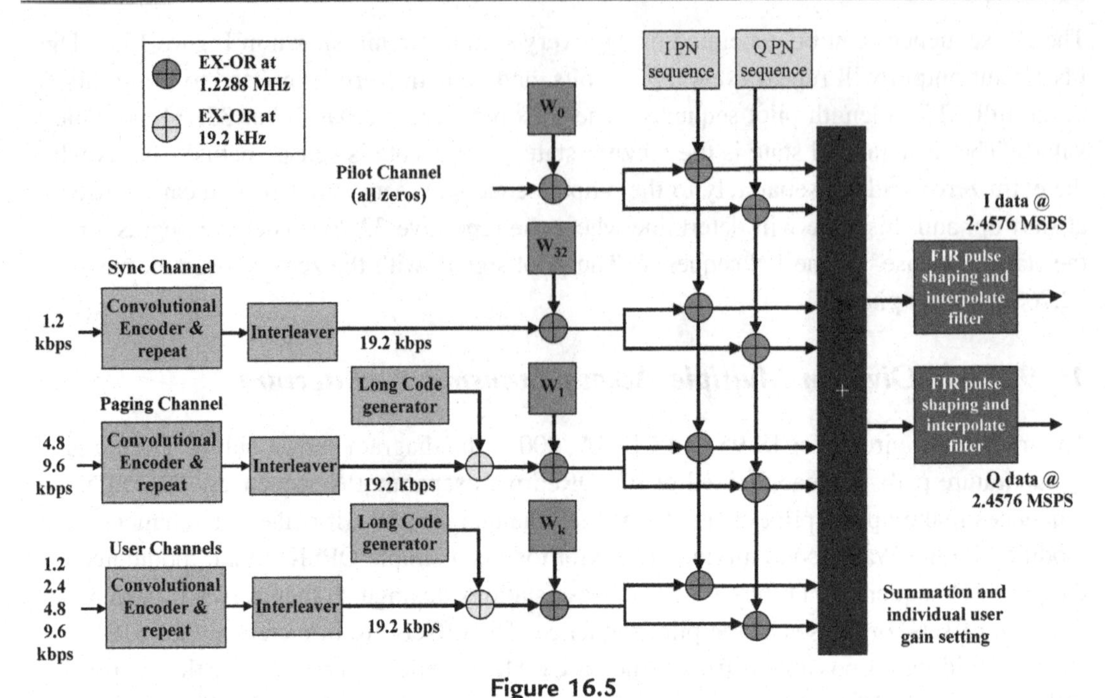

Figure 16.5
CDMA2000 downlink signal generation. *PN*, pseudorandom number.

to 55, can be allocated to users for voice or data calls. Due to mutual interference limitations, the number of simultaneous users is normally 40 or less.

All of the channels are then summed together. At this point, the signal is no longer a sequence of "1"s and "0"s, but is often 10–14 bits wide. In addition, gain is individually applied to each channel (or set to zero for unused channels). In CDMA, optimal capacity requires that each channel only be transmitted at the minimum power level necessary for low error rate reception. The exception is the pilot channel, which will be discussed further. After summation and gain setting stage, the transmit signal is then filtered and interpolated by the pulse-shaping filters, and then digitally upconverted to the IF frequency for conversion to analog form by the DAC. The analog RF circuitry will then mix and filter the signal to the carrier frequency, amplify using a high power RF amplifier, and transmit through the antennas. The RF signal bandwidth is approximately 1.25 MHz.

16.10 Variable Rate Vocoder

One of the many innovations introduced in the CDMA system was variable rate vocoder. Notice that the input data rate to each user or traffic channel can be one of four rates: 1.2, 2.4, 4.8, or 9.6 kbps. This was due to the use of a vocoder known as EVRC (enhanced variable rate coder), also known as the CTIA IS-127 standard. The EVRC vocoder exploits the nature of voice compression. Depending on the nature of the speech, or if there are pauses in conversation, the speech can be compressed to a much greater degree than the normal 8 kbps rate, without sacrificing voice quality. The EVRC was designed to do just that.

With a TDMA system, the allocated bandwidth in a slot is fixed. A variable rate vocoder would need to be allocated a channel with a data rate equal to the maximum possible rate. CDMA, however, can take advantage of variable rates. When EVRC is operating at a rate of 1.2 kbps, the data can be repeated four times. This does not directly help the system operation. But with this repetition, the correlation is stronger, as it is over four times the length. This, in turn, allows the data to be transmitted at lower power levels. This adjustment is possible because of the power, or gain, control adjustment for each individual channel. There is a very sophisticated power control algorithm to keep each mobile handset transmitting at the minimum power level to maintain a reasonable bit error rate. When EVRC is operating at a low data rate, the power level of that user channel can be reduced. Recall that in the uplink, each channel acts like noise all the other channels. The lower the power required for each channel, the less noiselike interference is experienced by the all of the channels. For that reason, CDMA is often referred to as an interference limited system. The level of mutual interference, or noise created by all of the channels with respect to each other is the major determining factor in the capacity of a CDMA system.

16.11 Soft Handoff

The CDMA system uses a common 1.25 MHz frequency band in each cell site and sector. By using different Walsh codes, interference is prevented between different users in the same cell or adjacent cells. For adjacent cells, even where the same Walsh code is in use, the different offsets of the short PN code will limit interference, by making the other users appear as noise. This use of the same frequency band allows "soft" handoff. In the last chapter, frequency handoff was discussed in analog and TDMA systems. This is called hard handoff, where the mobile phone must break communication with the existing base station and change frequency to communicate with the newly assigned base station. Soft handoff, in contrast, allows the mobile phone to maintain communication with the previous base station while simultaneously communicating with the next base station. Depending on the pilot strength thresholds set by the network, it is possible for a single mobile to be in communication with three or more base stations at once. The mobile phone has pilot signal strength thresholds to allow communication with a new base station, and thresholds to discontinue communication with a previous base station. These thresholds are under network control.

Soft handoff can provide greater reliability and voice quality, due to the ability to communicate with both or several base stations at or near cell boundaries, where the signals tend to be weak. There is also a network capacity tradeoff, for as the number of simultaneous base stations in the handoff process increases, or the duration of the soft handoff process increase, more network capacity is being consumed by that particular user.

Soft handoff is possible in CDMA networks because of the common frequency band used by all cells, and the network synchronization of all the base stations, neither of which exists in analog or TDMA systems. The network synchronization allows the voice or data to be simultaneously and synchronously routed to several base stations for transmission. The network can also synchronously combine voice or data traffic from a given mobile phone. This is an important advantage of CDMA voice systems over 2G systems.

16.12 Uplink Modulation

CDMA systems are so named for the downlink (base station to mobile) modulation method used. The uplink (mobile to base station) is somewhat different. The Walsh codes are orthogonal only if all the codes are sent and received synchronously. This is possible with the downlink as the base stations are all synchronized through GPS. In the uplink, the arrival time of the signals at the base station receiver will be delayed by the roundtrip transit time. There is a time or propagation delay from of the base station downlink signal to the mobile phone. It is the speed of light that works out to about one mile per microsecond (millionth of a second), which is close to the one chip delay in CDMA2000. The mobile phone then synchronizes to the received signal using the downlink sync channel. The mobile will, in turn, transmit the uplink signal, which will also experience a similar propagation delay to the base station. As this is dependent on the mobile phone position, and constantly changing, it is not possible to guarantee that the different mobile phone uplink signals are all aligned at the base station receiver. Recall in the discussion on Walsh code demodulation, we needed all of the 64-chip long Walsh code boundaries to be aligned to preserve the orthogonality. If one Walsh sequence is delayed by ½, 1, 2, or 5 chips, this demodulation process will not work. For this reason, CDMA modulation is not used in the uplink direction.

The uplink users are instead distinguished by different PN codes. Actually, they all use the same very long PN code, but are assigned different phases. A 2^{42} length PN code is used, which is more than 4 trillion chips in length before repeating. Due to the autocorrelation properties of PN sequences, every other user uplink signal appears like noise to a receiver tuned to the correct PN sequence phase.

Walsh codes are still used in the uplink, but not to separate the different users as in the downlink. In CDMA2000, there are 2^6 or 64 Walsh codes. In the uplink, after the convolutional encoder and interleaver, each set of 6 bits selects one of the 64 Walsh codes.

The 64 chips of that Walsh code are mixed with the selected phase of the 2^{42} length PN code sequence. By detecting which of 64 Walsh codes was sent, the base station receiver can recover the original 6 bits. This works similar to maximum likelihood in coding. There is a possible 2^{64} sequences with a 64-bit word. However, only the 64 Walsh codes sequences are valid transmit sequences. The base station receiver attempts to match the received sequence to the closest valid Walsh sequence.

After the Walsh encoding, the quadrature phases are mixed with I and Q PN short codes and pulse-shaped filtered, similar to the downlink circuit. In this case, however, offset QPSK modulation is used rather than QPSK. The advantage of this will be discussed in the next chapter.

The uplink is inherently the weaker link compared to the downlink. The Walsh codes in the downlink direction are perfectly synchronized and theoretically eliminated interference between users. But the uplink uses different PN phases to distinguish user channels. The autocorrelation properties will not eliminate each user's effect on each other, as the Walsh codes will, but will make the other users' signals appear as noise. This makes the uplink interference limited. The downlink, on the other hand, tends to be a power limited, as the base station power amplifier must provide sufficient power to all users.

16.13 Power Control

Transmit RF power control was one of the key challenges of CDMA, and many of the enhancements of CDMA2000 over the earlier IS-95 system involved power control. Power control is used in both the downlink and uplink. In the downlink, the prime consideration is to keep the pilot power constant, even as the user channel power varies with number of users and distance from base station. Consistent pilot power is important as the pilot signal strength is used by the mobile phones to determine cell boundaries and make handoff decisions.

Power control is even more critical in the uplink, as each uplink is interference limited. To minimize the interference, it is important that each mobile phone transmit an amount of power such that all the mobile transmit power levels are roughly equal *at* the base station antenna for equivalent user traffic bit rates (9.6 kbps).

Both open and closed loop power control methods are used. Open loop power control is performed by the mobile phone, with no assistance from the base station network. Essentially, it works as follows. The further from the base station, or more obstacles between, the lower the pilot and other downlink signal powers will appear to the mobile phone. The mobile phone can use this to estimate the uplink power level required. The weaker the pilot signal, the higher the loss and greater the distance to the base station, which requires a higher transmit power from the mobile to compensate. The converse is also true.

The rate of response of the open loop power control is nonlinear. If the received pilot signal suddenly increases, such as if the mobile emerges from behind a building, the mobile transmit power is immediately reduced. But if the pilot signal strength drops, then the mobile transmit power is increased slowly, to prevent inadvertent interference with other user's uplink signals.

In addition to open loop power control, there is also a closed loop power control loop operating. Closed loop power control is needed as estimating the uplink losses based on downlink losses can lead to errors. The downlink and uplink bands are different (separated by either 25 or 80 MHz depending on whether in the 800 or 1900 MHz bands), and different frequencies can fade independently.

Closed loop power control is more accurate because it is based upon the uplink power measurements at the base station. A key challenge is to close the loop quickly. For that reason, a special power control bit is allocated in each downlink data frame, which can incrementally increase or decrease the mobile transmit power every 1.25 ms. In addition, the base station is designed to use hardware circuits to drive this bit depending on receive correlation results. By not using a software-based control to the power control bit, very rapid response can be achieved, which leads to optimal uplink transmit power control.

16.14 Higher Data Rates

One advantage of CDMA is that multiple Walsh codes can be aggregated to a single user, to allow a higher data rate. Data rates in multiples of 9.6 kbps are available. Aggregating 16 channels together allows a data rate of 153.6 kbps. With 32 channels aggregated, the maximum as the pilot, sync, and paging channel are not available for user traffic, the data rate is 307.2 kbps. This is simply not possible for in a TDMA system.

However, the disadvantage is that a few data users can consume most of the downlink capacity, compromising the available capacity for voice users, who provide the bulk of the revenue to the wireless carriers.

Therefore, a companion technology was developed to support high-speed data users. The market for this type of service is primary business people who need and are willing to pay for mobile high-speed service to allow remote internet and email access. This service is known EVDO for Evolution for Data Only. It uses the same 1.25 MHz channel bandwidth and same uplink modulation techniques as CDMA2000. The downlink is completely redesigned and does not use CDMA techniques. It used a very high-speed form of QAM modulation, and the entire signal is devoted to one user at a time. The users share this high-speed link in a time division duplex fashion. This is suitable for high speed data access, where the data access is often intermittent and packet based. The uplink typically requires much lower data rates, so retains the original CDMA uplink technology. Where EVDO service is offered, it uses separate frequency spectrum as the CDMA2000 service.

16.15 Spectral Efficiency Considerations

So how does CDMA stack up against TDMA when it comes to spectral efficiency? The original goal was for digital systems to provide a 10-fold increase in spectral efficiency. The IS-136 TDMA efficiency provided a threefold increase.

AMPS: One voice call per 30 kHz channel bandwidth. With 7-cell frequency reuse, the effective spectrum per user was 210 kHz.
IS-136: Three voice calls per 30 kHz channel bandwidth. With 7-cell frequency reuse, the effective spectrum per user was 70 kHz.
PDC: Six voice calls (half rate vocoder) per 25 kHz channel bandwidth. With 7-cell frequency reuse, the effective spectrum per user was 29 kHz.
GSM: Eight voice calls per 200 kHz channel bandwidth. With 7-cell frequency reuse, the effective spectrum per user was 175 kHz.
CDMA2000: CDMA is tougher to calculate capacity, as this depends on network settings. In general, higher capacity is possible at the expense of voice quality, due to the mutual interference issues and soft handoff thresholds. In general, most CDMA systems operate with a maximum of about 40 voice calls per cell.

Assuming 40 voice calls per 1.25 MHz, with reuse every sector. The effective spectrum per user is 31 kHz.

To reach the 10-fold increase, the CDMA system would have to operate at the maximum of 55 voice channels per sector, which while possible, does not lead to satisfactory quality in practice. But CDMA does lead to about a factor of over twice the spectral efficiency of IS-136 TDMA. The GSM TDMA system is by far the least spectrally efficient system. It now tends to dominate where lowest cost of service is critical, and user density (and therefore spectrum requirements) is moderate, such as in third-world countries to provide basic phone service in rural areas.

16.16 Other Code Division Multiple Access Technologies

Alternate CDMA technologies were developed after CDMA2000. The most common CDMA system is known both as WCDMA, or Wideband CDMA, or UMTS. This form of CDMA was heavily based on IS-95 technologies. The biggest difference is a chip rate of 3.84 Mcps, a spectral bandwidth of 5 MHz, and higher count of 256 Walsh codes to allow more user or traffic channels. This system was developed principally by the European wireless OEMs, and designed to be a 3G upgrade from the GSM system. UMTS has basically become the world wide standard for CDMA, as CDMA2000 is largely limited to usage in North America, Japan, and South Korea.

Just as Qualcomm developed EVDO for high-speed data access, a companion service to UMTS for high-speed data users was developed. It is known as HSDPA.

Another higher rate version of CDMA2000 was developed by Qualcomm, called 3xRTT. This system had a chip rate of 3.68 Mcps, three times that of 1xRTT. It was designed to compete with the higher chip rate WCDMA. It was never really deployed, as similar capacity could be achieved using three separate 1xRTT systems.

A third CDMA system was developed in China. It is known as TD-SCDMA and is expected to be limited to deployment within China only.

CHAPTER 17

Orthogonal Frequency Division Multiple Access Wireless Communications

The latest mobile communication technology following code-division multiple access (CDMA) wireless systems is naturally called fourth generation, or "4G." The goals of fourth-generation system are yet even higher data capabilities, as to be able to support voice, internet, streaming video, and other services. It utilizes a completely different technology than 3G, known as orthogonal frequency division multiplexing or OFDM. OFDM has also been used in broadcast systems and for wireless LAN (Wi-Fi). These systems are point to multipoint systems. In the case of broadcast, it was basically one-way transmission. Using OFDM in mobile communications systems requires a multiple-access system, which requires some additional considerations. This is known as orthogonal frequency division multiple access, or OFDMA.

17.1 WiMax and Long-Term Evolution

There are two major standards for OFDMA mobile technology. The first is known as WiMax. It grew out of the Wi-Fi wireless LAN technology. Wi-Fi is now standard in virtually every PC laptop, DSL, or cable ISP gateway, providing private internet access in many homes and public coverage in most airports, coffee houses, and hotels. Wi-Fi is defined as IEEE standard 802.11. WiMax, which has received major additions to support mobility and multiple access, is defined as IEEE standard 802.16. Shortly after the standardization of WiMax, the mobile communications industry began definition of its own OFDMA standard, which is known by the acronym "LTE," for Long-Term Evolution. The technology path envisioned is a worldwide wireless technology roadmap for mobile service providers, starting with GSM, migrating to WCDMA, and eventually to LTE. As LTE is being promoted by most of the mobile wireless industry, it is expected to have much wider deployment than WiMax. WiMax is more likely to find use in wireless network backhaul, military communications, and wireless local loop (basic phone service for rural areas without landline phones).

OFDMA does utilize a common concept as CDMA, orthogonality. To maximize spectral efficiency, all of the user traffic channels will be made orthogonal to each other. In that way, there is no interference between users, even though they all share a large common frequency bandwidth. Yet in OFDMA, the different users do occupy different subsections

Digital Signal Processing 101. http://dx.doi.org/10.1016/B978-0-12-811453-7.00017-2

of the frequency band but are spaced much more closely together, compared to TDMA, by using a technique that prevents adjacent channel interference.

17.2 Orthogonal Frequency Division Multiple Access Advantages

It is natural to ask what advantages does OFDMA offer over CDMA mobile technology. First, OFDMA can be easily configured to support multiple bandwidths, and therefore system capacities. LTE, for example, is able to operate in frequency bands ranging from 1.25 MHz up to 20 MHz wide. CDMA, in contrast, has a fixed bandwidth, which is largely by the chip rate and filtering. This characteristic allows OFDMA system operators to deploy service initially in a small frequency band, and then expand that band as the number of users or customers increase. It also allows for higher data rates, as more RF bandwidth can be used (up to 16 times CDMA2000 or 4 times WCDMA). Secondly, the individual subchannel modulation method can change dynamically depending upon the quality of RF channel between the base station and mobile phone (which depends on distance to base station, obstacles between number of reflecting signals and elevation and motion of user). OFDMA has the ability to vary on the fly the modulation used to either provide more robust communications link or higher data rate, on a per user basis. This is done without changing the amount of frequency bandwidth or subchannels that an individual user is allocated. Another consideration is more business focused. Given the prominent role Qualcomm had in the design, commercialization and deployment of CDMA mobile technology, that company naturally enjoys dominance in the CDMA handset chip market, as well as the intellectual property rights and the associated royalties worldwide. OFDMA mobile technology, by comparison, is being more or less developed in parallel by multiple wireless and semiconductor companies, and therefore provides a more level business landscape. Actual OFDMA system user capacity, spectral efficiency, and quality of service relative to CDMA are not known at the time of this writing, as OFDMA wide-scale deployment is still in very early stages. Comparisons to OFDM used in broadcast industry may not be relevant, as this is a one to many broadcast system with little or no reverse link traffic. In OFDMA mobile communications, the uplink is expected to be the weaker link, due to its multiple access requirement. However, if a significant portion of the mobile user traffic becomes internet access or video streaming, then the downlink traffic loads will be much higher than the uplink and the uplink limitations may be less important than downlink capacity.

In an FDD system, each user channel occupies a separate frequency band. Other users are rejected at the receiver by downconverting and filtering out the desired signal. Since all filters require a transition region between the pass and stopbands, there must be a guardband or separation region between frequency band of the individual carriers. The larger the difference is signal levels of multiple carriers at the receiver, the more rejection required by the filtering. More rejection or attenuation of adjacent channels in the digital receive filters usually requires either a longer filter or a larger transition band.

With OFDM, we will pack the different frequency channels, or subcarriers, very close together, and both demodulate all the subcarriers simultaneously. The basis of OFDM is utilization of the property that a group of sinusoidal signals spaced at a specific frequency separation will be orthogonal, or independent to each other. Just as we saw in CDMA, orthogonal signals do not interfere with each other, and using proper techniques, the desired signal can be separated from the other signals. We will not be using traditional filtering to perform this separation of subcarriers.

17.3 Orthogonality of Periodic Signals

Orthogonality is defined as the product of two periodic signals (example—sinusoids) equals zero over an interval of a period is equal to zero. What this means is that if we cross multiply the two signals and add the results over a length of the time in which both signals repeat, the result is zero. We saw this happen with the CDMA2000 Walsh codes in the previous chapter. The period in that case was defined as 64 samples.

In continuous time, we can express the orthogonality relationship as

$$\int_0^T \cos(2\pi m f_0 t)\cdot\cos(2\pi n f_0 t)\,dt = 0, \text{ as long as m} \neq \text{n}$$

In this expression, m and n are integers, and $1/f_0 = T$, the period, which is the interval to integrate over.

We can see this graphically in Fig. 17.1. The fundamental frequency f_0 in the graph is 1 kHz, and the period T is 1000 microseconds (μs) or 1 millisecond (ms).

(A: 1 kHz, B: 2 kHz, C: 3 kHz, D: 4 kHz)

Next, the product or cross correlation of several cosines is graphed.

The results in Fig. 17.2 are

A: 1 kHz * 1 kHz will integrate to a positive value over the interval T
B: 1 kHz * 2 kHz will integrate or average to zero over the interval T
C: 1 kHz * 3 kHz will integrate or average to zero over the interval T
D: 1 kHz * 4 kHz will integrate or average to zero over the interval T

Similarly, the 2-kHz sinusoid multiplied by the 3-kHz sinusoid, and every other combination of different frequency cosines, will also integrate to zero over the integral T. Note that all the sinusoids are periodic over interval T, which simply means that they have

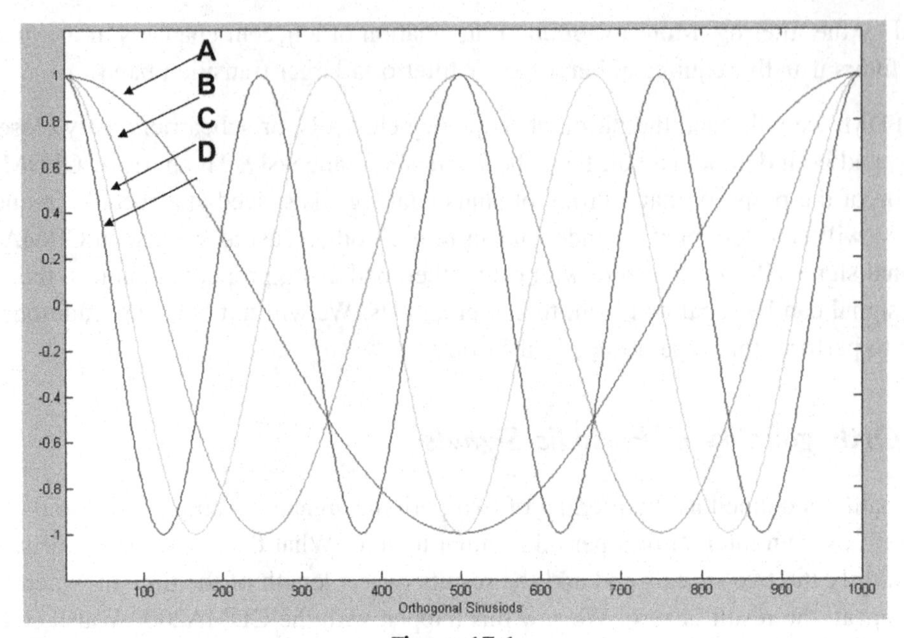

Figure 17.1
Orthogonal sinusiods.

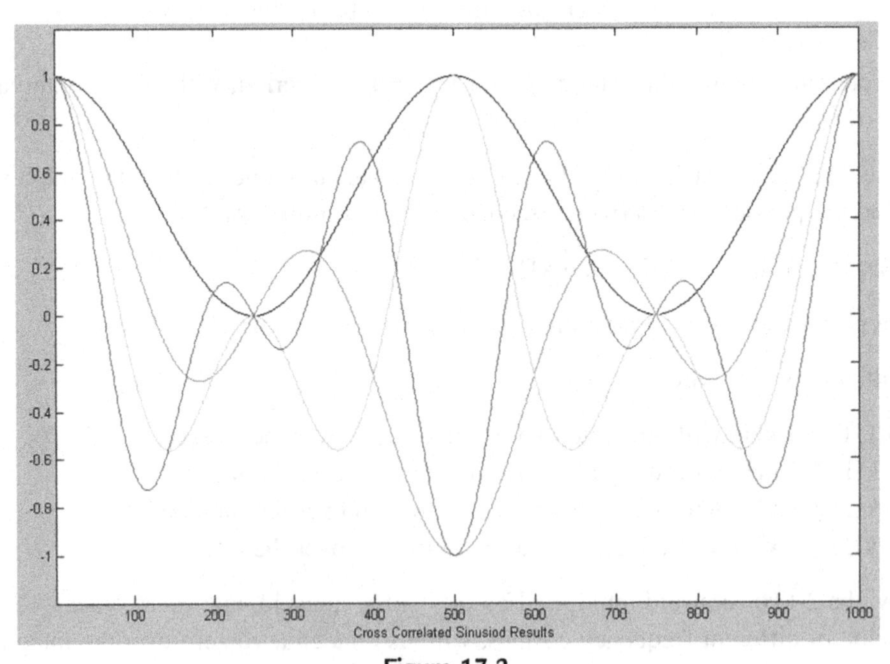

Figure 17.2
Multiplication of orthogonal sinusoid yields an orthogonal sinusoid.

an integer number of cycles in that interval. Notice in the first graph that since every sinusoid returns to its starting value of "1" at the end of the 1000-μs interval, they are all periodic over T, equal to 1000 μs in this example.

In digital signal processing, we will use sampled signals, rather than continuous signals. The orthogonality relationship can be expressed as

$$\sum_{k=0 \text{ to } N-1} \cos(2\pi mk/N) \cdot \cos(2\pi nk/N) = 0, \text{ as long as } m \neq n$$

For example, suppose we are sampling at 1 MHz. We can make the symbol period equal to 1000 samples, which equates to 1 k-symbol/second. Because of the orthogonality principle, we can have carriers at 1, 2 kHz... up to the Nyquist limit of 500 kHz. Each of these subcarriers can carry one symbol, perhaps using the phase of the subcarrier to carry the information. The demodulation process would require cross multiplying, or cross correlating, by each subcarrier frequency. This will exclude the other subcarriers and allows the recovery of the symbol information by determining the phase of the subcarrier. In this way, it is possible to transmit simultaneous data across multiple subcarriers, without interference.

17.4 Frequency Spectrum of Orthogonal Subcarrier

Each subcarrier is modulated, usually using either quadrature phase shift keying (QPSK) or quadrature amplitude modulation (QAM) modulation. Sinusoids have their frequency content all at the carrier frequency and appear as a vertical line in the frequency spectrum. A QPSK or QAM modulation will have nulls in the spectrum at the offset by the symbol frequency. In OFDM, these nulls line up perfectly with the adjacent subcarriers, resulting in minimal adjacent channel interference and efficient packing of the subcarriers in the frequency spectrum, as shown in Fig. 17.3.

One possible method of OFDM modulation and demodulation would be to multiply each complex baseband QPSK or QAM signal be a complex exponential corresponding to each subcarrier frequency, located at integer spacings of the symbol frequency. For N carriers, this would require N parallel circuits for both OFDM modulation and demodulation.

Let us review the discrete Fourier transform (DFT) and inverse discrete Fourier transform (IDFT) equations:

$$\text{DFT(time} \rightarrow \text{frequency)} \quad X_k = H(2\pi k/N) = \sum_{i=0 \text{ to } N-1} x_i e^{-j2\pi ki/N} \quad \text{for } k = \{0 \ldots N-1\}$$

$$\text{IDFT(frequency} \rightarrow \text{time)} \quad x_i = 1/N \cdot \sum_{k=0 \text{ to } N-1} X_k e^{+j2\pi ki/N} \quad \text{for } i = \{0 \ldots N-1\}$$

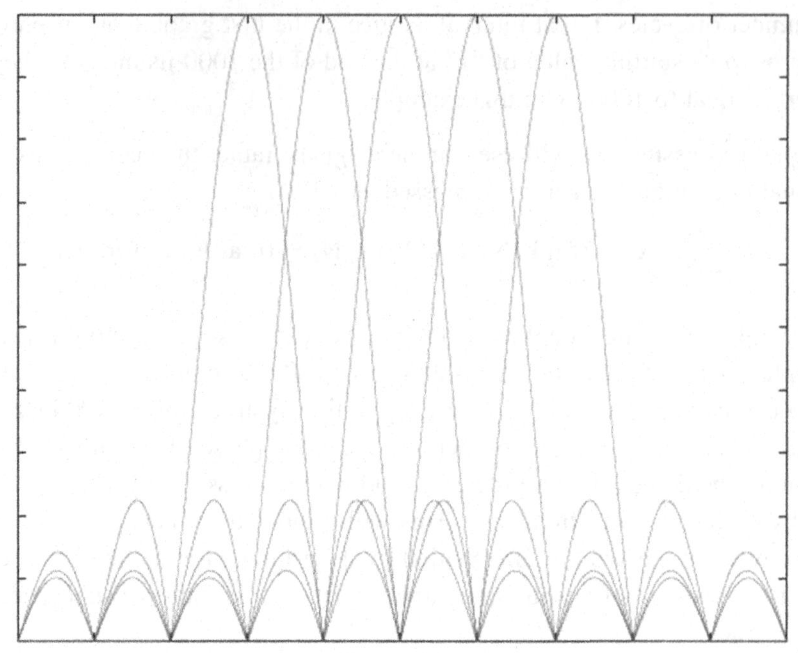

Figure 17.3
Orthogonal frequency division multiplexing modulation using fast Fourier transform.

Expanding the IDFT equation

$$x_0 = 1/N \cdot \sum_{k=0 \text{ to } N-1} X_k e^{+j0}$$
$$x_1 = 1/N \cdot \sum_{k=0 \text{ to } N-1} X_k e^{+j2\pi \cdot k/N}$$
$$x_2 = 1/N \cdot \sum_{k=0 \text{ to } N-1} X_k e^{+j2\pi \cdot 2k/N}$$
$$x_3 = 1/N \cdot \sum_{k=0 \text{ to } N-1} X_k e^{+j2\pi \cdot 3k/N}$$
$$\ldots$$
$$x_{N-1} = 1/N \cdot \sum_{k=0 \text{ to } N-1} X_k e^{+j2\pi \cdot (N-1)/N}$$

Notice that each complex exponential frequency is orthogonal to all the other. Each of these complex exponentials is periodic in N samples. That is because each has an integer number of cycles in N samples. To complete the orthogonality requirement, the symbol data rate should be equal to the frequency spacing, which is equivalent to the subcarrier spacing $e^{\ j2\pi 1/N}$, and will be $F_{sampling}/N$ symbols per second.

17.5 Orthogonal Frequency Division Multiplexing Modulation

The DFT can be used to perform OFDM modulation and demodulation. If we make N equal to 2^m, where y is an integer, then an even better way is to use the fast Fourier transform (FFT). For example, if m = 10, then N = 1024. In OFDM systems, the inverse fast Fourier transform (IFFT) is used for modulation in the transmit path, while the FFT is used for demodulation in the receiver. The OFDM modulation architecture is shown in Fig. 17.4 for 10-MHz LTE system.

The IFFT processes 1024 complex samples to form a single OFDM symbol. Each user is assigned to a separate subcarrier. For 10 MHz LTE, there about 600 allowable subcarriers, spaced 15 kHz apart. Each subcarrier has a QPSK or QAM modulator that takes each user's input data and creates a single constellation point. This forms one bin of the IFFT input buffer. Each bin will be mapped to an individual subcarrier in the frequency spectrum. Each OFDM symbol will contain a single constellation point from each of the subcarrier modulators. And each successive OFDM symbol will contain the successive constellation points from the modulators.

As shown in Fig. 17.4, there is a data source for each subcarrier. This data source will be modulated. If the chosen modulation for that subcarrier is QPSK, then every 2 bits of input data will be mapped to one of four possible complex points on the QPSK constellation. The QPSK symbol for our subcarrier becomes one of the complex input samples for the IFFT. In parallel, this is occurring with all the other subcarriers. They form the rest of the input samples for the IFFT. Additionally, certain samples in the IFFT input are always forced to zero. The IFFT output will form the OFDMA symbol. Therefore, the symbol rate of the OFDMA system is the same as the rate at which the IFFT is performed. Selection of the input bin of the IFFT input buffer will map to a specific frequency subcarrier in the IFFT output. A continuous transmit data signal can be formed by concatenating successive IFFT outputs.

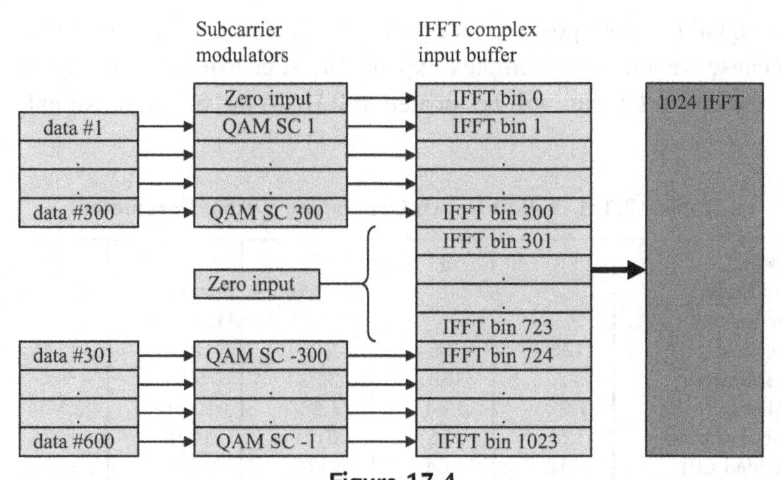

Figure 17.4
Formation of orthogonal frequency division multiplexing symbol.

This is the equivalent function as QSPK or QAM modulating hundreds of users data in parallel, and upconverting each using a different complex exponential with a frequency separation or difference equal to the QPSK or QAM symbol rate.

Note that each subcarrier can be independently modulated using either QPSK, 16-QAM or 64 QAM. The higher order modulation requires a higher signal to noise ratio at the receiver, which occurs when there is little degradation in signal path from transmitter to receiver. The benefit is higher amounts of data can be transmitted per subcarrier. Recall from Chapter 9 on modulation, QPSK carries 2 bits, 16-QAM carries 4 bits, and 64-QAM carries 6 bit of user data per symbol. This dynamic trade-off per user of data rate to signal link quality is one of the benefits of OFDMA.

The output of the IFFT forms the wide band transmit signal. Depending upon the IFFT size, the frequency spectrum can be made larger or smaller. The table below shows the configurable frequency bandwidths of the LTE system. In all cases, the subcarrier symbol rate and subcarrier frequency spacing is constant. Using a larger IFFT/FFT size allows for more subcarriers and occupies more bandwidth. This property allows a wireless service provide to initially deploy LTE in a smaller frequency spectrum, say 5 MHz. As customer usage grows, the LTE system can be reconfigured for up to 20 MHz. The mobile phones are dynamically configurable as well. A single subcarrier has high enough bandwidth to accommodate a compressed voice data rates. However, higher data rates are available by assigning multliple subcarriers to the same user (Table 17.1).

Of course, an OFDMA system is a bit more complicated than this. For example, not all subcarriers are available for users. Certain subcarriers are carry known, fixed data patterns, and are known as pilot subcarriers. These are used by the receiver to perform synchronization, frequency offset tracking, and equalization.

The baseband signal has both positive and negative frequency subcarriers (remember, this is possible because we are using complex exponential subcarriers). The middle subcarrier, which is also the first IFFT bin, will be located at 0 Hz in the baseband signal. It is always

Table 17.1: LTE Bandwidth Dependent System Parameters

BW (MHz)	1.4	3	5	10	15	20
Frame/slot length	10 ms/500 μs					
Subcarrier spacing	15 kHz					
FFT size used	128	256	512	1024	1536	2048
Number of subcarriers	72	180	300	600	900	1200
Sampling frequency (MHz)	1.92	3.84	7.68	15.36	23.04	30.72
OFDMA symbol samples (using extended CP)	128 +32	256 +64	512 +128	1024 +256	1535 +384	2048 +512

FFT, fast Fourier transform; *OFDMA*, orthogonal frequency division multiple access.

set to zero, to avoid introducing any DC into the baseband signal. And the outer subcarriers on both sides of the baseband signal are also set to zero, to provide a transition band for the low-pass filtering of the entire OFDMA signal from the last active subcarrier to the frequency channel edge (Fig. 17.5).

For example, the baseband sampling frequency of a 10-MHz LTE signal is 15.36 complex MSPS. So the Nyquist frequency is one half of this, or 7.68 MHz. The actual signal bandwidth is approximately +/− 300·15 kHz = +/− 4.5 MHz. This provides sufficient transition band for a low-pass filter (remember, there is another aliased image centered at each multiple of 15.36 MHz).

17.6 Intersymbol Interference and the Cyclic Prefix

Just as in the other wireless technologies, a direct line of sight is often impossible in mobile communications. This results in multiply reflected versions of the signal, of different signal strength, phase, and delay being received. This is sometimes referred to as "delay spread." The delay between different received reflections causes intersymbol interference or ISI. In TDMA, adaptative equalizers are used to cope with ISI, and in CDMA, the RAKE receiver is designed to operate in the presence of ISI. OFDM systems use a different method, called a cyclic prefix.

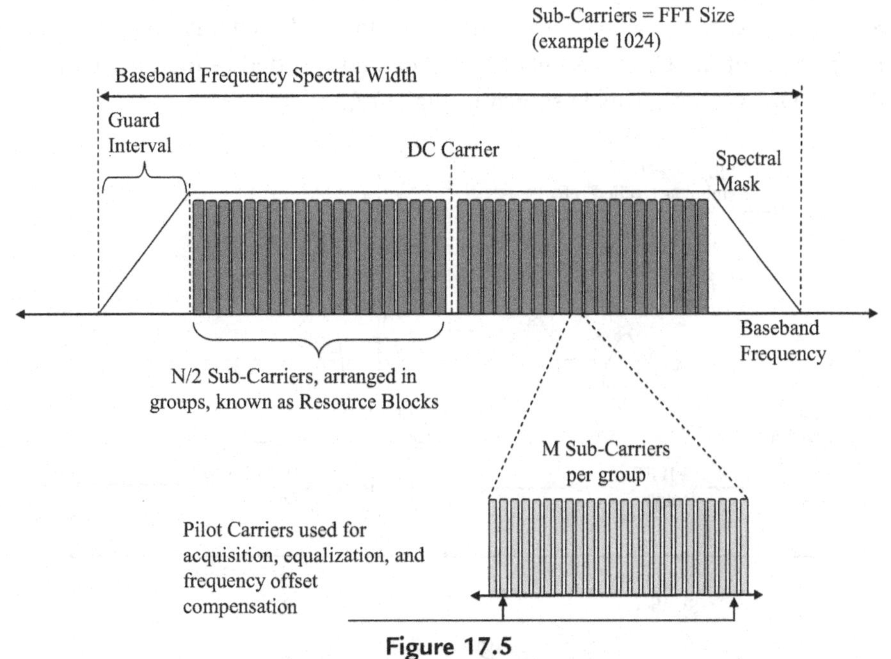

Figure 17.5
Orthogonal frequency division multiple access spectrum.

Of the three methods mentioned above, the cyclic prefix is the simplest. The easiest way to prevent ISI is to simply pause between symbols. Think about trying to talk to someone at the other end of a long cave. If you yell, the echoes will garble what the other person hears. But if you speak each word, separated by a pause of a few seconds, the echoes will die away before the next word, and the listener can hear each word distinctly. The cyclic prefix involves having a guard interval between symbols. This method works because the symbols are of long duration (for LTE the symbol length is about 67 μs, not counting guard interval), compared to TDMA (GSM symbol length is less than 4 μs) and especially CDMA (CDMA2000 chip rate is 1.2288 MHz, or less than 1 μs period). So having a guard interval of a fraction of a symbol is sufficient to prevent ISI, or in our analogy, let the echoes die away. For most mobile environments, 5–10 μs is sufficient for this purpose. For much longer time delays, the received signal echoes will have taken a much longer transmission path, and so be much lower amplitude (perhaps reflected from a mountain range outside your city), so can generally be ignored. This works because of the long OFDM symbol time. If we had to wait 10+ μs in a TDMA system, we would spend much more time waiting than transmitting and would have a very slow data throughput.

Still, even in OFDM, the guard interval does come at a penalty. During the guard interval, no user data can be transmitted, reducing the capacity and efficiency of the OFDMA system. And transmitting zero signal during the guard interval causes other issues. For example, the RF power amplifier would need to switch on and off each symbol, which can cause unwanted spectral output during these transitions. And this interval can still be used by the receiver, if a known signal is sent during this interval. What is done in practice is that the last portion of the OFDM symbol is copied and inserted in the guard interval just prior to that OFDM symbol. This is shown in Fig. 17.6.

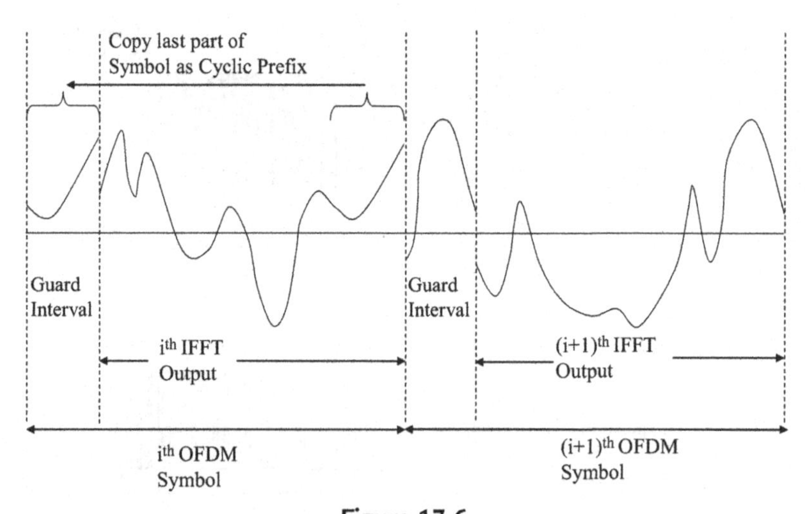

Figure 17.6

Time domain of orthogonal frequency division multiple access symbol.

The receiver will process the OFDM symbol period including guard interval. In LTE, for example, there is an extended cyclic prefix mode where the guard interval is set at 25% (the default mode is similar, but slightly more complex, as the guard interval varies in different symbols). In this mode, the IFFT symbol period is 66.67 μs, with guard interval of 16.67 μs, for a final symbol period of 83.33 μs. The receiver will process the OFDM symbol period of 83 μs including guard interval. During this interval, the symbol appears to be periodic or repeating. For this reason, the guard interval is called a cyclic prefix. Multiple OFDM symbols are aggregated to form slots and frames, which provide the structure needed to organize pilot subcarriers (use by mobile to synchronize timing and frequency to base station) as well as the user or data subcarriers.

Fig. 17.7 shows the organization of the LTE structure of frames, slots, and symbols, and how the samples are organized to allow for the cyclic prefix. This is shown for the 10-MHz bandwidth and extended cyclic prefix configuration. LTE systems are generally configurable to support all the different possible bandwidths and cyclic prefix options, as this allows the wireless service provider the ability to select the optimum configuration for their capacity and licensed spectrum allowances. The mobile phones are also able to support the various configurations.

In Fig. 17.8, a simplified block diagram of an example LTE transmitter is shown. A bank of modulators performs modulation for all active subcarriers, each of which provides data to a different bin on the IFFT. After the IFFT, the cyclic prefix is inserted as described above. The rest of the chain performs digital upconversion and interpolation. Analog RF circuitry (not shown) will further upconvert the signal to the actual transmit frequency and amplify it to a sufficient power level provide coverage over the cell sector and radius.

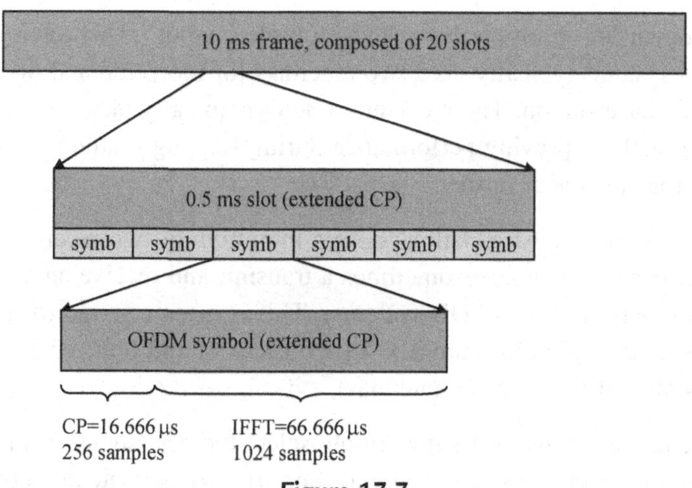

Figure 17.7
Orthogonal frequency division multiple access frame.

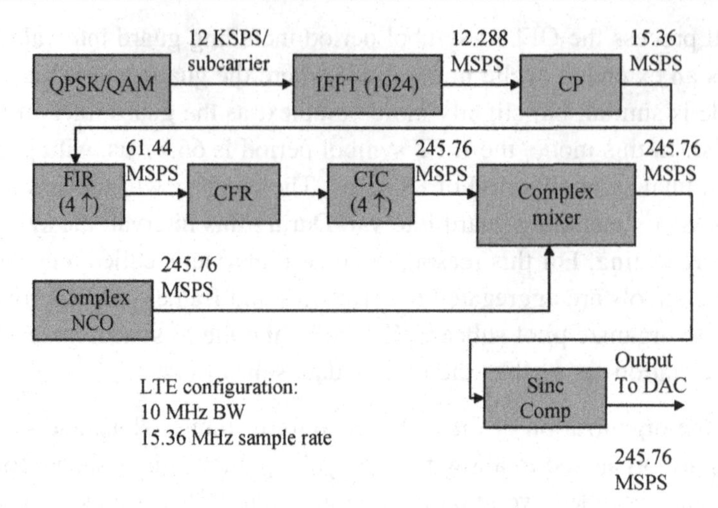

Figure 17.8
Long-Term Evolution (LTE) transmit circuit.

Two blocks in the diagram may be unfamiliar. The "CFR" block denotes crest factor reduction, which will be discussed in the next section. The "CIC" block denotes cascade integrate comb interpolation filter, which as a type of filter that does not require multipliers and so is inexpensive to build in hardware. The CIC will not be covered in this introductory book. Note the sinc compensation filter block, which compensated for the DAC response as described in Chapter 11 on Digital Up and Down Conversion.

17.7 Multiple Input and Multiple Output Equalization

MIMO is an acronym for "Multiple Input and Multiple Output." This refers to antennas. Previous wireless systems typically used two antennas for reception and one antenna for transmission at the base station. The receiver could dynamically pick between the two antenna signals, greatly improving performance during Rayleigh fading. In MIMO systems, this approach is taken further.

For 4G wireless systems, the base stations have a minimum of two receive and two transmit antennas per sector (note—sometimes a transmit and receive antenna are packaged together, particularly in FDD systems). This is often referred to as "2 × 2" MIMO. Other likely configurations are 2 × 4 (two transmit and four receive antennas) and 4 × 4 (four transmit and four receive antennas).

In a MIMO system, the receiver does not simply select the best signal from two or more antennas. Instead, the receiver uses both signals and tries to estimate and compensate for the degradation the signal experiences from in the separate paths to each antenna. This usually involves solving of multiple equations simultaneously and the use of matrix inversion algorithms to obtain the individual channel degradation estimates and perform the compensation or equalization of each receiver path.

In the uplink, the mobile signal will travel to multiple receive antennas, with each antenna having independent or uncorrelated path reflections, fading, and additive noise. In the downlink, the mobile has only one receive antenna. However, the use of multiple base station transmit antennas can be used in a similar manner. Often, each of the transmit antennas will transmit a slightly different version of the signal, using a technique called "space time encoding," which is known by the mobile receiver. Each version of the signal will experience uncorrelated path reflections, fading, and additive noise, which can be simultaneously processed to obtain the best estimate of the transmit signal.

These techniques involve sophisticated processing and statistical theory, and are not suitable for this introductory discussion.

The use 4×4 system will basically double the cost of the most expensive portion of the wireless base station compare with 2×2, but promises improved performance and system capacity. Whether this improvement is enough to justify the additional base station cost is still unknown at the time of this writing.

17.8 Orthogonal Frequency Division Multiple Access System Considerations

Before concluding this introduction to OFDMA, it may be worthwhile to summarize the relative merits and challenges of this fourth-generation mobile wireless technology.

In addition to the two benefits described earlier in the chapter of configurable RF BW and dynamically variable user data rates depending upon subcarrier SNR, there are some further benefits to OFDM technology.

OFDM efficiently deals with multipath, or ISI, using the cyclic prefix. This is much less complex and cheaper to implement than the adaptive equalizer in TDMA and RAKE receiver in CDMA.

OFDM is fairly insensitive to narrowband interference, as only a few subcarriers are affected. And given the dynamic monitoring of subcarrier SNR, this can be detected and users are allocated to subcarriers where interference is not present.

OFDM is however very sensitive to frequency offset. Both mobile and base station receivers must compensate for Doppler shift prior to demodulation, to preserve orthogonality of the subcarriers.

OFDM is also computationally efficient, when considering amount of circuitry and calculations required to support high data rates. Due to the high efficiency of FFT implementation, mobile phones will be able to receive high data rates with less DSP processing than comparable rates would require in a CDMA system, and would not be possible at all in a TDMA system.

17.9 Orthogonal Frequency Division Multiple Access Spectral Efficiency

Estimating spectral efficiency for OFDMA, particularly for voice capacity is difficult as no large, high capacity systems are in operation at the time of this writing. There are many parameters which can affect efficiency and capacity. For example, by use of dynamic modulation modes, multiple antenna receive and transmit (MIMO) and use of packetized voice (voice−over IP), proportion of voice to data traffic may all contribute to system optimization.

Similar to CDMA, OFDMA systems will use the same channel frequency in all cells (Fig. 17.9). This could be anywhere from a 1.25−20 MHz bandwidth. Interference between users is avoided by assigning different subcarriers to each base station. Groups of subcarriers are preassigned specific pilot subcarriers, so mobile phones are able to use known pilot subcarriers to acquire and synchronize to any assigned group of subcarriers. The frequency reuse of the subcarriers is not defined and may vary by wireless service provider. In TDMA and analog systems, the frequency reuse pattern was for every 7 cells. However, for OFDMA, it may be possible to have subcarrier frequency reuse of every 3 cells, due to the lower SNR requirements when using QPSK and MIMO techniques. Also, the channel spacing is tighter, a consequence of the OFDM modulation. For the LTE system, channel spacing is 15 kHz, much closer than any TDMA system.

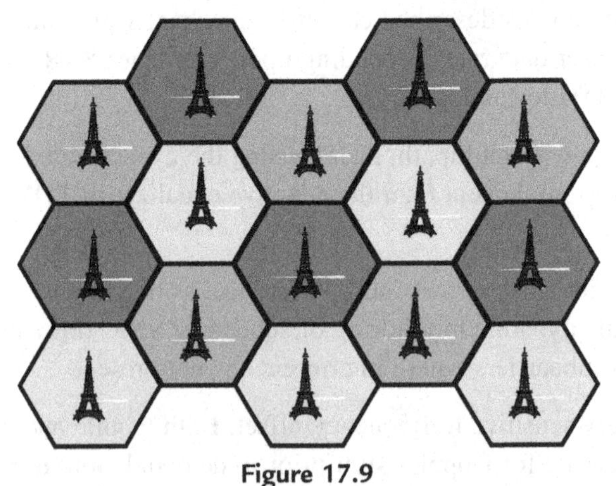

Figure 17.9
Frequency reuse across cells.

In summary, at this point it is very difficult to reliably predict actual capacity and spectral efficiency with so many variables. Similar to CDMA, the OFDMA network service providers will have a learning curve to go through in optimizing their systems for both high capacity and high quality of service.

17.10 Orthogonal Frequency Division Multiple Access Doppler Frequency Shift

All mobile communication systems must contend with Doppler frequency shift. However, OFDM is especially sensitive to Doppler shift, as it relies upon the precise alignment of subcarrier frequencies to provide orthogonality. The mobile phone velocity relative to the base station will cause Doppler frequency shift, which must be tracked and compensated for to provide proper demodulation using FFT processing. The following equation can be used to calculate the Doppler frequency shift.

$$\text{Doppler Frequency shift} = f_{\text{carrier}} \cdot \text{relative velocity/speed of light}$$

For example, assume a system operating at 2 GHz frequency band, with a mobile user traveling at a speed of 120 km/h (33.3 m/s).

$$\text{Doppler Frequency shift} = 2 \cdot 10^9 \cdot 33.3/(3 \cdot 108) = 220\ \text{Hz}$$

This is enough to cause a problem in OFDMA systems if not compensated for, by frequency shifting or rotating the input signal prior to demodulation.

17.11 Peak to Average Ratio

Both OFDM and CDMA have what is known as high "peak to average ratios" or PAR, on the order 10–20 times, or 10–13 in dB. Every signal has an average power. The "peak to average power" level, usually expressed in dB, is the power level of the highest instantaneous power compared to the average power level. A PAR of 1, or 0 dB, means the signal is of constant power, so the peak power is equal to the average power. A large PAR means that the signal power fluctuates occasionally to a very large value. Therefore a large PAR requires the linear transmit amplification circuits to operate over a wide power range, which tend to be both costly and inefficient. The plot in Fig. 17.10 shows a typical OFDM symbol amplitude over the duration of the symbol, with I and Q shown separately. This is a plot of the output of the IFFT in the modulator.

In mobile communication systems, the base station signals are amplified using fairly high power amplifiers (PA). The amplifiers must behave linearly over the output power range to avoid creating spectral energy outside the transmit band. The required power capacity and therefore power consumption and cost of the amplifier depend upon the PAR of the signal being amplified.

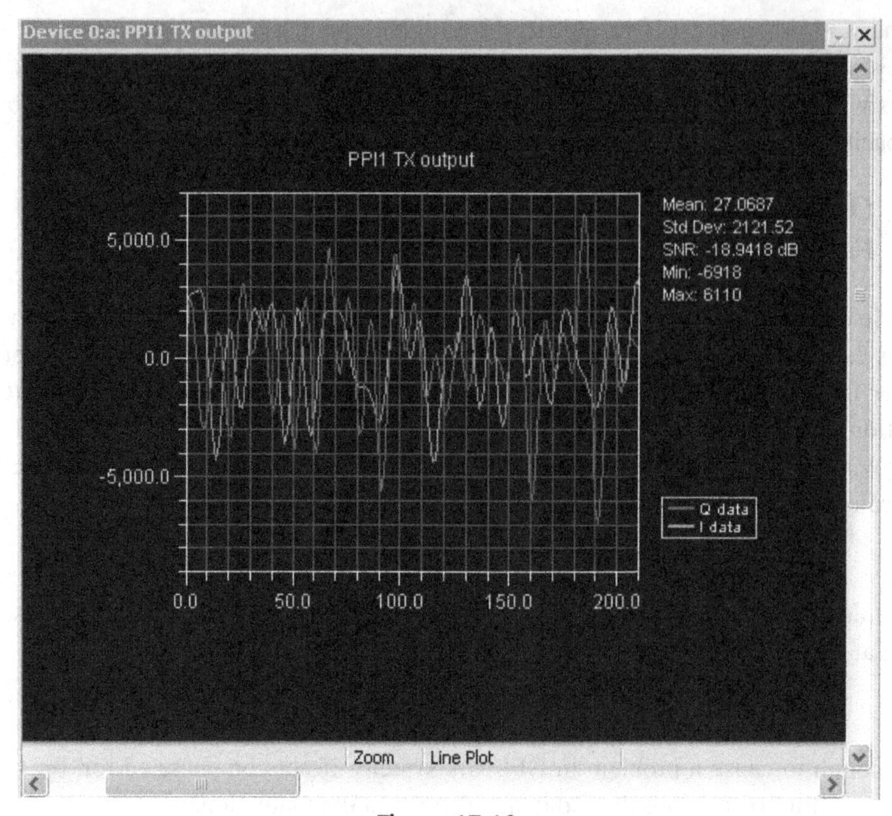

Figure 17.10
Actual orthogonal frequency division multiple access symbol (time domain).

Let us assume that we need to transmit a 50 W of power, on average, to provide RF coverage to a cell during high traffic hours. If the transmit signal has a PAR of 0 dB, we will need an amplifier with a power rating of 50 W (able to output 50 W while operating linearly). However, if the signal has PAR of 3 dB, we will ideally need an amplifier capable of 100 W, to linearly amplify peaks of the signal. And if the signal has a PAR of 10 dB, we will require an amplifier capable of 500-W amplifier, to linearly amplify the peaks of the signal. The amount of reduction in the input power to accommodate the peaks in the signal and still have linear performance in the PA is known in the industry as "backoff" and is expressed in dB. If no compensation techniques are used, the required PA backoff is approximated the same as the input signal PAR. This assumes use of a basic class AB power amplifier.

The PAR directly effects PA efficiency. The efficiency is defined as the RF power output divided by the DC power input. High PAR will tend to decrease efficiency, as a PA power consumption is roughly proportional to the peak RF output power capacity. High RF power output requires a high level of bias current at all times, which is reflected in the DC

Table 17.2: Power Amplifier Efficiency Variance by Technology

Wireless Technology	Wireless Standard	PAR (dB)	Typical PA Efficiency (%)
TDMA	GSM	0	60
TDMA	NADC	3	20
Multicarrier	Any	10+	10[a]
CDMA	CDMA2000 or UMTS	10–12	10[a]
OFDM	LTE or WiMax	10–13	10[a]

PA, power amplifiers; *PAR*, peak to average ratios.
[a]In order to achieve required linearity and meet adjacent channel spectral emission requirements, the PAs often included "feed forward compensation," which involved using analog techniques to cancel PA output distortion, improving linearity.

power consumption, even when the PA is not outputing high levels of output power (Table 17.2).

Note that GSM is 0 dB PAR. That is because the modulation technique used only affects the phase; there is no change in amplitude. This is one factor that makes GSM the lowest cost wireless system and why it is still commonly used in many cost-sensitive markets, despite its low spectral efficiency.

17.12 Crest Factor Reduction

New digital technology now allows for improvements in PA efficiency. The two technologies are commonly known as "crest factor reduction" and "digital predistortion" (DPD). A well-designed PA with CFR and DPD can achieve efficiency of about 30% in a typical OFDM application. This is a threefold increase in output power for the same PA circuit and power consumption. It results in major cost reductions and, due to lower power dissipation, higher reliability.

Crest factor reduction is a digital processing function which can reduce the PAR of a signal. A simple way to accomplish this is to simply limit or saturate the peaks in the digital signal amplitude. The problem with this is that it produces high-frequency spectrum components which will cause interference in the adjacent frequency spectrum. More sophisticated techniques are needed. There are usually three different conditions to satisfy, with measurements associated with each.

- Maximum reduction in PAR of signal. PAR is measured using complementary cumulative distribution function (CCDF) function on spectrum analyzer.
- Minimum distortion of transmitted signal. Signal quality measured using error vector measurement (EVM) mode of spectrum analyzer.
- Minimize adjacent channel frequency signals. This is measured using an emission mask function on the spectrum analyzer.

Spectrum analyzers designed for wireless system development come with personality modules, which are firmware-enabled functions to perform many different measurements required by a specific wireless standard, such as LTE for example.

The CCDF is used to measure PAR. A typical plot is shown in Fig. 17.11.

PAR is naturally statistical in nature. This is a plot of the percentage of time the signal power is a given amount of dB above the average power. For example, the plot here shows about 10% (1E-1) of the time, the signal is 4.5 dB higher than average power. The plot also shows that about 1% (1E-2) of the time, the signal is 8 dB higher than average power. The PAR is not just one value but varies according to the duty cycle of the time the signal is a given level above the average power. The maximum PAR depends how long the measurement system can wait, as the longer the measurement interval, the greater likelihood of a large peak in the signal to occur. Typically, for many wireless systems, the PAR is taken at about 0.001% to 0.0001%, or E-5 to E-6. In this case, the signal would have a PAR of about 13 dB. At percentages lower than this, higher peaks occur so infrequently to cause little effect in the other measurements, such as spectral mask or EVM measurements.

EVM stands for "error vector measurement." A standard way to measure distortion in a signal is to measure the actual constellation points compared with the ideal constellation. In a perfect constellation, each point will land exactly on one of the allowable constellation values. CFR techniques will introduce distortion. The EVM is defined as the distance between the actual demodulated point on the constellation and the nearest valid

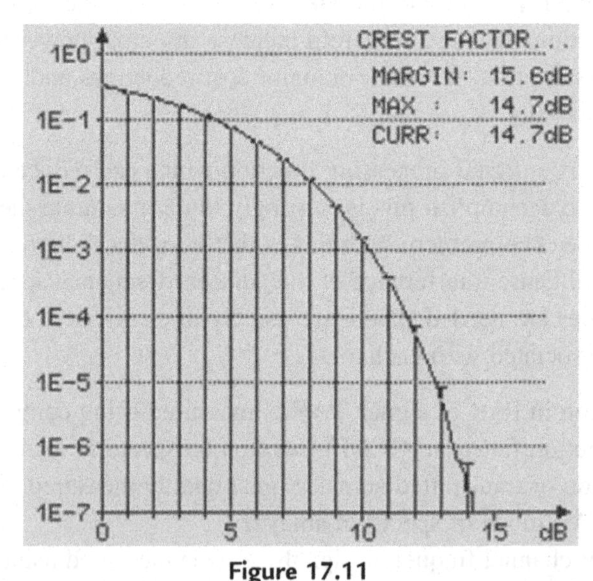

Figure 17.11
Complementary cumulative distribution function plot.

constellation, divided by the distance of the valid constellation point from the origin. Again, this is a statistical measurement, averaged over many points in a signal. It is normally expressed in percent. For example, for 64 QAM, the typical transmit quality must be 3% or less.

To prevent interference in nearby spectrum, all wireless standards have a transmit emission mask (GSM example in Fig. 17.12), which specifies how quickly the transmit power must fall off as a function of the distance from the center of the RF carrier frequency. The measurement is usually specified in dB, relative to the signal power at the carrier frequency. For example, in the GSM transmit mask requirement below, the RF power must drop by at least 30 dB at a 200 kHz offset to the carrier frequency.

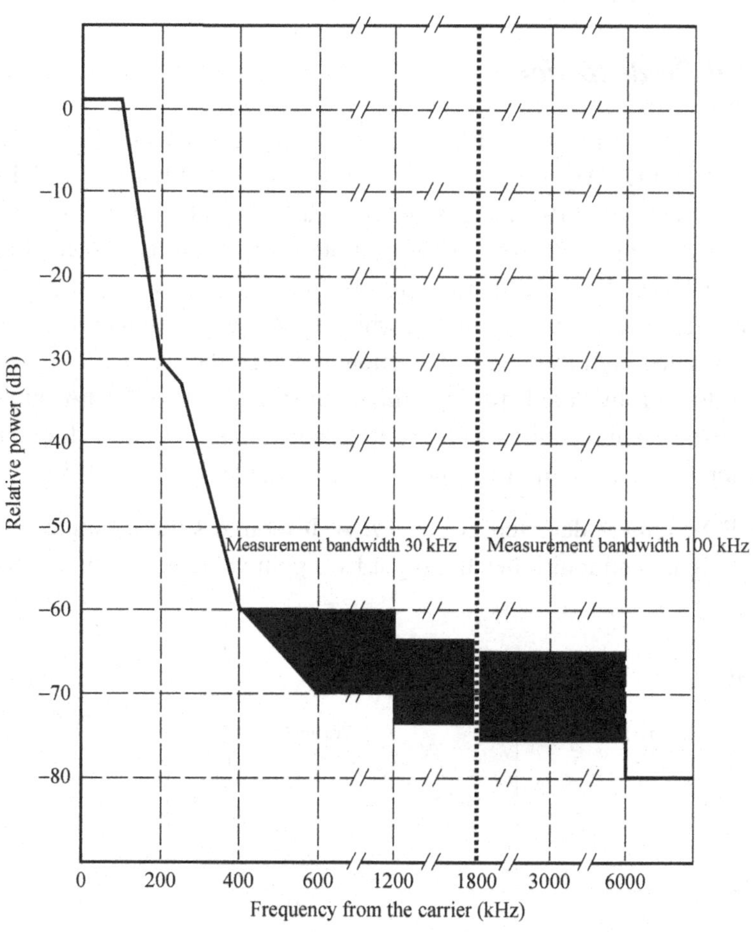

Figure 17.12
Spectral mask regulatory requirement.

Two common approaches are used in CFR, both applied to the digital baseband representation of the signal prior to upconversion. One is a time domain–based technique, where peaks in amplitude are monitored and reduced in amplitude in a smooth manner to present unwanted high frequency content being added to the signal. This approach has the advantage of being able to be applied to any signal.

Another approach is often used in OFDMA systems such as WiMax or LTE. In this approach, the signal is converted to the frequency domain, and the amplitude of various subcarriers are adjusted in such as way as to reduce the CFR of the complete composite signal. After this adjustment, the signal is reconverted back to the time domain. For both of these techniques, the signal must first be interpolated to a sample rate of four or more times faster, which also proportionally enlarges the baseband frequency band. This is required, as all CFR techniques introduce some higher frequency spectral content, which would be aliased into the baseband signal at a lower sample rate.

17.13 Digital Predistortion

A second technique used to increase PA efficiency is digital predistortion or DPD. While CFR works to reduce the PAR of the input signal to the PA, the purpose of DPD is to extend the linear range of the PA, thereby reducing the amount of backoff required. There are many proprietary techniques used to compensate for the nonlinear behavior of a PA at high output power levels. When a PA is overdriven, or operated at output power levels beyond the linear range, it tends to saturate, which causes undesirable effects, such as generation of unwanted high frequency distortion and degradation of EVM. PA saturation can be compensated for, by "predistorting" the input signal so that the PA output has the correct characteristics. This requires a closed loop operation, that is, feeding back an attenuated version of the PA output for monitoring by the DPD circuit (Fig. 17.13).

A very basic DPD circuit is depicted in Fig. 17.14. It monitors the instantaneous power of the digital input signal and multiplies the input by a gain value stored in a table. The table

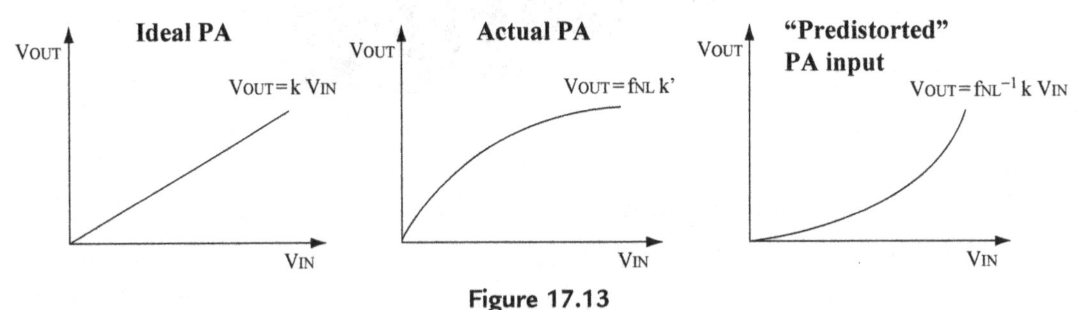

Figure 17.13
Power amplifiers (PA) characteristic responses.

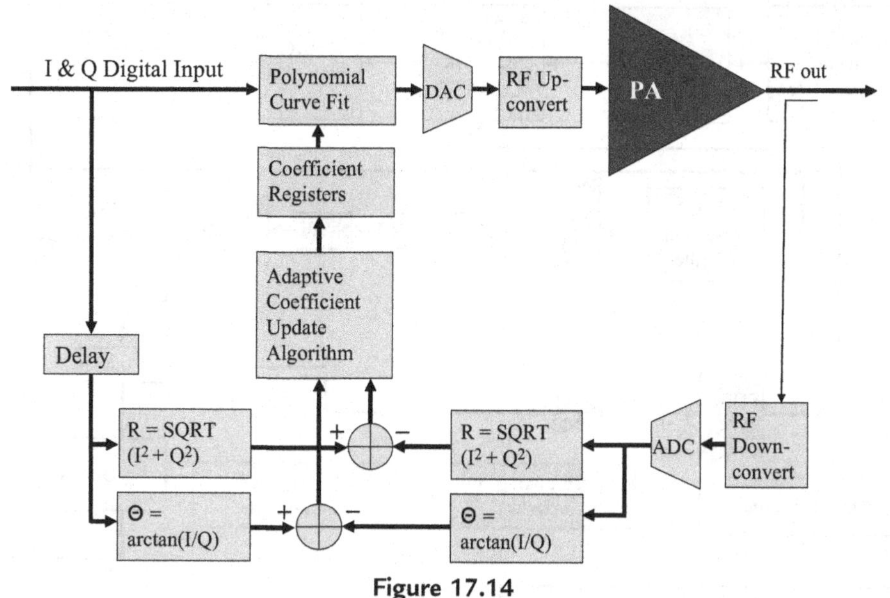

Figure 17.14
Digital predistortion block diagram.

has a value for different input power levels. A closed loop algorithm continually adjusts the table values to optimally compensate for the PA reduction in gain at high power levels, thereby predistorting the signal. Note that it cannot actually increase the RF power capacity of the amplifier but may increase the range over which the amplifier behaves linearly. One of the challenges of DPD is the multiple disciplines required. Digital signal processing, RF amplification, software, and hardware design knowledge are required.

By use of DPD and CFR techniques, PA efficiency can be increased threefold. But there is yet another method to further increase the overall efficiency of wireless base stations, which does not involve digital signal processing at all.

17.14 Remote Radio Head

For reasons for serviceability, cooling and equipment size reasons, a building or shed is used to house the base station at the base of the antenna tower. This typically required RF cables of 50 m or longer between the base station and the antennas mounted on the antenna mast tower. This will often result in a 2 dB for more reduction on transmit signal power, due to cable loss. This loss increases as the RF carrier frequency increases.

A 2 dB loss can mean almost two thirds of the RF amplifier power is dissipated in the cabling prior to reaching the antenna. To eliminate this lost, most modern wireless base stations are being designed with remote radio heads (RRH). The functions associated with

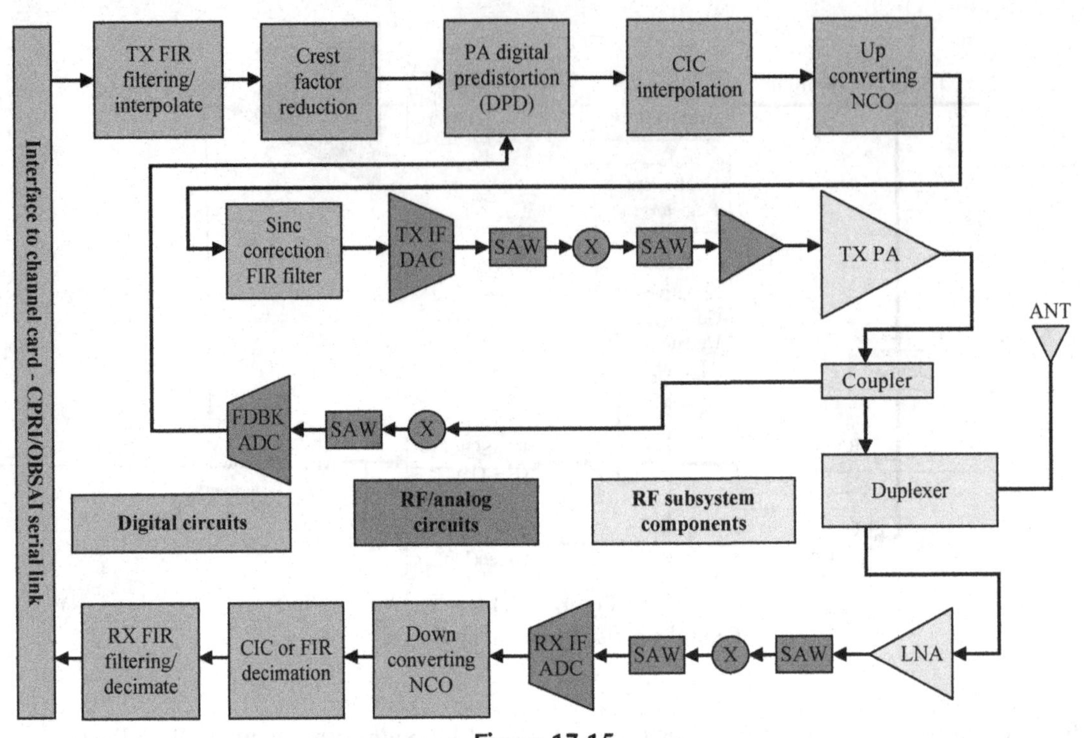

Figure 17.15
Remote radio heads block diagram.

the PA and antenna circuits are packed into an air-cooled module that is small and light enough to be mounted next to the antenna arrays, up to 50-m high in the antenna tower. This can reduce cost, primarily due to use of much smaller PA and elimination of RF cabling. It does, however, require very high reliability, as servicing RRH mounted 50-m high is costly and difficult. The industry has standardized upon two interface standards between the RRH and rest of the base station equipment mounted at the base of the tower or in a nearby builder. These RRH interface standards are known as open base station architecture initiative (OBSAI) or common public radio interface (CPRI). Eventually, CPRI became the dominant RRH interface.

A simplified block diagram of an RRH is shown in Fig. 17.15.

CHAPTER 18

Radar Basics

Radar stands for **R**adio **D**etection **a**nd **R**anging. It can be used to detect target range, direction, and motion. It was developed during World War II and has continually evolved since. Early radars were composed of analog circuits, but since the 1960s, radars have become increasingly digital. Today, radar systems contain some of the most sophisticated and powerful digital signal processing systems anywhere.

Radar has widespread use in both commercial and military applications. Air traffic control, mapping of ground contours, automotive traffic enforcement are just a few civilian applications. Radar is ubiquitous in military applications being used in air defense system, aircraft, missiles, ships, tanks, helicopters, and so forth.

18.1 Radar Frequency Bands

Radar systems transmit electromagnetic or radio waves. Most objects reflect radio waves, which can be detected by the radar system. The frequency of the radio waves used depends on the radar application. Radar systems are often designated by the wavelength or frequency band in which they operate using the band designations (Table 18.1).

The choice of frequency depends on the application requirements. The minimum antenna size is proportional to wavelength and inversely proportional to frequency. Airborne applications often are limited in the size of antenna that can be used, which will dictate a higher frequency and lower wavelength choice. Beamwidth, or the ability of the radar to focus the radiated and received energy in a narrow region, is also dependent on both

Table 18.1: Radar Frequency Bands

Radar Band	Frequency (GHz)	Wavelength (cm)
Millimeter	40–100	0.75–0.30
Ka	26.5–40	1.1–0.75
K	18–26.5	1.7–1.1
Ku	12.5–18	2.4–1.7
X	8–12.5	3.75–2.4
C	4–8	7.5–3.75
S	2–4	15–7.5
L	1–2	30–15
UHF	0.3–1	100–30

Digital Signal Processing 101. http://dx.doi.org/10.1016/B978-0-12-811453-7.00018-4

antenna size and frequency choice. Larger antennas allow the beam to be more tightly focused. Therefore, a higher frequency also allows the beam to be more tightly focused, for a given antenna size. The "focusing" ability of the antenna is often described using an antenna lobe diagram, which plots the directional gain of an antenna over the azimuth (side to side) and elevation (up and down).

The range of the radar system is also influenced by the choice of frequency. Higher frequency systems usually are of lower power due to electronic circuit limitations and experience greater atmospheric attenuation. The ambient electrical noise that can impair operation of analog circuitry also becomes more pronounced at higher frequencies. Most of the radar signal absorption and scattering is due to oxygen and water vapor. Water vapor, in particular, has high absorption in the "K" band. When this was discovered, the band was divided into Ka, for "above" and Ku for "under," the frequencies where radar operation is limited due to water vapor absorption. At higher frequencies in portions of the millimeter band, oxygen causes similar attenuation through absorption and scattering.

Another consideration, discussed more fully in the next chapter, is the effect of the radar operating frequency on Doppler frequency measurements. Doppler frequency shifts are proportional to both the relative velocity and the radar frequency. Doppler frequency shifts can provide important information to the radar system.

Most airborne radars operate between the L and Ka bands, also known as the microwave region. Many short range targeting radars, such as on a tank or helicopter, operate in the millimeter band. Many long range ground-based operations utilize UHF or lower frequencies, due to the ability to use large antennas and minimal atmospheric attenuation and ambient noise. At even lower frequencies, the ionosphere can become reflective, allowing very long range over-the-horizon operation.

18.2 Radar Antennas

A critical function in any radar system is the antenna. Early radars often used mechanical parabolic antennas. The antenna is capable of focusing both the receive and transmit energy in a given direction. The antenna could be moved mechanically using motors and aimed to search over different parts of the sky (Fig. 18.1).

The degree of directionality is often shown in azimuth and elevation gain diagrams. The diagram in Fig. 18.2 below shows an antenna that has a fairly wide or broad main lobe. Most radar antennas may have a much narrower main lobe, on the order of a few degrees. Frequently, the width of the main lobe is specified by the point at which the receive or transmit signals are attenuated by 3 dB or about one half. The antenna shown below has a lobe width of about 20 degrees. However, all antennas will receive some level of signal from undesired directions, even from behind. The antenna gain plots visually quantify the

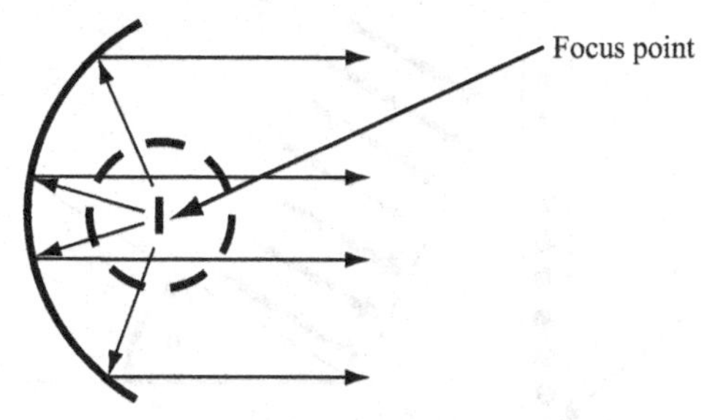

Figure 18.1
Parabolic antenna operation.

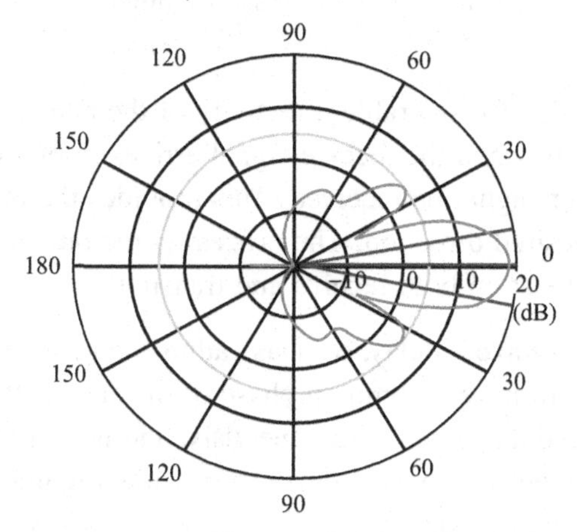

Figure 18.2
Antenna lobe diagram.

relative gain across both azimuths and elevations (usually a separate plot for each). In general, the narrower the main lobe, the higher the antenna gain will be.

The antenna design influences both the amount of energy the radar can transmit at the desired target space, as well as how much energy it can receive from the same direction. It also determines how much unwanted energy from other directions is attenuated (for example, reflections from the ground in an airborne search radar). Having a narrow or focused beam allows the energy to be more focused. To search across a wide area, the antenna must steer its beam across the entire search space. As just mentioned, this was done mechanically in early radars. However, more advanced radars, especially airborne, use electronically steerable antennas.

An electronically steerable antenna is built from many small antennas or individual elements. Each element can individually vary the phase of both receive and transmitted signals, as well as the signal strength using analog or digital electronic circuits. It is the

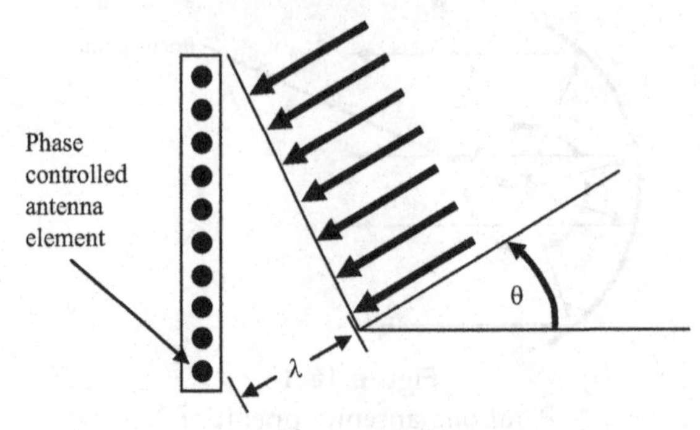

Figure 18.3
Planar antenna phase alignment.

changes in phase that provide for steerable directivity of the antenna beam over both azimuth and elevation. Only when the receive signal arrives in-phase across all the antenna elements will the maximum signal be received. This provides the ability to "aim" the main lobe of the antenna in a desired direction. The process is reciprocal, meaning that the same antenna lobe pattern will exist on both receive and transmit.

Each antenna element must have a delay, or phase adjustment, such that after this adjustment, all elements will have a common phase of the signal. If the angle $\theta = 0$, then all the elements will receive the signal simultaneously, and no adjustment is necessary. At a nonzero angle, each element will have a delay to provide alignment of the wavefront across the antenna array (Fig. 18.3).

This has several advantages. It can be steered very rapidly, which can allow fast searching as well as tracking of objects. Using a technique called "lobing," the radar beam can be rapidly steered on either side of a target. By noting where the stronger return is, the target location can be tracked. Further, different regions of the antenna can be aimed in different directions to scan or track multiple regions or targets, albeit with a reduced transmit power and receive gain. A disadvantage of an electrically steered antenna is the reduced aperture at larger incident angles. The aperture is one factor in the antenna gain and will decrease by the cosine θ, where θ is the angle of the steering direction, relative to the perpendicular vector from the antenna (Fig. 18.4).

18.3 Radar Range Equation

Detection of objects using radar involved sophisticated signal processing. However, all of this is first dependent on the amount of energy received from the target echo.

$$\text{Receiver Power } P_{receive} = P_t G_t A_r \sigma F^4 (t_{pulse}/T) \Big/ \left((4\pi)^2 R^4\right)$$

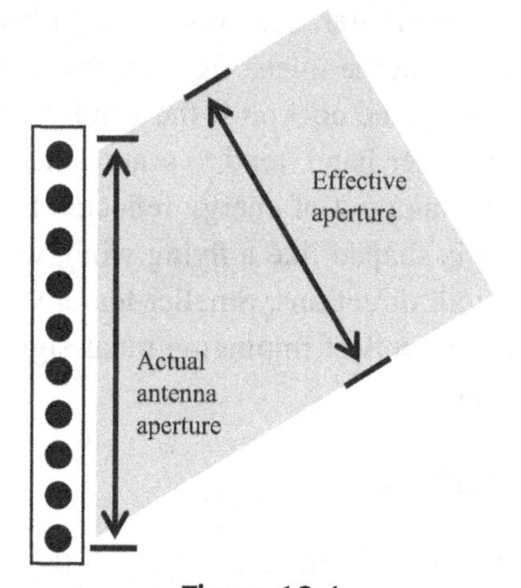

Figure 18.4
Antenna aperture.

where P_t = transmitted power, G_t = antenna transmit gain, A_r = Receive antenna aperture area, σ = radar cross section (function of target geometric cross section, reflectivity of surface, and directivity of reflections), F = pattern propagation factor (unity in vacuum, accounts for multipath, shadowing, and other factors), t_{pulse} = duration of receive pulse, T = duration of transmit interval (the inverse of the PRI), and R = range between radar and target.

Notice that the received power drops with the fourth power of the range, so radar systems must cope with very large dynamic ranges in the receive signal processing. The radar energy seen by the target drops proportional to the range squared. The reflected energy seen by the radar receive antenna further drops by a factor of the range squared. The ability to detect very small signals is crucial to operate at longer ranges.

18.4 Stealth Aircraft

Military planes have been developed with "stealth" characteristics. This means that such a plane has a very small σ, or radar cross section, relative to other aircraft of similar size. It can still be detected by a sufficiently powerful radar or at sufficiently close ranges. Because the size of stealth aircraft is similar to other military aircraft, the stealth characteristic is achieved by reducing the amount of radar signal power that is reflected back from the stealth aircraft to the transmitting radar. There are two fundamental methods to reduce the reflected energy: either absorb the radar signal or deflect it in a different direction that the radar transmitter. Special radar absorbant materials are used in stealth aircraft. The shape and contours of the aircraft greatly influence effective radar cross section. A concave surface tends to reflect radar waves in the general direction of the direction of arrival, back to the transmitter. This is to be avoided in stealth aircraft. Examples of concave surfaces are engine inlets, right angles where wings join the fuselage, open bomb bays, and even the cockpit if the windscreens are transparent to radar signals. Convex surfaces, on the other hand, tend to scatter the radar waves in widely separated directions, reducing the amount of energy reflected back to the source. For example, the B2 stealth bomber is shaped like a flying wing, which is basically a convex shape when viewed from nearly all directions. Smaller features, such as the engine air inlets, have a geometry designed to reflect impinging radar signals in a direction other than that of the illuminating radar.

18.5 Pulsed Radar Operation

Most radar systems are pulsed, which means the radar will transmit a pulse and then listen for receive signals or echoes. This avoids the problem of a sensitive receiver trying to operate simultaneously with a high power transmitter. The pulse width or duration is an important factor. The shorter the pulse width, the easier it is to determine range, as the receive signal is of short duration also. Radars operate by "binning" the receive signals. The receive signal returns are sorted into a set of bins by time of arrival relative to the transmit pulse. This is proportional to the round-trip distance to the object(s) reflecting the radar waves. By analyzing the receive signal strength in the bins, the radar can sort the objects by radar cross section size and across different ranges. This is performed over for all desired azimuths and elevations.

Having many range bins allows more precise range determinations. A short duration pulse is likely to be detected and mapped into only one or two range bins, rather than being spread over many bins. However, a longer pulse duration or width allows for greater amount of signal energy to be transmitted, and a longer time for the receiver to integrate the energy. This means longer detection range. To optimize for both fine range resolution and long range detection, radars use a technique called pulse compression.

18.6 Pulse Compression

The goal of pulse compression is to transmit a long duration pulse of high energy, but to detect a short duration pulse to localize the receive filter output response to one or at most two radar range bins. Early radars accomplished this by transmitting a signal with linear frequency modulation. The pulse would start at a low frequency sinusoid, and increase the frequency over the duration of the radar pulse. This is referred to a "chirp." A special analog filter is used at the receive end, with nonlinear phase response. This filter has a time lag that decreases with frequency. When this rate of time lag decrease is matched to the rate of increase in the chirp, the result is a very short, high amplitude output from the filter. The response of the pulse detection has been "compressed."

All digital radars can also perform pulse compression, but using a different method. Recall the matched filter in the chapter on Complex Modulation and Demodulation. The matched filter will perform the same effect as the analog pulse compression technique just described. If the transmitted radar pulse uses a pseudo random sequence of phase modulations and is detected using a matched filter, then the resulting output will be of high amplitude for only when the receive signal sequence matches up in phase (or delay) to the transmitted pulse sequence. This can be used to precisely identify delay or time of arrival of the receive pulse. The sequence used for radar transmit pulses must have strong autocorrelation properties (sequence of length N correlates to value N with zero offset, and to 0 for any nonzero sequence offset). In radar systems, sequences known as Barker sequences are sometimes used.

18.7 Pulse Repetition Frequency

A high pulse repetition frequency (PRF) has several advantages. First, the higher the PRF, the greater the average power the radar is transmitting (assuming the peak power of each pulse is limited by the transmit circuitry), and the better the chance of detection of targets. A high or fast PRF also allows for more rapid detection and tracking of objects, as range measurements at a given azimuth and elevation can be performed during each PRF interval. A high PRF also allows easier discrimination of the Doppler frequency, a topic discussed in the next chapter. But a low PRF also has an important advantage, which is to allow unambiguous determination of range over longer distances. This is our next topic.

Range to target is measured by round-trip delay in the received echo. It is the speed of light multiplied by the time delay and divided by two to account for the roundtrip.

$$R_{measured} = v_{light} t_{delay}/2$$

The maximum range that can be unambiguously detected is limited by the PRF. This is more easily seen by an example. If the PRF is 10 kHz, then we have 100 μs between pulses. Therefore, all return echoes should ideally be received before the next transmit pulse. This range is simply found by multiplying the echo delay time by the speed of light and dividing by two to account for the roundtrip.

$$R_{maximum} = (3 \times 10^8 \text{ m/s})(100 \times 10^{-6} \ s)/2 = 15 \text{ km}$$

Let us suppose the radar system sorts the returns into 100 range bins, based on the time delay of reception. The range resolution of this radar system is then 0.15 km or 150 m. However, there may be returns from distances beyond 15 km. Suppose that a target aircraft 1 is 5 km distant, and a target aircraft 2 is 21 km distant. Target aircraft 1 will have a delay of

$$t_{delay} = 2R_{measured}/v_{light} = 2(5 \times 10^3)/3 \times 10^8 = 33 \ \mu s$$

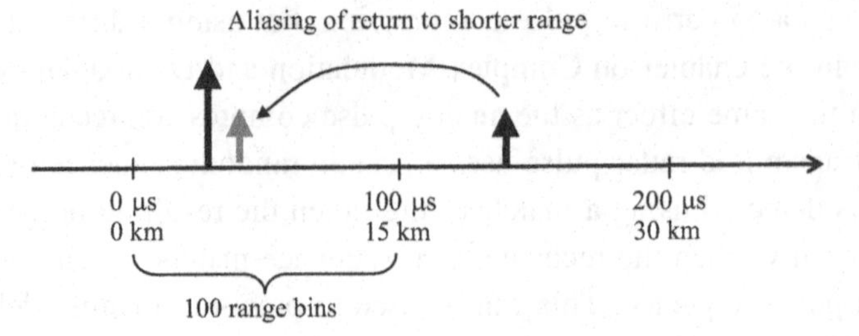

Figure 18.5
Range aliasing.

Target aircraft 2 will have a delay of

$$t_{delay} = 2R_{measured}/v_{light} = 2(21 \times 10^3)/3 \times 10^8 = 140 \ \mu s$$

The first target return will be mapped into the 33rd out of 100-range bin, and the second target to 40th range bin. This is called a range ambiguity. The target(s) that are within the 15 km are said to be in the unambiguous range. This is analogous to the sampling rate. The range ambiguity is analogous to aliasing during the sampling process (Fig. 18.5).

One solution to this problem is to transmit different pulses at each PRF interval. However, this has the downside of complicating the receiver, as it must now use multiple matched filters at each range bin, and at each azimuth and elevation. This will effectively double the rate of digital signal processing required for each separate transmit pulse and matched filter pair used.

Another approach can be used instead. If we periodically change the PRF slightly, we will find that the returns in the unambiguous range do not move. However, those beyond that range will have shifted in their apparent ranges. This can be illustrated using an example.

Suppose we switch the PRF to 11 kHz, from 10 kHz, or 90.9 μs. The maximum unambiguous range is

$$R_{maximum} = (3 \times 10^8 \ m/s)(90.9 \times 10^{-6} s)/2 = 13.6 \ km$$

The target aircraft at 5 km distance will still have a 33-μs delay. Using 100 bins like before, the target return will appear in bin number 100 × 33/90.9 = 36th bin. The target aircraft 2 at 21 km will have a target return delay of 140 μs. This will appear as a return at 140 − 90.9 = 49.1 us, and in 100 × 49.1/90.9 = 54th bin.

So by switching PRFs, we are able to determine that at least one of our targets is beyond the ambiguous range.

PRF = 10 kHz: Target aircraft returns in bin 33, and in bin 40
PRF = 11 kHz: Target aircraft returns in bin 36, and in bin 54

We assume Scenario A:

Target1 moved from bin 33 → 36 when we changed PRF
Target2 moved from bin 40 → 54 when we changed PRF

Instead, what if Scenario B:

Target1 moved from bin 33 → 54
Target2 moved from bin 40 → 36

From this information, we cannot be sure that the first target was at 5 km, and the second at 21 km. We will not work this out here, but if Target1 is at 34.8 km (or 232 μs) and Target2 is at 141 km (or 940 μs), the result would be scenario B.

The way to tell which scenario is in fact occurring is to use a third PRF, perhaps at 9 kHz, or 111 μs.

$$R_{\text{maximum}} = \left(3 \times 10^8 \text{m/s}\right)\left(111 \times 10^{-6} s\right)/2 = 16.7 \text{ km}$$

The target aircraft at 5 km will appear in bin number 100 × 33/111 = 30th bin. The target aircraft 2 at 21 km will appear in bin number 100 × (140 − 111)/111 = 26th bin.

This additional information allows us to know that scenario A is the true one.

In reality, there may be many target returns, and they may also be obscured by noise or clutter in the return. The higher the PRF, the more ambiguity will be present in the range returns. For these reasons, radar detection is at best a statistical process, with calculated probability of detections (at a given range for a given radar cross section). There is also a probability of false detection, which must also be considered when setting detection thresholds.

18.8 Detection Processing

Most radars have thousands of range bins. They may scan wide sweeps of azimuth and elevation. Or in tracking mode, may be focused in narrow regions containing targets, which have been detected. In either case, the rate of digital signal processing can be very high.

A matched filter can be used to detect incoming radar pulses. The radar will focus at a particular azimuth and elevation for one or many PRFs. For each PRF, the incoming data is filtered using an FIR filter with an impulse response that is the complex conjugate of radar transmit pulse. This will produce a large peak in the filter output at the point where the incoming data stream contains radar pulse, which will correspond to a particular range bin.

The computation load in modern radars can be very high. To reduce the amount of computations, filtering, or convolving the receive signals by using matched filters usually not performed using an FIR filter.

An alternative is to perform a fast Fourier transform (FFT) transform of the receive signal sequence from each PRF. The spectral representation of the receive signal can then be multiplied by the frequency response of the radar pulse. After this multiplication, the result can then be transformed back into the time domain using an inverse fast Fourier transform. This will perform the equivalent function as FIR filtering. This sounds counterintuitive, but due to the efficiency of the FFT algorithm, this is often a less computationally intensive process than a large FIR filter. In any case, once the receive pulse has been processed, the amplitude of the matched filter operation is compared to a threshold to determine if this is a valid radar pulse return. Use of results over multiple PRFs can be used to discard or confirm valid target radar returns, maximize the probability of detection and minimize the probability of false detections.

In the next chapter, Doppler processing will be discussed. This also requires use of the FFT algorithm. In radar systems, the FFT is the most common digital signal processing algorithm, and efficient implementation of FFTs is critical for any digital radar system.

CHAPTER 19

Pulse Doppler Radar

We have mentioned Doppler frequency shift in our discussions of wireless systems. In general, these frequency shifts degrade wireless receiver performance and must be compensated for. In radar, however, Doppler shifts are a key part of the detection and tracking of objects. For this reason, nearly all radar systems incorporate Doppler processing.

By measuring the Doppler rate, the radar is able to measure the relative velocity of all objects returning echoes to the radar system—whether planes, vehicles, or ground features. For targeting radars, estimating the targets' velocity is equally important as determining its location. And for all radars, Doppler filtering can be used to discriminate between objects moving at different relative velocities. This can be especially important when there is a high level of clutter obscuring the target return. An example of this might be an airborne radar trying to track a moving vehicle on the ground. Since the ground returns will be at the same range as the vehicle, the difference in velocity will be the means of discrimination using Doppler measurements.

19.1 Doppler Effect

Because sensing Doppler frequency shifts is so important, it is worth reviewing the cause of Doppler frequency shifts. A common example we have all experienced is standing beside a train track or highway. As a train or truck approaches, we hear a certain frequency sound. As a high speed train or truck passes, the sound immediately drops several octaves. This is caused by a frequency shift caused by the Doppler effect. Although we cannot sense this, the light waves are affected in the same way as sound waves. In fact, the realization that our universe is expanding was determined by making very fine Doppler measurements of the light from stars in the night sky.

The relationship between wavelength and frequency is as follows:

$$\lambda = v/f$$

where f = wave frequency (Hz or cycles per second), λ = wavelength (meters), v = speed of light (approximately 3×10^8 m/s).

The speed of light is constant—Einstein proved this. Technically this is true only in a vacuum, but the effect of the medium such as our atmosphere can be ignored in radar

Digital Signal Processing 101. http://dx.doi.org/10.1016/B978-0-12-811453-7.00019-6

discussions. What is happening in a radar system is that the frequency is modified by the process of being reflected by a moving object. Consider the transmission of a sinusoidal wave. The distance from the crest of each wave to the next is the wavelength, which is inversely proportional to the frequency. Each successive wave is reflected from the target object of interest. When this object is moving toward the radar system, the next wave crest reflected has a shorter round-trip distance to travel, from the radar to the target and back to the radar. This is because the target has moved closer in the interval of time between the previous and current wave crest. As long as this motion continues, the distance between the arriving wave crests is shorter than the distance between the transmitted wave crests. Since frequency is inversely proportional to wavelength, the frequency of the sinusoidal wave appears to have increased. If the target object is moving away from the radar system, then the opposite happens. Each successive wave crest has a longer round-trip distance to travel, so the time between arrival of receive wave crests is lengthened, resulting in a longer (larger) wavelength and a lower frequency. This effect becomes more pronounced when the frequency of the transmitted sinusoid is high (short wavelength). Then the effect of the receive wavelength being shorted or lengthened due to the Doppler effect is more noticeable. Therefore, Doppler frequency shifts are more easily detected when using higher frequency waves, as the percentage change in the frequency will be larger.

This effect only applies to the motion relative to the radar and the target object. If the object is moving at right angles to the radar, there will be no Doppler frequency shift. An example of this would be an airborne radar directed at the ground immediately below the aircraft. Assuming level terrain and the aircraft is at a constant altitude, the Doppler shift will be zero, even though the plane is moving relative to the ground. There is no change in the distance between the plane and ground.

If the radar is ground based, then all Doppler frequency shifts will be due to the target object motion. If the radar is vehicle or airborne based, then the Doppler frequency shifts will be due to the relative motion between the radar and target object. For example, if you are driving on the highway at 70 mph and an approaching police car is traveling at 50 mph, the radar will show a Doppler shift corresponding to 120 mph. The police radar will need to subtract the speed of the police car to display your speed.

This can be of great advantage in a radar system. By binning the receive echoes both over range and Doppler frequency offset, target speed as well as range can be determined. Also, this allows easy discrimination between moving objects, such as an aircraft, and the background clutter, which is generally stationary.

For example, imagine we have a radar operating in the X band at 10 GHz ($\lambda = 0.03$ m or 3 cm). The radar that is airborne, traveling at 500 mph, is tracking a target ahead moving at 800 mph in the same direction. In this case, the speed differential is -300 mph, or -134 m/s.

Another target is traveling head on toward the airborne radar at 400 mph. This gives a speed differential of 900 mph, or 402 m/s The Doppler frequency shift can be calculated as follows:

$$f_{Doppler} = 2v_{relative}/\lambda$$
$$\text{First target Doppler shift} = 2(-134\ \text{m/s})/(0.03\ \text{m}) = -8.93\ \text{kHz}$$
$$\text{Second target Doppler shift} = 2(402\ \text{m/s})/(0.03\ \text{m}) = 26.8\ \text{kHz}$$

The receive signal will be offset from 10 GHz by the Doppler frequency. Notice that the Doppler shift is negative when the object is moving away (opening range) from the radar and is positive when the object is moving toward the radar (closing range).

19.2 Pulsed Frequency Spectrum

Measuring Doppler shift is complicated by the fact that the radar is transmitting pulses. This has an effect on the spectrum of the radar transmit signal. To understand this, we need to start with the frequency or spectral representation of a pulse. Previously, in the chapter on digital upconversion, the frequency response of DACs was discussed. The time response of a DAC is also pulse, and we saw that the frequency response in the sin(x)/x or sinc function (Fig. 19.1). If the pulse has the sharp edges removed, we can reduce the sidelobes, although this will broaden the mainlobe.

The frequency response of an infinite train of pulses is similar, except that it is composed of discrete spectral lines in the envelope of the sinc function (Fig. 19.2). Also, the spectrum repeats at intervals of the pulse repetition frequency (PRF). We will forgo the mathematical derivation of this, but this is available in any engineering text on radar. This is not unlike a sampled signal, in which the frequency representation repeats at the sampling frequency interval (Fig. 19.3).

The important point is that this will impose restrictions on Doppler frequency shifts. To unambiguously identify the Doppler frequency shift, it must be less than the PRF

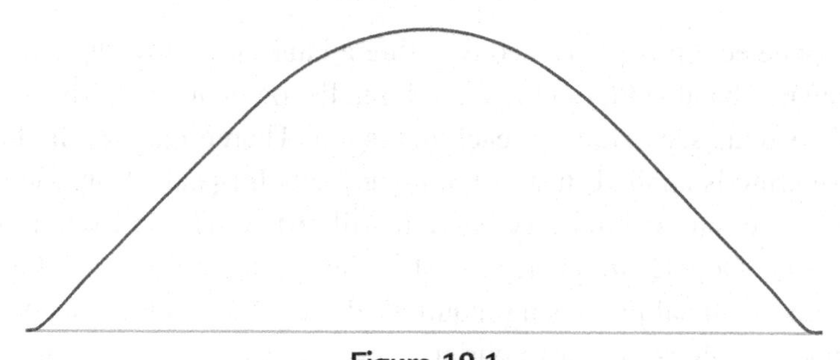

Figure 19.1
Spectrum of single pulse.

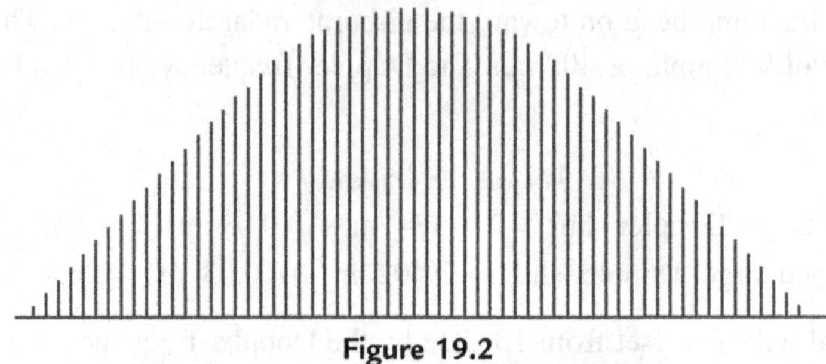

Figure 19.2
Spectrum of pulse train repeating slowly.

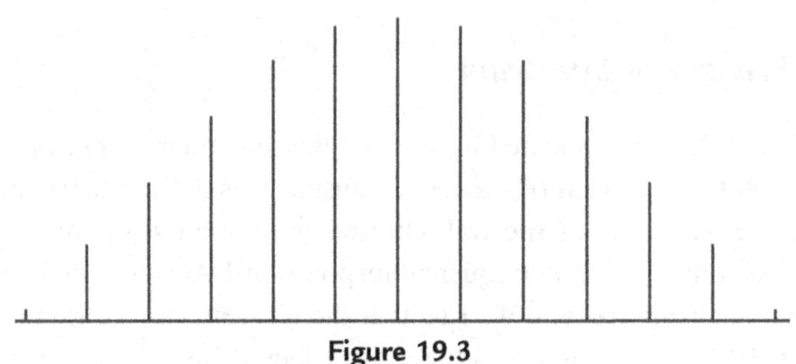

Figure 19.3
Spectrum of pulse train repeating rapidly.

frequency. Doppler frequency shifts greater than this will alias to a lower Doppler frequency and be ambiguous just as radar returns beyond the range of the PRF interval time are ambiguous, as they alias into lower range bins.

The Doppler frequency range placement will be some what arbitrarily determined by the digital downconversion of the received radar high-frequency carrier to baseband. Assuming downconversion of the carrier to 0 Hz, then the Doppler frequency effect will cause the target return signal to have a positive or negative offset, as computed below.

$$f_{Doppler} = 2v_{relative}/\lambda$$

Doppler frequency detection is performed by using a bank of narrow digital filters, with overlapping frequency bandwidth (so there are no nulls, or frequencies that could go undetected). This is done separately for each range bin. Therefore, at each allowable range, Doppler filtering is applied. Just as the radar looks for peaks from the matched filter detector at every range bin, within every range it will test across the Doppler frequency band to determine the Doppler frequency offset in the receive pulse. This dramatically expands the amount of signal processing required. Rather than building many individual narrow-band frequency filters, the fast Fourier transform is used to perform spectral filtering across the spectral bandwidth of the PRF signal.

19.3 Doppler Ambiguities

Doppler ambiguities can occur if the Doppler range is larger than the PRF. The maximum Doppler requirement of a given radar can be estimated. Using a military airborne radar example, the fastest closing rates will be with targets approaching, as both speeds of the radar-bearing aircraft and the target aircraft are summed. This should assume the maximum speed of both aircraft. The highest opening rates might be when a target is flying away from the radar-bearing aircraft. Here, we should assume the radar-bearing aircraft is traveling at minimum speed, and the target aircraft is flying at maximum speed. We should also assume the target aircraft is flying at a large angle θ from the radar-bearing aircraft flight path, which further reduces the radar-bearing aircraft speed in the direction of the target.

$$\begin{aligned}
&\text{Maximum positive Doppler frequency(fastest closing rate)at 10 GHz/3 cm} \\
&\text{Radar} - \text{bearing aircraft maximum speed: } 1200 \text{ mph} = 536 \text{ m/s} \\
&\text{Target aircraft maximum speed: } 1200 \text{ mph} = 536 \text{ m/s} \\
&\text{Maximum positive Doppler} = 2(1072 \text{ m/s})/(0.03 \text{ m}) = 71.5 \text{ kHz}
\end{aligned}$$

$$\begin{aligned}
&\text{Maximum negative Doppler frequency(fastest opening rate)at 10 GHz/3 cm} \\
&\text{Radar} - \text{bearing aircraft maximum speed: } 300 \text{ mph} = 134 \text{ m/s} \\
&\text{Effective radar} - \text{bearing aircraft minimum speed with } \theta \\
&\quad = 60 \text{ degree angle from target track is } \sin(60) = 0.5\text{: } 150 \text{ mph} = 67 \text{ m/s} \\
&\text{Target aircraft maximum speed: } 1200 \text{ mph} = 536 \text{ m/s} \\
&\text{Maximum negative Doppler} = 2(67 - 536 \text{ m/s})/(0.03\text{m}) = -31.3 \text{ kHz}
\end{aligned}$$

This gives a total Doppler range of $71.5 + 31.3 = 102.8$ kHz. Unless the PRF exceeds 102.8 kHz, there will be aliasing of the detected Doppler rates and the associated ambiguities.

If we assume a PRF of 10 kHz from the previous chapter's example, we will clearly have Doppler ambiguities. Doppler ambiguities can be resolved using a number of methods.

In Fig. 19.4, the aliasing resulting in Doppler ambiguity is shown for a higher PRF of 80 kHz. If the PRF was 10 kHz, there would be many more Doppler ambiguities in the spectrum.

We have already discussed range and Doppler ambiguities. The PRF directly affects the size of the unambiguous zone. But ambiguities are not the only issue. Just as a range or Doppler measurement return can be outside the unambiguous zone and is aliased into the primary zone, so is all other returns and radar clutter. This can raise the noise floor of the radar to a degree that lower amplitude returns become obscured.

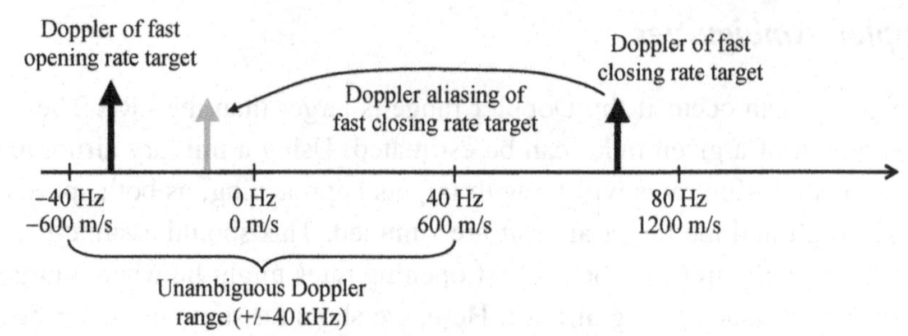

Figure 19.4
Doppler aliasing. *PRF*, pulse repetition frequency.

- *Range differentiation*: Using range measurements over a period of time, the difference in range can be measured over the time interval. Using this, the radar can estimate the change in range, which is the relative velocity between the radar and the target. This method is less precise than Doppler-based measurements but can provide an estimate to use in resolving the Doppler ambiguity.
- *Multiple or offset PRFs*: This is very similar to resolving range ambiguities. Multiple PRFs with slightly different values can be used, and the ambiguities resolved by analysis of how the aliased Doppler frequency measurements move within the unambiguous range.
- *Variable PRF*: The PRF need not be constant, particularly in a digitally programmable system. The PRF can be varied. The PRFs are generally grouped into low, medium, or high ranges. A low PRF is generally from 2 to 8 kHz. A medium PRF is generally from 8 to 30 kHz. And a high PRF is generally from about 30 to 250 kHz. Each PRF zone has its advantages and disadvantages.

19.4 Radar Clutter

There are two categories of radar clutter. There are mainlobe clutter and sidelobe clutter. Mainlobe clutter occurs when there are undesirable returns in the mainlobe or within the radar beamwidth. This usually occurs when the mainlobe intersects the ground. This can occur because the radar is aimed downward (negative elevation), there is higher ground such as mountains in the radar path, or even if the radar beam is aimed level and as the beam spreads it hits the ground. Because the area of ground in the radar beam is often large, the return can be much larger than targets.

Sidelobe clutter is unwanted returns that are coming from a direction outside the mainlobe. Since the radar is not pointed in this direction, it is never a desired radar return signal. Sidelobe clutter is usually attenuated by 50 dB or more, due to the antenna directional selectivity or directional radiation pattern. A very common source of sidelobe clutter is ground return. When a radar is pointed toward the horizon, there is a very large area of ground area covered by the sidelobes in the negative elevation region. The large reflective area covered by the sidelobe can cause significant sidelobe returns despite the antenna attenuation.

Different types of terrain will have a different "reflectivity" that is a measure of how much radar energy is reflected back. It also depends on the angle of the radar energy relative to the ground surface. A related parameter, known the backscattering coefficient, has the angle incorporated into the coefficient and is therefore normalized over all angles. Some surfaces, such as smooth water, reflect most of the radar energy away from the radar transmitter, particularly at shallow angles. A desert would reflect more of the energy back to the radar, and wooded terrain would reflect even more. Man-made surfaces, such as in urban areas, would reflect the most of the energy back to the radar system.

This is one reason why Doppler processing is so important. Most targets are moving, and this is an effective method to distinguish them from the background clutter of the ground. Remember, the Doppler frequency of the ground will usually be nonzero if the radar is in motion. In fact, sidelobe Doppler clutter will vary by the elevation and azimuth angle, as the relative velocity will be equal to cosine θ, where θ is equal to the angle between direction on aircraft flight line and the ground. Different points on the ground will give different Doppler values, depending on how far ahead or off to the side of the radar-bearing aircraft that particular patch of ground is located. So Doppler sidelobe clutter will be present over a wide range of Doppler frequencies.

Mainlobe clutter is more likely to be concentrated at a specific frequency, since the mainlobe is far more concentrated (typically 3–6 degrees of beamwidth), so the patch of ground illuminated is likely to be far smaller and all the returns are at or near the same relative velocity.

A simple example can help illustrate how the radar can combine range and Doppler returns to obtain a more complete picture of the target environment (Fig. 19.5).

The diagram above illustrates unambiguous range and Doppler returns. This assumes the PRF is low enough to receive all the returns in a single PRF interval, and the PRF is high enough to include all Doppler return frequencies.

The ground return comes though the antenna sidelobe, known as sidelobe clutter. The reason ground return is often high is due to the amount of reflective area at close range, which results in a strong return despite the sidelobe attenuation of the antenna. The ground return will be at short range, essentially the altitude of the aircraft. In the mainlobe, the range return of the mountains and closing target are close together, due to similar ranges. It is easy to see how if just using the range return, it is easy for a target return to be lost in high-terrain returns, known as mainlobe clutter.

The Doppler return gives a different picture. First of all, the ground return is more spread out, around 0 Hz. The ground slightly ahead of the radar-bearing plane is at slightly

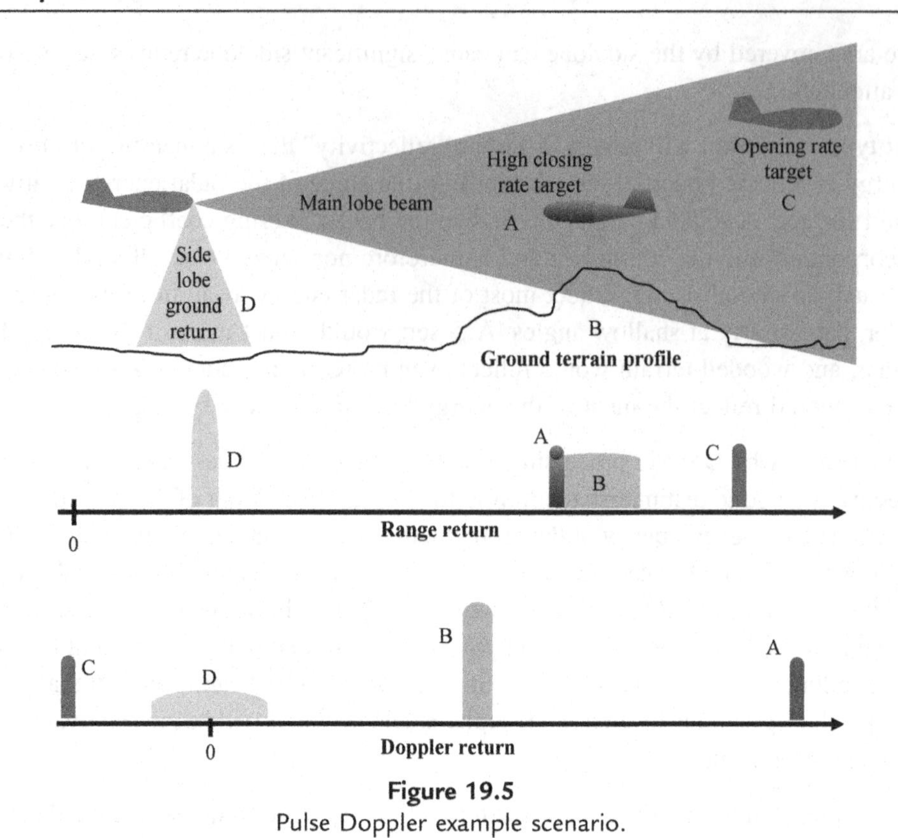

Figure 19.5
Pulse Doppler example scenario.

positive relative velocity, and the ground behind the plane is at slightly negative relative velocity. As the horizontal distance from the radar-bearing plane increases, the ground return weakens due to increased range.

Notice that the Doppler return from the mountain terrain is now very distinct from the nearby closing aircraft target. The mountain terrain is moving at a relative velocity equal to the radar-bearing plane's velocity. The closing aircraft relative velocity is the sum of both aircraft velocity, which is much higher, producing a Doppler return at high velocity. The other target aircraft, which is slowly opening the range with radar-bearing aircraft, is represented as a negative Doppler frequency return.

19.5 Pulse Repetition Frequency Trade-Offs

Different PRF frequencies have different advantages and disadvantages. The following discussion summarizes the trade-offs.

Low PRF operation is generally used for maximum range detection. It usually requires a high-power transmit power, to receive returns of sufficient power for detection at a long range (remember, receive echo power levels are proportional to the range to the fourth power). To get the highest power, long transmit pulses are sent, and correspondingly long-matched filter processing (or pulse compression) is used. This mode is useful for precise range determination. Strong sidelobe returns can often be determined by their relatively close ranges (ground area near radar system) and filtered out. Disadvantages are that Doppler processing is relatively ineffective due to so many overlapping Doppler frequency ranges. This limits the ability to detect moving objects in the presence of heavy background clutter, such as moving objects on the ground. This can also be a problem for detecting low-flying aircraft because of ground terrain clutter at similar ranges in the mainlobe of the radar.

High PRF operation spreads out the frequency spectrum of the receive pulse, allowing a full Doppler spectrum without aliasing or ambiguous Doppler measurements. The clutter that is present in the spectrum is not folding or aliased from higher frequencies, which lowers the noise floor of the receive spectrum. A high PRF can be used to determine Doppler frequency and therefore relative velocity for all targets. It can also be used when a moving object of interest is obscured by a stationary mass, such as the ground or a mountain, in the radar return. The unambiguous Doppler measurements will make a moving target stand out from a stationary background. This is called mainlobe clutter rejection, or filtering. Another benefit is that since more pulses are transmitted in a given interval of time, higher average transmit power levels can be achieved. This can help improve the detection range of a radar system in high PRF mode.

Pulse delay–based ranging performance becomes very compromised in high-PRF operation. One solution is to use the high-PRF mode to identify moving targets, especially fast moving targets, and then switch to a low-PRF operation to determine range. Another alternative is to use a technique called FM ranging (Fig 19.6). In this mode, the transmit duty cycle becomes 100%—the radar transmits continuously. But it transmits a continuously increasing frequency signal and then, at the maximum frequency, abruptly begins to transmit at a continuously decreasing frequency until it reaches the minimum frequency resets to begin another cycle of increasing frequency. The frequency over time looks like a "sawtooth wave." The receiver can operate during transmit operation, as the receiver is detecting time-delayed versions of the transmit signal, which are at a different frequency than current transmit operation. Therefore, the receiver is not desensitized by the transmitter's high power at the received signal frequency. Detection of what frequency is detected, and knowing the transmitter frequency ramp timing, can be used to detect round-trip delay time and therefore range. It is not as accurate as pulse-delay ranging using a matched filter but can provide ranging information nonetheless. Of course, the receive frequency will be affected by the Doppler frequency. On a rapidly closing target, the receive frequencies will be all offset by a positive $f_{Doppler}$, which can be measured by the receiver once the peak receive frequency is detected. The Doppler addition can be found as the receiver knows the peak frequency of the transmitter.

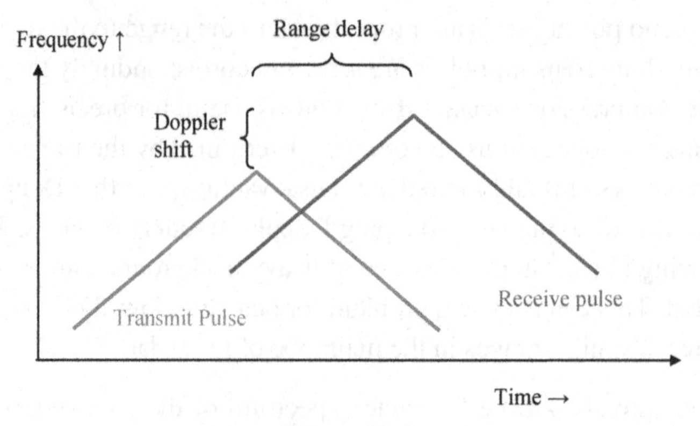

Figure 19.6
FM ranging operation.

Medium-PRF operation is a compromise. Both range and Doppler measurements are ambiguous, but each will not be aliased or folded as severely as the more extreme low- or high-PRF modes. This can provide a good overall capability for detecting both range and moving targets. However, the folding of the ambiguous regions can also bring a lot of clutter into both range and Doppler measurements. Small shifts in PRFs can be used to resolve ambiguities, as has been discussed, but if there is too much clutter, the signals may be undetectable or obscured in both range and Doppler.

19.6 Target Tracking

So far, we have been discussing how a radar system performs measurements of range and velocity of potential targets. After measuring, the targets can be identified using some of the methods already described. A target may have a specific return amplitude, azimuth, elevation, range, and relative velocity. Since measurements are repeated continuously, this allows for tracking of targets.

To track an identified target, repeated measurements are used over time. These measurements can be filtered to reduce measurement error, and the results of the filtering are fed back to control the measurement process. The radar system can respond by aiming of the mainlobe, by changing PRF, and by using measurements to anticipate future behavior of the target. For example, by estimating the target velocity and knowing the lag or latency in measurements, the radar can estimate the next position of target and have the mainlobe lead the target motion. This can also be done for the range binning and Doppler filtering. Also, if the radar itself is on a moving platform, such as an aircraft, the motion of the radar-bearing platform needs to be taken into account. This is referred to as platform stabilization.

Filtering of target measurements can be much more complex than the basic digital filtering discussed in previous chapters. The filtering may be recursive, where previous filter outputs are fed back, and will be adaptive, with gains and frequency cutoffs be varied in response to the measurement accuracy, degree of clutter, angle of antenna main beam, and other factors. There may be a number of independent or dependent filtering loops in operation. One loop may be tracking the range of target, by monitoring the range bins by detecting the comparative changes in adjacent range bin results, known as range gating. By doing this, the rate of change of the range can be coarsely estimated, which can lead to decision on when to switch to a high PRF to confirm with Doppler measurements. The antenna mainlobe may be electronically steered by making measurements at elevation and azimuths slightly above/below or side to side to estimate the azimuth and elevation of the highest return, to keep the mainlobe centered on a target. This is known as angle tracking, and this process also must account for motion of the radar platform. Note that this tracking activity may be a portion of the time, while another portion of time can be used for scanning or tracking of other targets.

In addition to tracking, there are often software-implemented algorithms to correlate the various measurements to particular targets. Target ranges, velocities, azimuths, elevations may cross over each other. These changes need to be interpolated into trajectory that can be matched to a specific target. Using radar digital signal processing followed by software-enabled target identification, tracking, monitoring, and classification, these functions can all be automated by the radar system. Higher level tracking by software can also allow for improvements in probability of detection and minimization of false alarms, as behavior of potential targets can be correlated and analyzed over longer time intervals than the radar measurement functions typically perform. This can allow the operator or pilot to more quickly understand the situation and spend more time on deciding how to respond.

CHAPTER 20

Automotive Radar*

Radar is becoming an important automotive technology. Automotive radar systems are the primary sensor used in adaptive cruise control and are a critical sensor system in autonomous driving assistance systems (ADAS). In ADAS, automotive radar is one of the several sensor systems for collision avoidance, pedestrian and cyclist detection, and complements vision-based camera-sensing systems. The radar technology generally used is frequency-modulated continuous-wave or FMCW radar, which is quite different than the pulse-Doppler radar. The analog and RF hardware in FMCW is considerably less complex than that of pulse-Doppler radar. In addition, the digital processing requirements are generally modest and can be performed in low-cost field programmable gate arrays, microprocessors with specialized acceleration engines, or specialized application-specific integrated circuits.

Pulse-Doppler radars are used in longer range radars where the range can be one kilometer to hundreds of kilometers. The transmitter operates for a short duration, then the system switches to receive mode until the next transmit pulse. The pulse-Doppler radar sends successive pulses at specific intervals or a pulse repetition interval. As the radar returns, the reflections are processed coherently to extract range and relative motion of detected objects. Sophisticated processing methods are employed to further process radar returns and extract target data even when heavily obscured by ground clutter or background returns surrounding the object(s) of interest. However, with automotive radar, the range can be as short as a few meters to as much as a few hundred meters. For a range of 2 m, the round-trip transit time of the radar pulse is 13 ns. This short range requires that the transmitter and receiver operate simultaneously, which requires separate antennas. The pulse-Doppler radar sends a pulse periodically, and the ratio of the time the transmitter is active to the total time elapsed is the duty cycle. Since duty cycles are typically small, this ratio limits the total transmit power. The power, in turn limits the range of detection. Achieving a 1- to 2-m range resolution also requires very high-speed analog pulsing circuitry, high digital sample rates, high digital signal processing capability, which is not feasible in low cost automotive systems.

20.1 Frequency-Modulated Continuous-Wave Theory

FMCW radar is a much more optimal method for short range radar. FMCW does not sent out pulses and then monitor the returns or radar echo. Rather, a carrier frequency transmits

* With contributions by Ben Esposito.

Digital Signal Processing 101. http://dx.doi.org/10.1016/B978-0-12-811453-7.00020-2

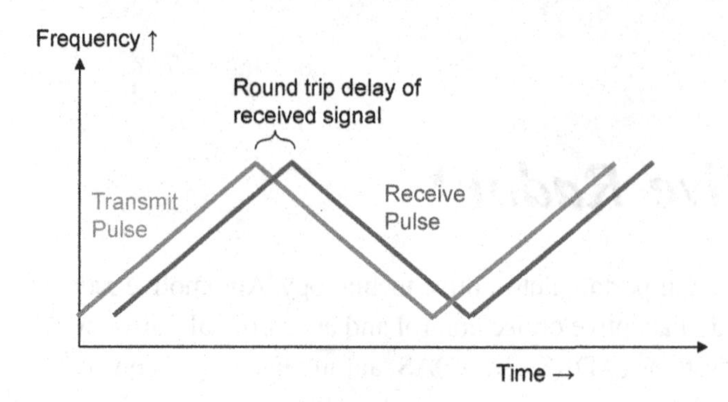

Figure 20.1
Frequency-modulated continuous-wave transmit and receive waveforms.

continuously. To extract useful information from the continuous return echo, the carrier frequency is increased steadily over time, and then decreased, as shown in Fig. 20.1. Both the transmitter and receiver operate continuously. To prevent the transmit signal from leaking into the receiver, separate transmit and receive antennas are used.

20.2 Frequency-Modulated Continuous-Wave Range Detection

The radar must determine the range of objects detected. In FMCW, the range is accomplished by measuring the instantaneous receive frequency difference, or delta, from the transmit frequency. During the positive frequency ramp portion of the transmit cycle, the receive frequency will be somewhat less than the transmit frequency, depending on the time delay. During the negative frequency ramp portion of the transmit cycle, the receive frequency will be greater than the transmit frequency, again depending on the time delay. These frequency differences, or offsets, will be proportional to the round-trip delay time, which provides a means to measure the range. The greater the range, the more time delay there is from the transmitter to the receiver. Since the transmit frequency is constantly changing, the difference between transmit and receive frequencies at any given instant will be proportional to the elapsed time for the transmit signal to travel from the radar to the target and back.

An example block diagram of an FMCW radar is shown in Fig. 20.2. Automotive radars operate in the millimeter range, meaning the wavelength of the transmit signal is a few millimeters. Common frequencies used are 24 GHz ($\lambda = 12.5$ mm) and 77 GHz ($\lambda = 3.9$ mm). Millimeter wave frequencies are used because of the small antenna size needed, relative availability of spectrum, and rapid attenuation of radio signals (automotive radar ranges are limited to hundreds of meters). By using FMCW, there is no amplitude modulation, and the transmitter only varies in frequency. Use of FM allows for the transmit circuit to operate in saturation, which is the most efficient mode for any RF amplifier.

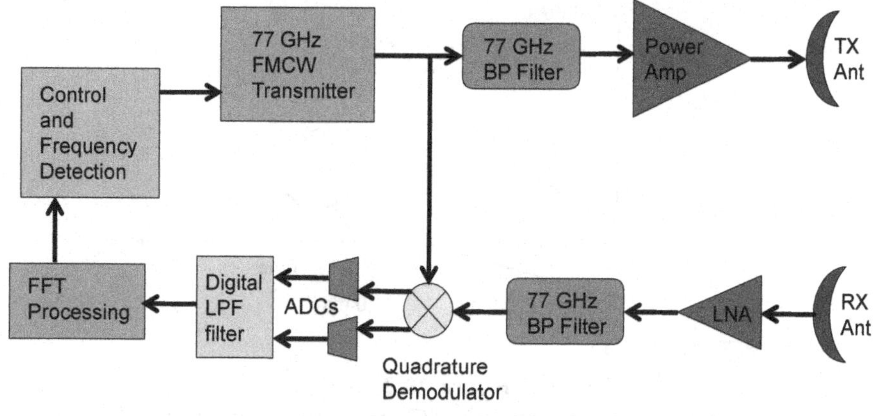

Figure 20.2
Automotive radar block diagram. *ADC*, analog-to-digital; *LPF*, low-pass filter.

Due to the analog mixer circuit, the receive low-pass filter only needs to pass the difference between the receive and transmit signals. It does not need to pass the 500 MHz bandwidth of the receive signal over the transmit cycle. This passing of the difference in signals is most easily illustrated by example. Assuming receive reflections at the minimum and maximum ranges of distances 1 and 300 m.

With a frequency ramp of 500 MHz in 0.5 ms, or 1 kHz per ns, the frequency of the receive signal will be as follows, using the speed of light at 3×10^8 m/s:

$$1\ \text{m distance} = 2\ \text{m round trip delay} = 2\ \text{m}/\left(3 \times 10^8\ \text{m/s}\right) = 7\ \text{ns}$$
$$300\ \text{m distance} = 600\ \text{m round trip delay} = 600\ \text{m}/\left(3 \times 10^8\ \text{m/s}\right) = 2\ \mu\text{s}$$

During the positive frequency ramp, the return from the object at a 1 m distance will be a −7 kHz offset. During the negative frequency ramp, the return from the object at a 1 m distance will be a +7 kHz offset.

During the positive frequency ramp, the return from the object at a 300 m distance will be a −2 MHz offset. During the negative frequency ramp, the return from the object at a 1 m distance will be +2 MHz offset.

These offsets indicate that the receiver input will have frequencies in the range of +/−2 MHz, depending on the range of the target generating the return. This frequency can be detected by operating a fast Fourier transform (FFT) for the time intervals shown in Fig. 20.3. If the receiver samples at 5 MSPS (mega samples per second), then for a receiver FFT sampling interval of about 0.4 ms, a 2048-point FFT can be used, which will give a frequency resolution of 1 kHz, sufficient for a fraction-of-a-meter-range resolution. Further resolution is possible by performing an interpolation of the FFT output.

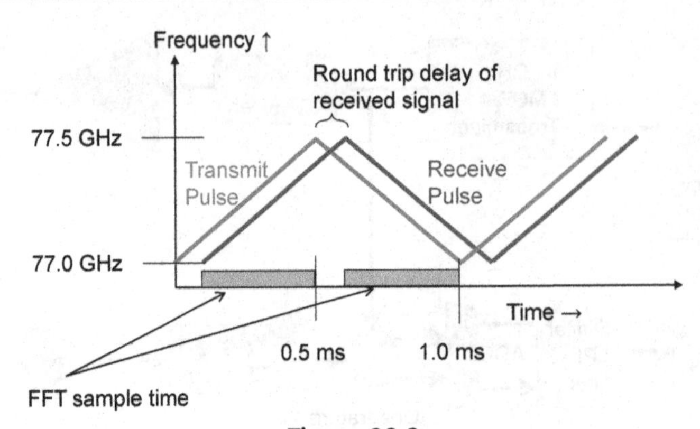

Figure 20.3
Receiver sampling intervals. *FFT*, fast Fourier transform.

This observed frequency offset provides the receiver range returns but cannot be used to discriminate between oncoming traffic, traffic traveling in same direction, or stationary objects at similar ranges. To make this discrimination, the Doppler frequency shift of the return must be exploited.

20.3 Frequency-Modulated Continuous-Wave Doppler Detection

A simple example can illustrate how the Doppler frequency shift can be detected. Assume the 77 GHz radar-equipped automobile is traveling at 80 km/h, or 22.2 m/s, and consider three targets at 30 m distance—one an oncoming auto at 50 km/h, another traveling in same direction at 100 km/h, and a stationary object. When the objects are closing range, the Doppler frequency shift will be positive, meaning the return signal will have a higher frequency than that transmitted. Intuitively, this behavior is due to the wave crests of the signal appearing closer together due to the closing range. Conversely, when the range between the radar and the target is opening, or getting larger, the result will be a negative Doppler frequency shift. The amount of Doppler shift can be calculated as shown in Eq. (20.1).

$$\begin{aligned}
&\text{Doppler frequency shift} = (2\cdot\text{velocity difference})/\text{wavelength}\\
&\text{Oncoming auto at } 50\ \text{km/h, radar auto at } 80\ \text{km/h, closing rate}\\
&\quad\text{of } 130\ \text{km/h or } 36.1\ \text{m/s}\\
&\text{Doppler frequency shift} = 2(36.1\ \text{m/s})/(.0039\ \text{m}) = 18.5\ \text{kHz}\\
&\text{Stationary object, radar auto at } 80\ \text{km/h, closing rate of } 80\ \text{km/h or } 22.2\ \text{m/s}\\
&\text{Doppler frequency shift} = 2(22.2\ \text{m/s})/(.0039\ \text{m}) = 11.4\ \text{kHz}\\
&\text{Auto ahead at } 100\ \text{km/h, radar auto at } 80\ \text{km/h, opening rate}\\
&\quad\text{of } 20\ \text{km/h or } 5.56\ \text{m/s}\\
&\text{Doppler frequency shift} = -2(5.56\ \text{m/s})/(.0039\ \text{m}) = -2.8\ \text{kHz}
\end{aligned} \tag{20.1}$$

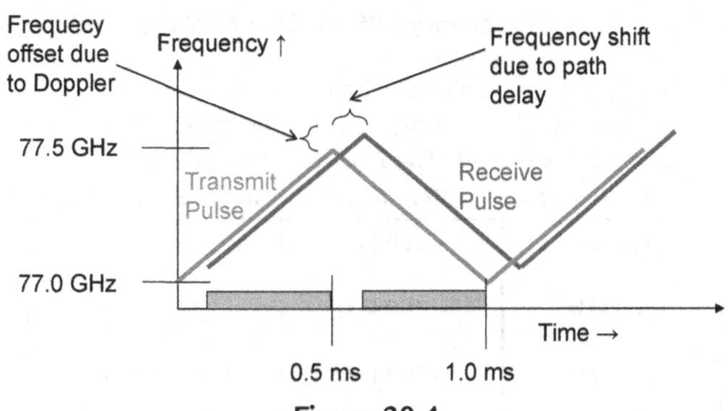

Figure 20.4
Doppler frequency detection.

These Doppler shifts will offset the detected frequency differences on both the positive and negative frequency ramps. When there is no relative motion, and no Doppler shift, the difference in the receive frequency will be equal, but of opposite sign, in the positive and negative frequency ramps. So the Doppler shift can be found by comparing the receive frequency offsets during the positive and negative portions of the frequency ramp of the transmit signal. The relationship is described in Eq. (20.2), where the range and the relative velocity can be determined by the output of the FFT using the results during both positive and negative frequency ramps of the transmitter.

$$\text{Relative velocity(target to radar)} = (\text{wavelength}/2) \cdot (\text{positive ramp detected frequency} - \text{negative ramp detected frequency})/2 \tag{20.2}$$

The frequencies seen by the receiver and processed by the FFT are shown in Fig. 20.4. The receiver is demodulated, or down converted, by using the transmit signal as the local oscillator (LO), so the FFT will process the frequency difference between the transmit and receive waveforms. Fig. 20.4 shows only one target pulse return; however, there could be multiple (three in our example) different targets generating multiple frequencies for the FFT to detect.

Returning to the example, there are three targets at 30 m distance—an oncoming automobile at 50 km/h, another traveling in same direction at 100 km/h, and a stationary object. The automobile mounting the radar is traveling at 80 km/h, or 22.2 m/s. For all three targets at 30 m, the receive frequency offset will be as follows:

$$30\ \text{m range} = 60\ \text{m round trip delay} = 60\ \text{m}/(3 \times 10^8\ \text{m/s}) = 200\ \text{ns}$$

Frequency offset: +/−200 kHz on the negative and positive frequency ramp, respectively.

The Doppler offset must be added to the frequency offset caused by the range delay. The values are summarized in Table 20.1.

Table 20.1: Target Frequency Offsets Due to Range and Doppler

30 m Range, 80 km/h Radar Vehicle Speed	Negative Ramp Frequency Offset Due to 30 m Range	Positive Ramp Frequency Offset Due to 30 m Range	Frequency Offset Due to Doppler (Negative and Positive Ramp)	Total Observed Negative/Positive Frequency Offset
Oncoming at 50 km/hr	200 kHz	−200 kHz	18.5 KH	218.5 kHz/ −181.5 kHz
Stationary object	200 kHz	−200 kHz	11.4 kHz	211.4 kHz/ −188.6 kHz
Same direction at 100 km/h	200 kHz	−200 kHz	−2.8 KHz	197.2 kHz/ −202.8 kHz

Using only the observed negative and positive frequency offsets, both the range and relative Doppler shift of the targets can be determined using Eqs. 20.3–20.5. Since the forward velocity of the radar-equipped auto is known, the velocities of the targets can be easily computed. Similarly, the frequency ramp rate of the radar is also a known quantity.

$$\text{Range} = (\text{speed of light}/4 \cdot \text{frequency ramp rate}) \cdot (\text{negative frequency offset} - \text{positive frequency offset}) \tag{20.3}$$

$$\text{Relative velocity} = (\text{speed of light}/4 \cdot \text{carrier frequency}) \cdot (\text{negative frequency offset} + \text{positive frequency offset}) \tag{20.4}$$

$$\text{Absolute target velocity} = \text{Relative velocity} - \text{radar-equipped vehicle velocity} \tag{20.5}$$

The frequency ramp rate in this case is 500 MHz over 0.5 ms, or rate of 1000 GHz per second. Plugging in our example values, we find:

For vehicle approaching 50 km/h:

$$\text{Range} = \left(3 \cdot 10^8/(4 \cdot 1000 \cdot 10^9)\right) \cdot \left(218.5 \cdot 10^3 - (-181.5 \cdot 10^3)\right) = 30\ \text{m}$$

$$\text{Relative velocity} = \left(3 \cdot 10^8/(4 \cdot 77 \cdot 10^9)\right) \cdot \left(218.5 \cdot 10^3 + (-181.5 \cdot 10^3)\right) = 36\ \text{m/s} = 130\ \text{km/h}$$

$$\text{Absolute target velocity} = 130 - 80 = 50\ \text{km/h closing rate}$$

The other target calculations are similar.

However, real-world systems often have many targets, some of which are of interest and some of which are not. Along with clutter, it may not be possible to pair up many frequencies in the negative and positive ramp period. This problem is known as ambiguities in radar jargon. One approach to the problem is varying the duration and frequency of the ramps and evaluating how the detected frequencies move in the spectrum with different steepness of frequency ramps.

More commonly, the range and velocity detection processing is performed separately, using multiple sample intervals. This is a combination of pulse-Doppler and FMCW techniques. Essentially, it is a pulsed system that utilizes FMCW modulation. But first a quick review of the FMCW RF and antennas.

20.4 Frequency-Modulated Continuous-Wave Radar Link Budget

Radar performance is governed by the link budget equations, which determine what level of receive signal are available for detection. A simplified version of the radar link budget is expressed in Eq. (20.6).

$$P_{rcv} = P_{trx} \cdot G^2 \cdot \sigma \cdot \lambda^2 \cdot \tau \Big/ \left((4 \cdot \pi)^3 \cdot R^4\right) \qquad (20.6)$$

where: P_{trx} is the peak transmit power; G is the transmit and receive antenna gain; σ is the radar cross section or area of the target; λ is the radar wavelength; τ is the duty cycle of transmitter; R is the range to the target.

Often, parameters are specified on a logarithmic scale, using decibels (dB), or dBm (dB referenced to 1 mW). This scale will be used in this case, but figures in Watts will also be given. We must make a few assumptions. With an achievable receiver noise figure of 5 dB, the receiver sensitivity should be about -120 dBm (10^{-15} W). Assuming about 20 dB signal-to-noise ratio to achieve reasonable frequency detection, results in a requirement that P_{rcv} should be at least 100 dBm (10^{-13} W) under worst-case circumstances.

We will use an antenna gain of 30 dB, or 1000. For a parabolic antenna, the gain in the direction of the bore sight of the antenna can be calculated as shown in Eq. (20.7).

$$G = 4 \cdot \pi \cdot A_{eff}/\lambda \qquad (20.7)$$

Working backward, we find the area required for the antenna at 77 GHz is 0.0012 m^2, which works out to a diameter of 0.04 m, or 4 cm, which is a very reasonable size to mount on the front of a vehicle. (Note, we need at least one transmit and one receiver antenna.)

The duty cycle τ in FMCW is 100%, or 1. If we assume a transmit power of 0.1 W (20 dBm), a maximum range of 300 m, and a target vehicle reflecting area of 1 m^2, we can find the worst-case receive power under reasonable circumstances.

$$P_{rcv} = 0.1 \cdot 1000^2 \cdot 1 \cdot .0039^2 \cdot 1 \Big/ \left((4 \cdot \pi)^3 \cdot 300^4\right) = 9.4 \cdot 10^{-14}\,\text{W or} -100\ \text{dBm}$$

Using a very close range, say 2 m, and same cross-sectional area of 10 m^2 (the back of a tractor trailer), we can calculate the maximum P_{rcv} we can expect under reasonable circumstances.

$$P_{rcv} = 0.1 \cdot 1000^2 \cdot 1 \cdot .0039^2 \cdot 10 \Big/ \left((4 \cdot \pi)^3 \cdot 2^4\right) = 4.8 \cdot 10^{-4}\,\text{W or} -3.2\ \text{dBm}$$

These calculations tell us that our system requires a very high-dynamic—range receiver, on the order of 120 dB. The high-dynamic range imposes a severe requirement on linearity of the receiver and the analog-to-digital converter (ADC). However, a large target only 2 m away will obscure the radar line of sight for other targets. Therefore, an analog automatic gain control (AGC) loop can be used to reduce the dynamic range of the receiver and ADC by desensitizing the receiver with attenuation in the presence of extremely large returns.

Full sensitivity could be desired in a slightly less demanding situation, such as a 1 m^2 target at 4 m distance, and still used to detect other targets at much longer range. In this case, the large return will have a receive power:

$$P_{rcv} = 0.1 \cdot 1000^2 \cdot 1 \cdot .0039^2 \cdot 1 \Big/ \left((4 \cdot \pi)^3 \cdot 4^4 \right) = 3.0 \cdot 10^{-6}\,\mathrm{W} \text{ or } -25 \ \mathrm{dBm}$$

The use of AGC reduces the dynamic range requirement to about 95 dBm, achievable using a 16-bit ADC. To give some margin, the ADC could be operated with 32X oversampling (beyond Nyquist requirements) to achieve another effective 3 bits, or reducing the quantization noise floor a further 18 dB. Alternatively, an 18-bit ADC could be used, but would likely be much more costly.

20.5 Frequency-Modulated Continuous-Wave Implementation Considerations

On the analog side, the transmitter can be implemented using a direct digital synthesizer (DDS) with a standard reference crystal. The DDS generates an analog frequency ramp reference for the phase-locked loop (PLL) to generate the desired transmit frequency modulation. For example, if the PLL has a divider of 1000, then in our example, the reference would be centered at 77 MHz, with a 5 MHz frequency ramp. This analog ramp signal drives the reference of a PLL, which disciplines a 77 GHz oscillator. The oscillator output of the circuit is amplified and produces the continuous-wave signal ramping up and down over 500 MHz with a center frequency of 77 GHz. Filtering and matching circuits at 77 GHz can be accomplished using passive components etched into high-epsilon R dielectric circuit cards, minimizing the components required. Fig. 20.5 illustrates an analog circuit block diagram.

In the receiver, the front end requires filtering and a low-noise amplifier (LNA), followed by a quadrature demodulator. The quadrature demodulator mixes the 77 GHz receive signal with the ramping transmit signal, outputting a complex baseband signal that contains the difference between the transit and receive waveforms at any given instant. The ramping has been canceled out, as we see fixed frequencies depending on the range and Doppler shift of the target returns. Again, the high-frequency filtering at 77 GHz can be implemented using etched passive components. The output of the quadrature

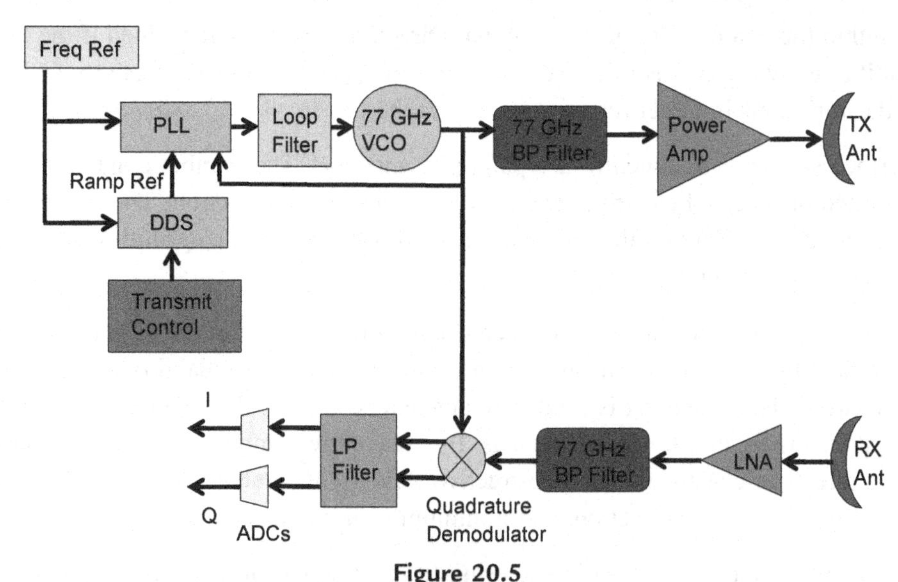

Figure 20.5
Analog circuit block diagram. *LP*, low pass.

demodulator will be at low frequency, up to +/− 2 MHz at maximum range. Therefore, traditional passive components and operational amplifiers can be used to provide antialiasing low-pass filtering prior to the in-phase (I) and quadrature (Q) ADCs. Alternately, an intermediate frequency architecture could be used, but would require an offset receive LO generation circuit.

20.6 Frequency-Modulated Continuous-Wave Interference

Eventually, there may be many vehicles equipped with radars operating at 77 or 24 GHz. The radar transmitter of an oncoming vehicle will likely produce a much stronger signal than most target reflections. However, the transmitter is operating over hundreds of MHz, specifically 500 MHz in this example. The receiver input bandwidth is on the order of 5 MHz, or about 1% of transmit bandwidth. Interference can only occur if the oncoming transmitter is sweeping through this 1% of its bandwidth at the same time the other receiver happens to also be sweeping through that particular bandwidth. Statistically, this overlap is not likely to occur very often, and when it does, it can be eliminated by randomly adjusting the transmit ramp timing. This problem is common in systems where many devices infrequently communicate over a shared channel using random access techniques.

20.7 Frequency-Modulated Continuous-Wave Beamforming

The radar system described so far can detect range and velocity or targets, but cannot provide any information on the direction of the target, other than it is in front of the vehicle, within the beam width of the antenna. Directivity can be determined if the system has the ability to sweep or steer the radar transmit or receive antenna directivity and monitor the variations in target return echo across the sweep.

The described system is assumed to use parabolic antennas. The parabolic antenna focuses the transmitted or received electromagnetic wave in a specific direction. The degree of focusing depends primarily on the antenna area and wavelength. Using millimeter wave radar allows for small antennas.

A parabolic antenna can be "aimed" by mechanically orientating it in the desired direction, which is limited by the speed of mechanical movement, as well as reliability and cost issues. Instead, electronic beam steering is used. The antenna becomes either a linear or rectangular array of separate receive or transmit antennas. By coherently combining the separate antenna signals, the effects of constructive and destructive wave front combining will result in maximum gain in a specific direction, and minimum gain on other directions.

In the automotive radar case, elevation steering (up and down) of the radar is generally not required, so a two-dimensional antenna array is not required. A linear array, or line of antennas, will permit azimuth (side to side) steering of the antenna. The trade-off is cost and complexity. In this case, steering of the receive direction is more straightforward, due to the digital processing of the receive signal. Each receiver must individually vary the phase of receive signal.

This phase adjustment provides for steerable directivity of the antenna beam. Only when the receive signal arrives in-phase across all the antenna elements will the maximum signal strength occur. The array antenna provides the ability to "aim" the main lobe of the antenna in a desired direction. Each antenna element must have a delay, or phase adjustment, such that after this adjustment, all elements will have a common phase of the signal. If the angle $\theta = 0$, then all the elements will receive the signal simultaneously, and no phase adjustment is necessary. At a nonzero angle, each element will have a delay to provide alignment of the wave front across the antenna array, as shown in Fig. 20.6.

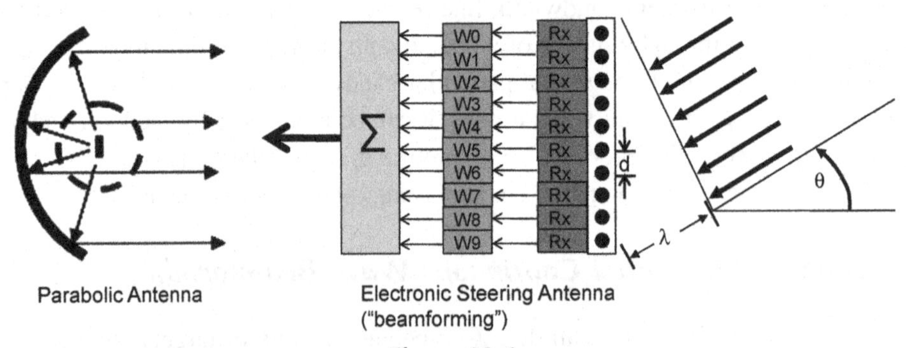

Figure 20.6
Electronic steerable antennas.

The electronic steerable antenna requires replicating the analog receiver circuits for each of the N antenna receive nodes. Fortunately for millimeter radars, much of the circuitry, including antenna patches, filters, and matching circuits can be implemented directly on the printed circuit board. The LNA, quadrature demodulators, and ADCs must also be replicated for each of the N nodes.

Digitally, each set of I and Q inputs from the ADC pair of each antenna node must be delayed in-phase. This delay is accomplished by a complex multiplier with N separate complex coefficients W_i for each of the N receive nodes. The control processor "sweeps" the receive antenna by updating N complex coefficients periodically and monitoring the changes in target-return amplitudes.

In a forward-looking automotive radar, the degree of desired azimuth steering may only be 5–10 degrees off the automobile centerline. In terms of cost effectiveness, it may be possible to use a parabolic transmit antenna with sufficient antenna lobe beam width and utilize a narrower lobe beam steering receive antenna to provide the ability to distinguish targets across different azimuths. Alternately, a more complex transmitter and transmit beamforming antenna can also be used to provide more gain in the desired transmit azimuth direction, but at greater cost and complexity.

20.8 Frequency-Modulated Continuous-Wave Range-Doppler Processing

Returning to the signal processing, the method combining pulse-Doppler and FMCW modulation is described. This includes multiple receive antennas, range FFT processing, Doppler FFT processing, noncoherent magnitude detection, constant false alarm rate (CFAR) processing, even angle of arrival (AoA) estimation (Fig. 20.7). Much of this processing can be performed in parallel, with hardware-based architecture (FPGA or ASIC).

Automotive radar systems require low latency processing. To meet latency requirements, much of the processing needs to be performed in parallel. For example, a typical radar frame

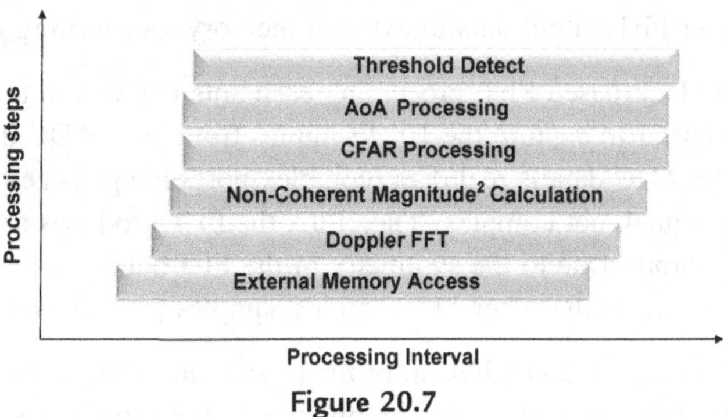

Figure 20.7
Parallel radar processing.

time is 40 ms, meaning that the entire radar processing must be completed and results generated in that time, at a rate 25 times per second. The signal processing would need to be performed in significantly less time, to allow for subsequent host CPU processing to organize and sort the detection results. In addition, memory access times must be considered, as the amount of data being processed requires storage of intermediate results, for the radar "corner turn" and to store the radar "data cube," which is generated during each frame.

Several system key parameters will determine the FMCW radar capability and are needed to size the signal processing and memory capabilities. For the purposes of discussion, the following will be assumed:

- 40 m, frame time, with 25 ms for sampling and signal processing,
- Eight receive antennas,
- 20 MSPS sample rate per ADC,
- 1024 samples per pulse (real data),
- 256 pulses per frame, and
- 800 MBps or greater external SRAM memory access rate.

20.9 Frequency-Modulated Continuous-Wave Radar Front-End Processing

Radar front-end processing includes the following steps:

- Sample the data during the each pulse across all antennas—requires buffering 1024 samples per ADC input (or per antenna) k,
- Apply window function to each block of date from each antenna;
- Performing range 1024 FFT across the pulse interval;
- Storing range FFT data to external memory;
- After all pulses have been processed and stored to external memory, read out data ("corner turn"), apply window function;
- Perform Doppler FFT processing; and
- Save each Doppler FFT output data to external memory, overwriting Doppler input data.

The FFT engine is a block-based FFT, processing each antenna data in turn, which necessitates storage buffer to capture the 1024 samples from each ADC during the pulse interval. Note that the ADC data is real data (complex portion equals zero), since FM input signal is a real signal, not complex. Therefore, the FFT produces a 1024 sample complex symmetric output. Due to the symmetry of the FFT output, only 512 complex samples need to be saved, as the other 512 complex samples are redundant and discarded.

A timing diagram illustrates the parallelism of the processing (Fig. 20.8). The range of FFT processing time is determined by the sampling rate. The range active sampling time

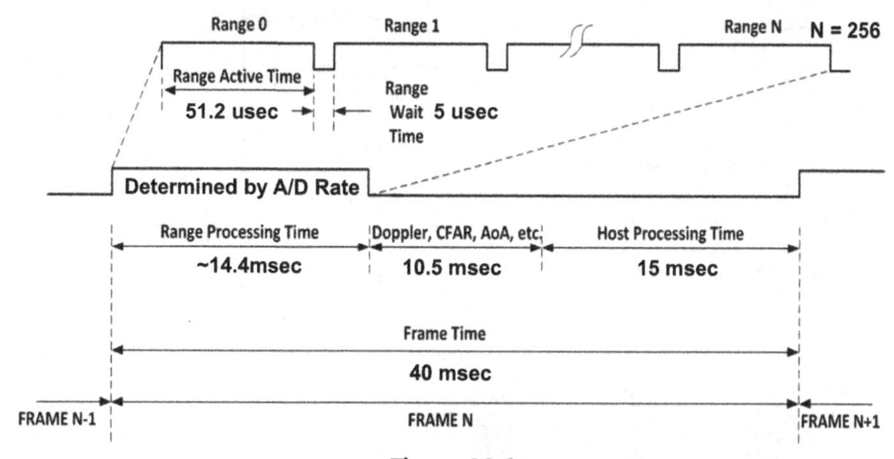

Figure 20.8
Radar timing diagram.

is 50 ns (at 20 MSPS) × 1024 real samples, which gives 51.2 μs. There is a small interval between pulse intervals, set here at 5 μs. This provides enough time for all receive radar returns to arrive, so will not overlap in the next pulse interval. The total sampling time is therefore (51.2 μs + 5 μs) × 256 (the number of pulses), which is about 14.4 ms.

During this 51.2 μs pulse interval time, the transmitter is operating with a linear ramping FM waveform. The transmit and receive frequencies are mixed, and this mixer output, or frequency difference, is what is sampled by the ADC. Since the transmit frequency ramp is only positive during the pulse interval, only a real frequency difference is sampled. Therefore, quadrature sampling (used for complex baseband signals) is not required.

After the range sampling time of 14.4 ms the back-end processing including Doppler, complex magnitude, CFAR, and AoA estimation is estimated to take about 10.5 ms. These functions can be largely done with parallel processing capabilities to allow for a substantial reduction in processing time for these complex functions and allows the remaining 15 ms for host processing of the output data (assuming a 40 ms period).

A single FFT can process all eight antenna range data. Each antenna data is double buffered to allow for flexible, nonsynchronous clocking of ADCs. Assuming the FFT can process one input sample per clock cycle and is operating at 200 MHz. Therefore, one FFT can keep up with the aggregate input sample rate of eight times 20 or 160 MSPS. The FFT processes each block of 1024 antenna samples. This will repeat across all eight antennas, creating the data structure on the lower left of Fig. 20.9. The FFT output data results will be written to the external memory, interleaving the data between eight antennas. This whole process repeats 256 times, which is the number of transmitter pulses and creates the radar data cube of post-FFT range data shown in the middle of Fig. 20.9.

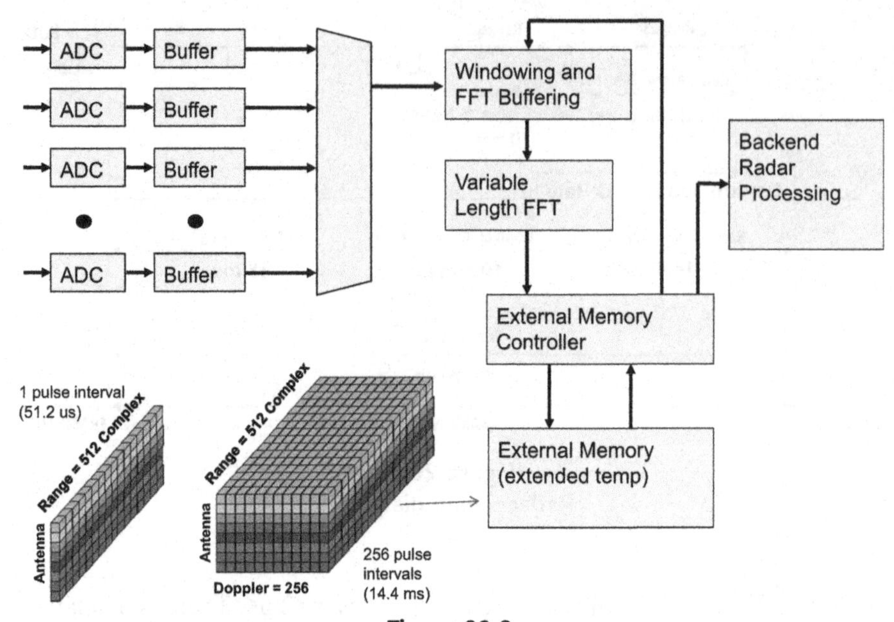

Figure 20.9
Radar range processing. *ADC*, analog-to-digital.

The FFT processing and memory writes occur in parallel with the receive data sampling, so at the end of the 14.4 ms, the range data cube is contained in the external memory. This memory also requires access in bursts of consecutive 8 bytes (or 9 bit bytes), which is four complex samples. The sequential burst access requirement must be observed to maintain maximum memory bandwidth, and the data ordering must be considered for Doppler and CFAR processing.

The FFT processing and memory writes occur in parallel with the receive data sampling, so at the end of the 14.4 ms, the range data cube is contained in the external memory. The memory size is 8 (number of antennas) × 512 (number of range FFT samples) × 256 (number of pulses) × 2 (complex data) × 2 (16 or 18 bit data), which requires a 4 MB memory. Often a burst size must be observed, and also sequential burst access memory requirements needed to maintain maximum memory bandwidth and the data ordering for subsequent Doppler and CFAR processing.

20.10 Frequency-Modulated Continuous-Wave Pulse-Doppler Processing

The radar data cube now need to be read back into the FPGA for Doppler processing. Recall in the FMCW radar, it is possible to perform both range and Doppler estimation by using triangular frequency pulses, with both positive and negative frequency ramps. This technique may be suitable for simple detection, with few targets, such as blind spot detection. However, in a forward-looking radar, there will be myriad of targets, and this more sophisticated processing is needed.

The Doppler FFT will be performed for each range bin, across a specific range for all 256 pulses. This requires a total of 512 FFTs (as there are 512 range bins) to be processed for each antenna, each of length 256. A single FFT engine will perform the 8 × 512 FFTs, requiring 8 × 512 × 256 clock cycles at 200 MHz, or 5.25 ms. The limitation is the memory bandwidth, as the Doppler FFT data must be both read in from the external memory, and then written back out, overlaying the original 256 block of complex samples. A 5 ns per 4-byte memory transaction, this requires 10.5 ms in total.

A single FFT engine can perform both range and Doppler FFTs, as these happen at different times. During the range FFT processing phase, the FFT operates as a 1024-point FFT. In Doppler processing, the same FFT operates at 256 points. If different combinations of range and Doppler sizes are needed, the FFT can be dynamically configured for the appropriate 2^N size transform.

The indexing of the memory is important in the organization of the radar dataflow. In pulse-Doppler radars, the range data is normally written into an external memory in sequential order with regard to the range samples. Each radar pulse range data is written in blocks, one block after the other. For Doppler processing, the data must be read in a different order. This reordering is known as the corner turn in radar terminology, and in DDR memories, internal buffering is needed to manage this with consideration of the burst length access requirements.

The data must be read from the memory for Doppler FFT processing (Fig. 20.10). Due to the CFAR processing requirements in the back end, the antenna data interleaved in the external memory. However, the burst data reads for the Doppler processing are in a different order than the postrange FFT writes. While the antenna data is still interleaved in the burst accesses, the burst addresses now index along the 256 long pulse counts, rather than the 512 long range counts. Each block of 256 complex samples is read in for a specific range, across the 256 pulses. Eight of the FFT data blocks are buffered, as the antenna data must be deinterleaved as it is read from the external memory.

The Doppler FFT processing will replace the individual range bin data for each pulse interval with a frequency distribution for individual range bin, allowing determination of radial velocity. At this point, target detections can be made on the basis of range and relative velocity to the radar system. This has been performed for each of the eight antenna receive inputs.

A plot of range-Doppler data is depicted (Fig. 20.11).

In Fig. 20.10, the first three of the memory accesses have been discussed. The fourth access occurs during AoA estimation, which will be covered shortly. An internal data

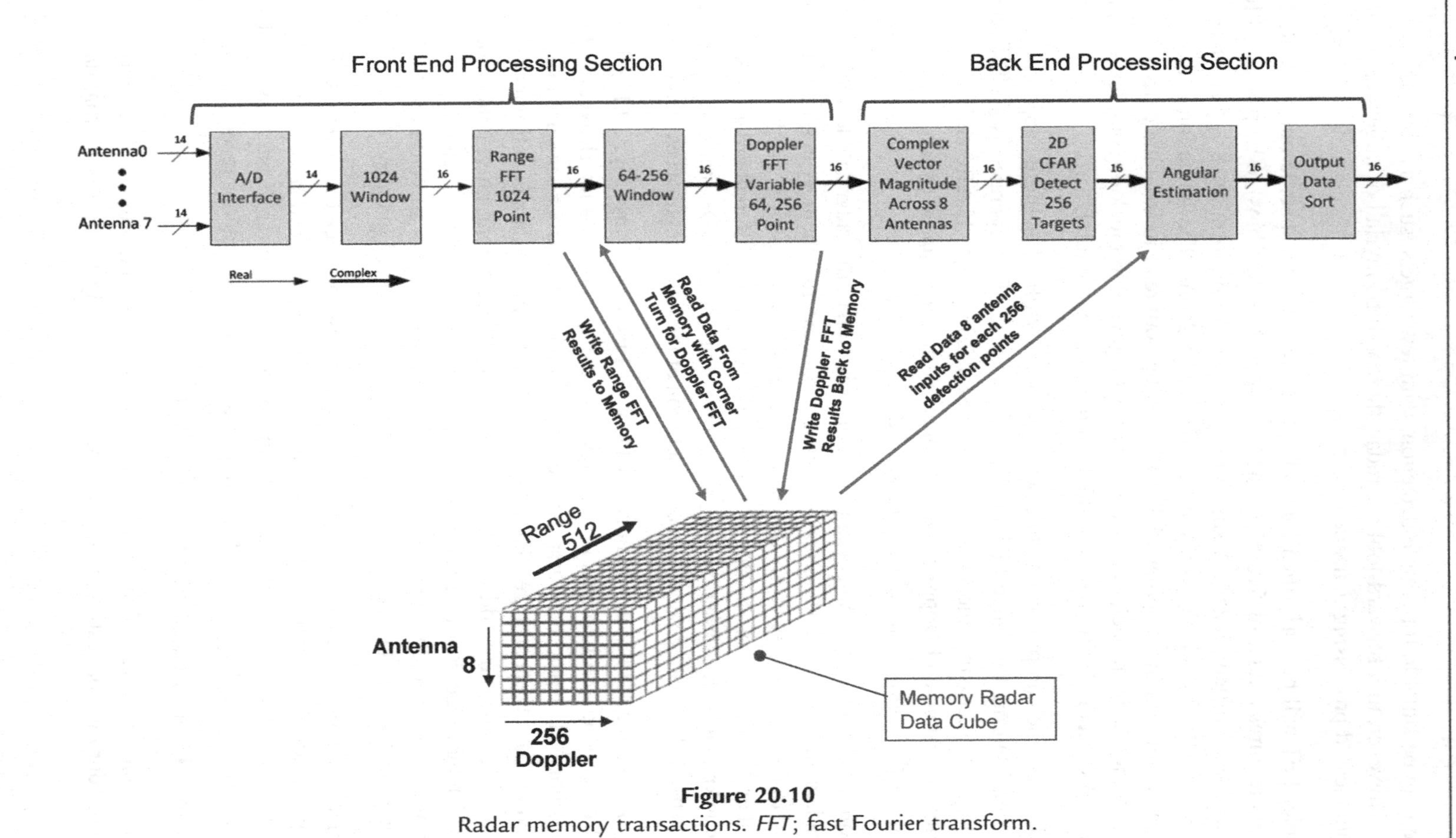

Figure 20.10
Radar memory transactions. *FFT*; fast Fourier transform.

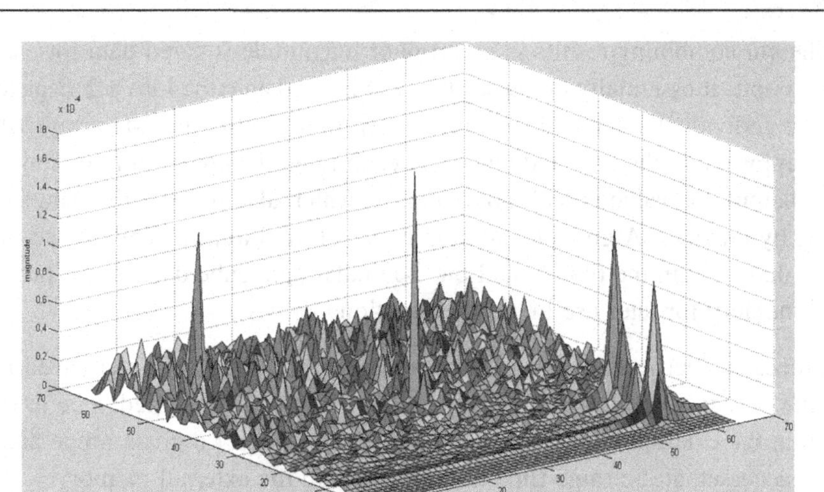

Figure 20.11
Range-Doppler plot.

transfer also occurs, whereas the Doppler FFT results are processed real time as they are generated by the detection process, which operates largely in parallel.

20.11 Frequency-Modulated Continuous-Wave Radar Back-End Processing

The back-end processing is focused on detection. Due to the stochastic nature and often high levels of noise and clutter, detection is far from simple. Most radar systems use some sort of CFAR detection algorithm to determine whether it is a legitimate target. This involves comparing the magnitude, or magnitude squared of a given range and Doppler cell to an estimate of the clutter and noise in nearby cells. When this ratio exceeds a given threshold, then detection is assumed. By using a variable and local estimate of clutter, the rate of false alarms should, in theory, remain at a constant level.

20.12 Noncoherent Antenna Magnitude Summation

For the CFAR detection process, the input energy from all antennas must be used. To facilitate this, the outputs of the eight FFTs for a give range and Doppler are combined, across the eight antennas, using the equation below.

$$y_{rd} = \sum_{a=0}^{7} I_{r,d,a}^2 + Q_{r,d,a}^2 \quad \begin{matrix} \text{Range, r: from}\{0\ldots511\} \\ \text{Doppler, d: from}\{0\ldots255\} \\ \text{Antenna, a: from}\{0\ldots7\} \end{matrix}$$

This noncoherent combining results in an array of magnitude squared data for each range and Doppler point. It essentially collapses the 3-D radar data cube into a 2-D array, by combining the individual antenna data (Fig. 20.12). In this case, the 2-D array will be 512×256, representing the detected energy at each range-Doppler combination. Since this is noncoherent, the value is real, not complex, which also reduces the data storage requirements by one half. A given location is often referred to as a "cell." Note this collapsing is done within the FPGA, and the 3-D radar data cube data is still preserved in the external memory for later use in AoA estimation.

The front processing across for the eight antennas is processed in an interleaved manner. Therefore, the output of the Doppler data is processed as it is generated. Due to the organization in the external memory, all eight antennas data for a given range and Doppler value can be accessed at the same time it is written into the external memory.

Two basic variations of CFAR detection are commonly used. Cell averaging, or CA-CFAR, performs an average calculation of the magnitude (or magnitude squared) of the cells surrounding the cell under test (CUT). Since every cell in the range-Doppler array must be tested, this results in a sliding window-type calculation. The other basic type of CFAR is ordered statistic, or OS-CFAR. In this case, magnitude (or magnitude squared) of the cells surrounding the CUT are ranked, and a median, or kth order value is used as the clutter estimate. OS-CFAR can be more difficult to process, but it has proven in many scenarios to provide a superior clutter estimate, and less sensitivity to regions of strong clutter obscuring target detection.

The front-end processing generates the range-Doppler array at a rate of one cell every 100 ns, or at a 10 MSPS rate. Therefore, the detection processing must be able to operate at the same rate.

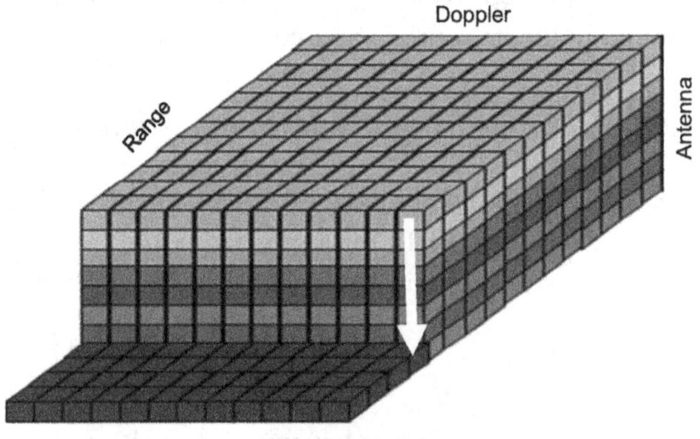

Figure 20.12
Magnitude squared radar array.

20.13 Cell Averaging—Constant False Alarm Rate

As each set of 256 Doppler row data is generated, it can be stored in a single block memory. For an N × N CFAR window, N + 4 rows need to be stored. As the CUT moves through the array, the next row of data can overwrite the oldest row. This process is depicted in Fig. 20.13, where the 7 × 7 block of cells is the sliding window. The upper cells are the stored row data, and lower cells are in the process of being computed and written into the array for CFAR processing.

A variation of CA-CFAR is to use weighted averaging, rather than simple averaging. Using multipliers, the cells can be weight the cells in the sliding window by coefficients stored in local memories. This allows for a "tapering" of the cell values near the sliding window boundaries, rather than a sharp cutoff. Weighting can also be used to deemphasize the guard cells immediately adjacent to the CUT, to avoid including possible target energy in the clutter estimate (left side of Fig. 20.14). The coefficient can also be modified near the array boundaries; for example, doubling the weighting on one side when the cells on the other side of CUT do not exist. Other options would be to perhaps expand the guard cells in regions of high Doppler, when there tends to be more spreading of target return energy across multiple cells.

Filtering or weighting can also better concentrate CUT energy from immediately adjacent cells. This would be to provide better CUT estimates when the target energy is split across two or more cells (Fig. 20.14).

Multiple sliding windows can be used to expand the regions used to estimate clutter. In this case, two regions, one leading and one lagging the CUT, are used, with the results

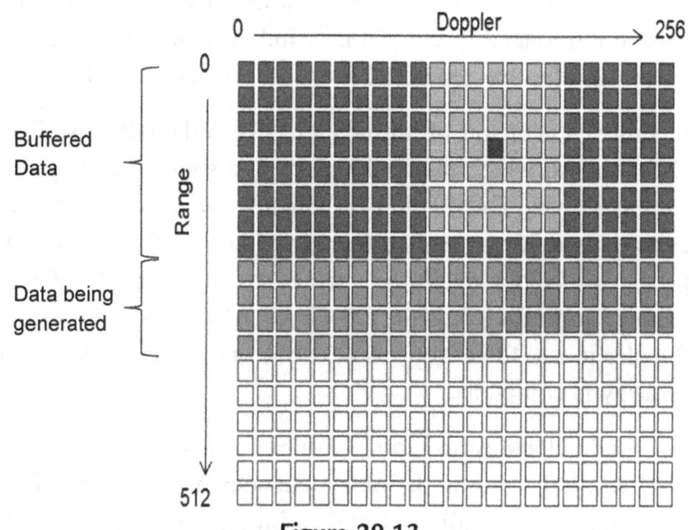

Figure 20.13
Sliding window data management.

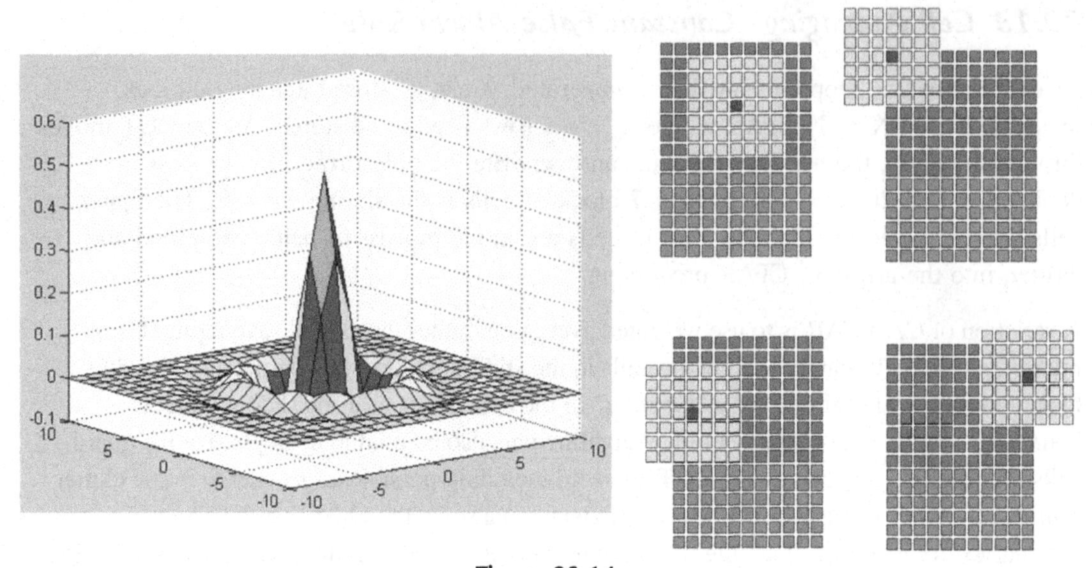

Figure 20.14
Weighted average interference estimator and boundary weighting.

averaged to provide the final clutter interference estimate. This could be computed at the same rate but requires more computational resources (Fig. 20.15).

20.14 Ordered Sort—Constant False Alarm Rate

The other popular CFAR algorithm is ordered sort (OS). This is more a logic intensive algorithm, as it requires sorting. As the N × N window moves across the array, N new magnitude^2 samples are introduced to the sliding window, and N old samples. As mentioned above, new samples are generated every 100 ns, or at a 10 MSPS rate.

Each new sample requires a new sort of all the elements in the window. This can be accomplished for N elements, which must take place in a single clock cycle. Here a window size of 7 × 7 is shown. The 49 element window size resorted seven times for the seven new samples as the window is shifted to a new CUT location. The window slides every 100 ns, so provides 10 clock cycles on average, as shown for OS-CFAR in Fig. 20.16. To maximize sort efficiency, the OS-CFAR window slides to right one sample at a time across Doppler index, then drops to next row and cycles in the opposite direction repeating until 2-D array is completed.

To build an efficient single clock cycle–sort algorithm, it is necessary to add only one new element per sort, while discarding the oldest element. When the CUT increments to a new row, a complete new window will be started, with all new cells, necessitating many clock cycles to complete the sort. A revised method for selecting the next CUT and

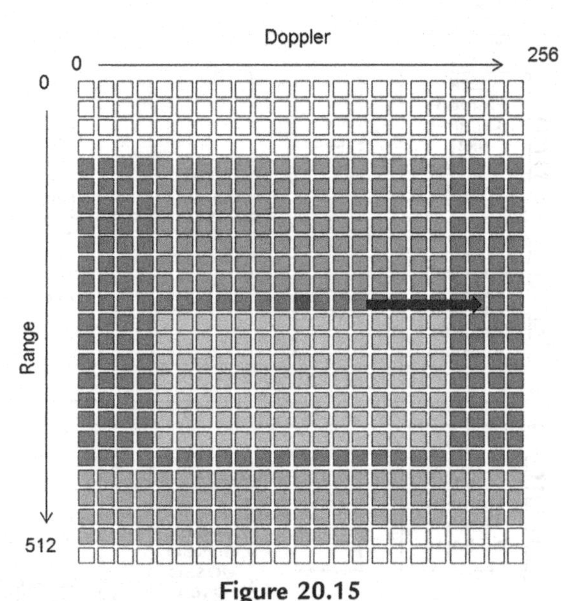

Figure 20.15
Lagging and leading cell averaging—constant false alarm rate.

therefore direction of sliding window movement is using a "zig-zag" pattern where the new row has begun just below where that last row sort was completed.

Multiple or leading and lagging sliding windows may be employed in OS-CFAR, using parallel processing circuits. This also provides opportunity for additional sorting variants. For example, the clutter estimate can be estimated by using the mean of the median magnitude^2 value of multiple sorts, or the mean of the kth-order value, or the mean of the

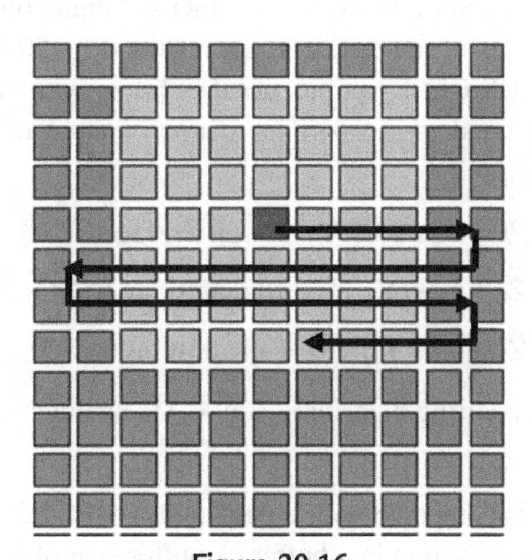

Figure 20.16
Ordered sort—constant false alarm rate processing order.

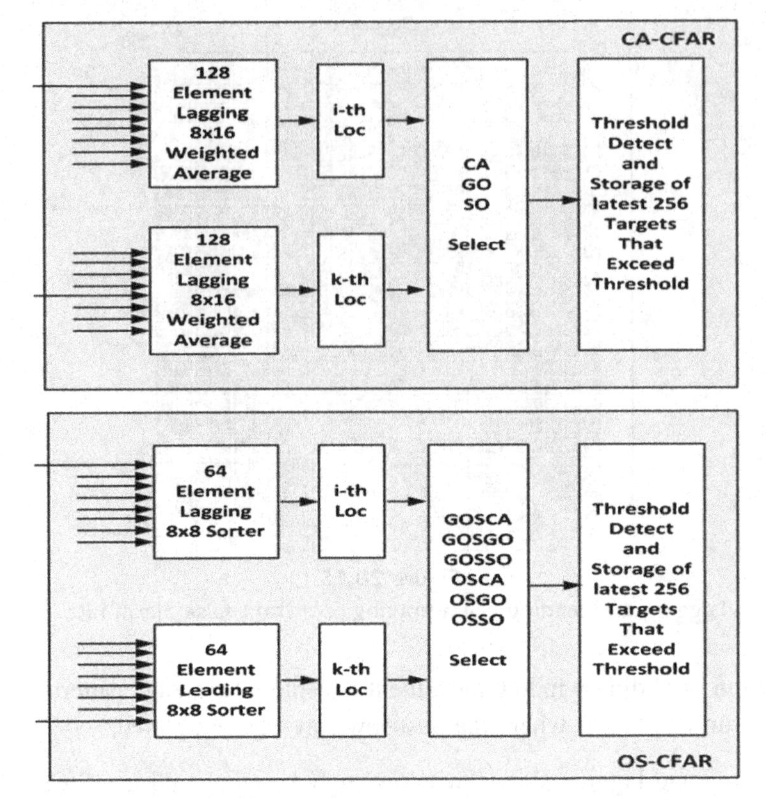

Figure 20.17
Cell averaging—and ordered sort—constant false alarm rate variations.

kth- and jth-order samples of multiple windows. Other combinations are also possible, such as taking the min or max of the different sort results (Fig. 20.17). Empirical testing is often used to optimize the OS-CFAR options. As the sliding window sorts can be completed for each sliding window in a parallel manner, many variations of CA-CFAR and OS-CFAR are possible.

$$Z = \tfrac{1}{2}(Y_1 + Y_2) \rightarrow \text{cell averaged(CA)}$$

$$Z = \max(Y_1, Y_2) \rightarrow \text{greatest order(GO)}$$

$$Z = \min(Y_1, Y_2) \rightarrow \text{smallest order(SO)}$$

Due to the logic needed for sorting algorithms, OS-CAR methods use considerably more resources than CA-CFAR.

Each of the CFAR results is compared with a threshold, and if it does, the value, as well as the range-Doppler index, is stored in a buffer. A buffer size of M depth is used. By having a host controller, adjust that depth M at the beginning of each 40 ms radar frame,

Mag	Rng	Dopp	AoA
Mag	Rng	Dopp	AoA
Mag	Rng	Dopp	AoA
Mag	Rng	Dopp	AoA
Mag	Rng	Dopp	AoA
Mag	Rng	Dopp	AoA
Mag	Rng	Dopp	AoA
Mag	Rng	Dopp	AoA
Mag	Rng	Dopp	AoA
Mag	Rng	Dopp	AoA
Mag	Rng	Dopp	AoA
Mag	Rng	Dopp	AoA
Mag	Rng	Dopp	AoA

Figure 20.18
Constant false alarm rate detection result storage.

the number of target detections can be kept to a reasonable number for further processing, as well as prevent overflows. It is also suggested by starting the CFAR processing with the longest range, and finishing with the shortest, that if any results are overwritten and lost, they will be the longer range results. (Fig. 20.18) This technique also eliminates the need for a resource-intensive bubble sort at each detection.

Once the detection results are available, the next step is to determine the AoA. This can be done by accessing the individual antenna data, which is coherent, from the external memory for the range-Doppler–indexed locations.

20.15 Angle of Arrival Estimation

The AoA can be estimated by using a complex-valued matrix of beamforming coefficients. Using the eight antenna inputs from a given range-Doppler location, the energy can be estimated from different directions and the maximum magnitude is used to estimate the AoA. A single complex multiplier can be used to perform this function across all the detection results. The $d_{0...7}$ represent the eight antenna values at specific range-Doppler locations corresponding to peaks from the detected targets. The $w_{0...7}$ represent the coefficient sets for eight angular directions. The $y_{0...7}$ represent value of the detected energy from the eight angular directions, and the direction of the largest value is of interest.

Once the AoA is determined, it can be included in the information reported to the host processor, as shown in Fig. 20.19. This is useful for "sensor fusion," where multiple sensor

$$\begin{bmatrix} y_0 \\ y_1 \\ \cdot \\ \cdot \\ y_n \end{bmatrix} = \begin{bmatrix} w0_0 & w0_1 & w0_2 & w0_3 & w0_4 & w0_5 & w0_6 & w0_7 \\ w1_0 & w1_1 & w1_2 & w1_3 & w1_4 & w1_5 & w1_6 & w1_7 \\ \cdot & \cdot & \cdot & \cdot & \cdot & \cdot & \cdot & \cdot \\ wn_0 & wn_1 & wn_2 & wn_3 & wn_4 & wn_5 & wn_6 & wn_7 \end{bmatrix} \begin{bmatrix} d_0 \\ d_1 \\ \cdot \\ \cdot \\ d_7 \end{bmatrix}$$

Figure 20.19
Angle of arrival estimation.

systems work together. For example, vision- or camera-based systems are more capable in terms of identifying objects and precisely determining azimuth angles. Radar systems are more capable of measuring range, as well as relative velocity and are not degraded in poor light or weather conditions. Having approximate AoA from the radar system can better map detected objects to those identified by the camera sensors.

Space Time Adaptive Processing (STAP) Radar

In previous chapters, the use of Doppler processing was discussed as a key method to discriminate both in distance and velocity. Discrimination in direction (or angle of arrival to the antennas) is provided by aiming the radar signal, using either traditional parabolic or more advanced electronic steering of array antennas.

Under certain conditions, other methods are required. For example, jammers are sometimes used to prevent detection by radar. Jammers often emit a powerful signal over the entire frequency range of the radar. In other cases, a moving target has such a slow motion that Doppler processing is unable to detect against stationary background clutter—such as a person or vehicle moving at walking speed. A technique called space time adaptive processing (STAP) can be used to find targets that could otherwise not be detected.

Because the jammer is transmitted continuously, its energy is present in all the range bins. And as shown in Fig. 21.1, the jammer cuts across all Doppler frequency bins, due to its wideband, noiselike nature. It does appear at a distinct angle of arrival, however. Fig. 21.1 also depicts the ground degree of clutter in a side-looking airborne radar, due to the Doppler of the ground relative to the aircraft motion. A slow-moving target return can easily blend into the background clutter.

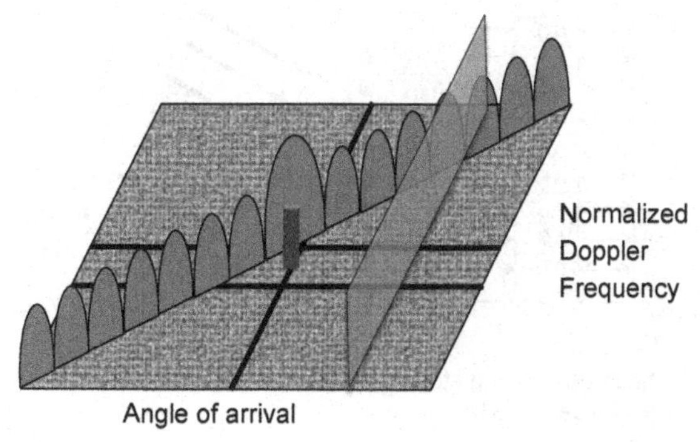

Figure 21.1
Clutter and jammer effects in Doppler space.

Digital Signal Processing 101. http://dx.doi.org/10.1016/B978-0-12-811453-7.00021-4

STAP radar could provide benefits in automotive applications. Metallic vehicles provide strong reflections of the radar pulses. Doppler processing allows for the radar to easily distinguish the radar returns from moving vehicles compared to the background clutter. One issue with traditional radar systems is that desensitization by interfering signals, perhaps even from another automotive radar system. Detecting nonmetallic and slow moving subjects such as pedestrians, bicyclists, and animals against stationary background clutter is difficult for traditional radar systems, and normally camera-based systems are used for this purpose. However, a camera-based system can be impaired by low-light conditions, weather, or any other circumstances that cause poor visibility. A possible solution to all of these issues is to integrate STAP into the radar system.

Linear algebra and matrix operations are used in STAP. For those unfamiliar with this, please refer Chapter 13 on matrix operations and inversion.

21.1 Space Time Adaptive Processing Radar Concept

STAP radar processing is combining temporal and spatial filtering that can be used to both null jammers and detect slow-moving targets. It requires very high numerical processing rates, as well as low latency processing, with dynamic range requirements that generally require floating-point numerical representation.

STAP processing requires use of an array antenna. However, in contrast to the active electronically scanned array (AESA), for STAP the antenna receive pattern is not electronically steered as with traditional beamforming arrays. In this case, the array antenna provides the raw data to the STAP radar processor, whereas the antenna processor does not perform the beam steering, phase rotation, or combining steps, as indicated in Fig. 21.2. The directional processing is done in a later stage as part of the STAP algorithm. Also, while the AESA is depicted in one dimension, this array can and often is

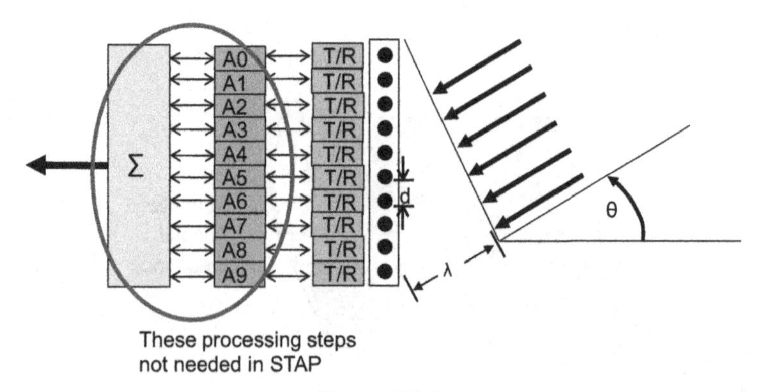

Figure 21.2
Antenna array allowing directional processing. *STAP*; space time adaptive processing.

two dimensional, in both elevation (up and down) and azimuth (side to side). In this way, the antenna receive pattern can be steered or aimed in both elevation and azimuth.

The radar processing can occur over N consecutive pulses, as long as they lie within the coherent processing interval, considered "slow" time. In the case of the automotive radar system considered here, N = 256. Performing STAP processing over all 256 pulses simultaneously would result in hundreds of TFLOPs of processing power. A common technique is to break the radar data cube into smaller sections, perform STAP on them individually, and then integrate the detection results from the separate computations.

In this case, 16 pulses will be used for STAP to bring the processing requirement to a reasonable level, and this will divide the radar data cube into 16 sections (as there are 256 pulses in the radar frame). To ensure the maximum Doppler sensitivity, the selected pulses will not be contiguous but rather will be data from every 16th pulse. Then the first STAP dataset would be over pulses {0, 16, 32, 48, 64, 80, 96, 112, 128, 144, 160, 176, 192, 208, 224, 240}. Then a second STAP dataset would be performed over pulses {1, 17, 33, 49, 65, 81, 97, 113, 129, 145, 161, 177, 193, 209, 225, 241}. Using datasets containing pulses spread over a longer interval (14 ms) means there will be more relative moment resulting in greater Doppler shift, which should provide better detection.

With airborne radar trying to detect slow-moving targets on ground, the range to the suspected targets of interest to perform STAP is known. This is not the case for automotive radar. Therefore, it will be suggested that the processing be performed at ranges of 20, 40, 60, and 80 m, which should allow timing detection of pedestrians, animals, or cyclist. However, detailed simulations and field testing will be required to fully optimize these radar processing configurations.

The L = 512 radar samples collected during the pulse interval are binned, and after fast Fourier transform (FFT) processing, corresponds to the range. The range sampling rate is referred to as "fast" time in radar jargon, whereas processing across the pulses is referred as "slow" time.

STAP radar will operate on the postrange-FFT processed cube of data, shown in Fig. 21.3. The M dimension corresponds to the number of inputs from the array antenna. The resulting radar data cube will be of dimensions M (number of array antenna inputs) by L (number of range bins in fast time) by N (number of pulses in slow time). Doppler processing, which is not part of STAP, occurs over the data slice across the L and N dimensions. In STAP, arrays of data (or slices) across the M and N dimensions are processed.

STAP is basically an adaptive filter that can filter over the spatial and temporal (or time) domain. The goal of STAP is to take a hypothesis that there is a target at a given location and velocity and create a filter that has high gain for that specific location and velocity

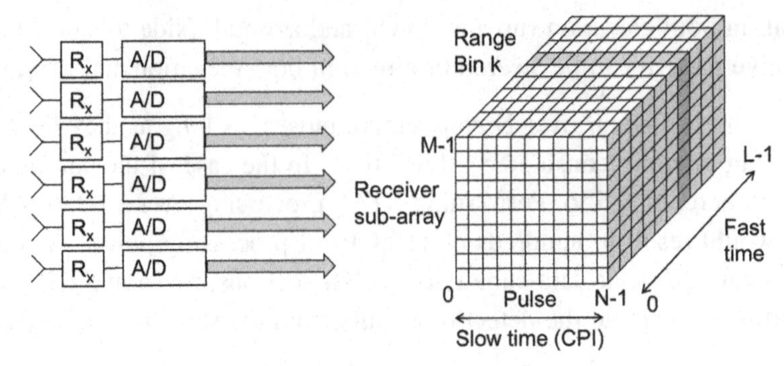

Figure 21.3
Radar data cube used in Space Time Adaptive Processing radar.

and proportional attenuation of all signals (clutters, interferers, and any other unwanted returns). There can be many returns of interest to generate location and velocity hypothesis for, and these are all normally processed together in real time. This produces very high processing and throughput requirements on the STAP radar processor.

This brings up the issue of how the spatian and Doppler hypotheses for the weak returns are identified for subsequent STAP processing. This can come from weak detections found in normal pulse-Doppler processing or information from other sensor systems; in an auto radar, from ADAS camera-based sensors or IR sensors on the vehicle.

STAP has the capability to pull targets that are below the clutter into a range that can be reliably detected. A good analogy is a magnifying glass. Conventional methods are used to view the big picture, but if something of interest is noted, STAP can be used to act as a magnifying glass to zoom into a specific area and see things that would be otherwise undetectable.

21.2 Steering Vector

For each suspected target, a target steering vector must be computed. This target steering vector is formed by the cross product of the vector representing the Doppler frequency and the vector representing the antenna angle of elevation and azimuth. For simplicity, we will assume only azimuth angles are used.

The Doppler frequency offset vector is a complex phase rotation:

$$\mathbf{F_d} = e^{-2\pi \cdot n \cdot F_{dopp}} \quad \text{for } n = 1 \ldots N-1$$

The spatial angle vector is also a phase rotation vector:

$$\mathbf{A_\theta} = e^{-2\pi d \cdot m \cdot \sin(\theta/\lambda)} \quad \text{for } m = 1 \ldots M-1, \text{ for given angle of arrival } \theta \text{ and wavelength } \lambda$$

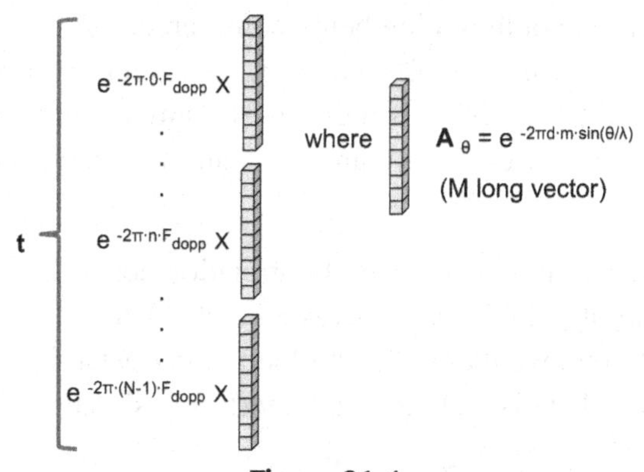

Figure 21.4
Target Steering Vector $\mathbf{t} = f$ (Angle, Doppler).

The target steering vector **t** is the cross product vector $\mathbf{F_d}$ and $\mathbf{A_\theta}$ as shown in Fig. 21.4, and **t** is vector of length N·M. This must be computed for every target of interest. Assuming 12 antennas, and that every 16th of the 256 pulse data is used, the vector **t** is $16 \times 12 = 192$ complex samples long.

21.3 Interference Covariance Matrix

The interference covariance matrix $\mathbf{S_I}$ must be estimated. A common method is to just compute and average this for many range bins surrounding the range of interest. To compute this, a column vector **y** is built from a slice of the radar data cube at a given range bin k. The covariance matrix by definition will be the vector cross product:

$$\mathbf{S_I} = \mathbf{y}^* \cdot \mathbf{y}^T$$

Here the vector **y** is conjugated and then multiplied by its transpose. As **y** is of length N·M, the covariance matrix $\mathbf{S_I}$ is of size [(N·M) × (N·M)] as shown in Fig. 21.5. This is

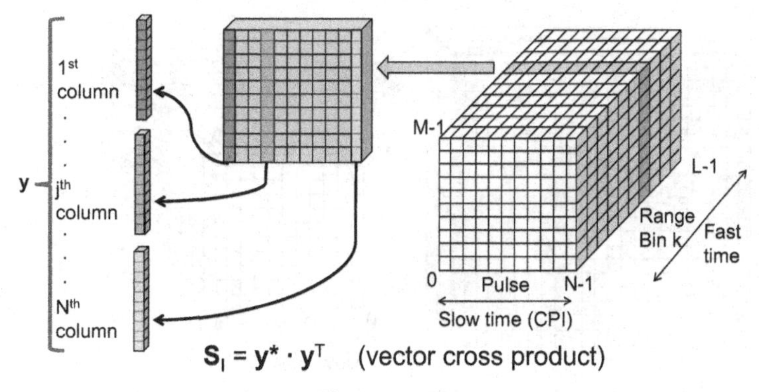

Figure 21.5
Computing interference covariance matrix.

a matrix of [192,192] for the configuration being considered. All the data and computations are being done with complex numbers, representing both magnitude and phase. An important characteristic of $\mathbf{S_I}$ is that it is Hermitian, which means that $\mathbf{S_I} = \mathbf{S_I^{*T}}$ or equal to its conjugate transpose. This symmetry is an inherent property of covariance matrices.

The covariance matrix represents the degree of correlation across both antenna array inputs and over the group of pulses. The process is to characterize undesired signals, and create an optimal filter to remove them, thereby facilitating detection of the target. The undesired signals can include noise, clutter, and interfering signals.

$$\mathbf{S}_{\text{Interference}} = \mathbf{S}_{\text{noise}} + \mathbf{S}_{\text{interfering signal}} + \mathbf{S}_{\text{clutter}}$$

The covariance matrix is very difficult to calculate or model; therefore, it is estimated. Since the covariance matrix will be used to compute the optimal filter, it should not contain the target data. Therefore, it is not computed using the range data right where the target is expected to be located. Rather, it uses an average of the covariance matrices at many range bins surrounding but not at the target location range as shown in Fig. 21.6. This average is an element by element average for each entry in the covariance matrix, across these ranges. This also means that many covariance matrices need to be computed from the radar data cube. The assumption is that the clutter and other unwanted signals are highly correlated to that at the target range, if the difference in range is reasonably small.

The estimated covariance matrix can used to build the optimal filter that involves inversion of the covariance matrix, which is computationally very expensive, and generally requires the dynamic range of floating-point numerical representation. The matrix is of size $[(N \cdot M) \times (N \cdot M)]$ and can be quite large.

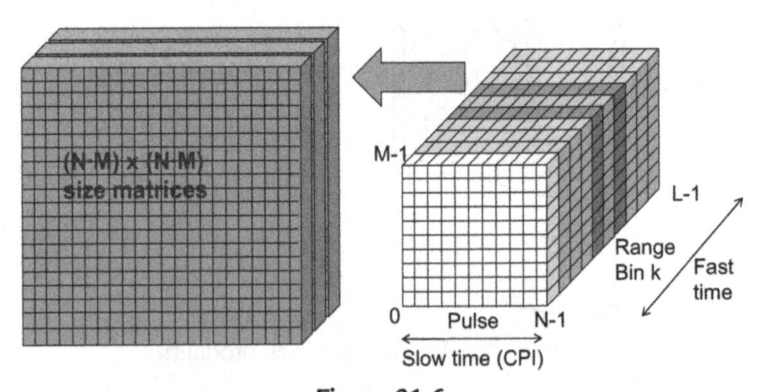

Figure 21.6
Estimating covariance matrix using neighboring range bin data.

21.4 Space Time Adaptive Processing Optimal Filter

Fortunately, the matrix inversion result can be used with multiple targets at the same range. The steps are as follows:

$$\mathbf{S_I}\cdot\mathbf{u} = \mathbf{t}^*,\ \text{or}\ \mathbf{u} = \mathbf{S_I}^{-1}\cdot\mathbf{t}^*$$

One method for solving for $\mathbf{S_I}$ is known as QR decomposition, which will be used here. Another popular method is the Cholesky decomposition, as the interference covariance matrix is Hermitian symmetric.

Perform the substitution $\mathbf{S_I} = \mathbf{Q}\cdot\mathbf{R}$, or product of two matrices.

Q and **R** can be computed from $\mathbf{S_I}$ using one of several methods, such as Gram–Schmidt, Householder transformation, or Givens rotation. The nature of the decomposition into two matrices is that **R** will turn out to be an upper triangular matrix, and **Q** will be an orthonormal matrix, or a matrix composed of orthogonal vectors of unity length. Orthnonormal matrices have the key property as follows:

$$\mathbf{Q}\cdot\mathbf{Q}^{\mathrm{H}} = \mathbf{I}\ \text{or}\ \mathbf{Q}^{-1} = \mathbf{Q}^{\mathrm{H}}$$

Therefore it is trivial to invert **Q**. Please refer to the chapter on matrix inversion (Chapter 13) for more detail on QR decomposition.

$$\mathbf{Q}\cdot\mathbf{R}\cdot\mathbf{u} = \mathbf{t}^*\text{then multiply both sides by } \mathbf{Q}^{\mathrm{H}}$$
$$\mathbf{R}\cdot\mathbf{u} = \mathbf{Q}^{\mathrm{H}}\cdot\ \mathbf{t}^*$$

Since **R** is an upper triangular matrix, **u** can be solved by a process known as "back substitution." This started with the bottom row that has one nonzero element and solving for the bottom element in **u**. This result can be back-substituted for the second to bottom row with two nonzero elements in the **R** matrix, and the second to bottom element of **u** solved for. This continues until the vector **u** is completely solved. Notice that since the steering vector **t** is unique for each target, the back-substitution computation must be done for each steering vector.

Then solve for the actual weighting vector **h**.

$\mathbf{h} = \mathbf{u}/(\mathbf{t}^{\mathrm{H}}\cdot\mathbf{u}^*)$, where dot product $(\mathbf{t}^{\mathrm{H}}\cdot\mathbf{u}^*)$ is a weighting factor (this is a complex scaler, not vector).

Finally solve for the final detection result z by the dot product of **h** and the vector **y** from the range bin of interest.

$$z = \mathbf{h}^{\mathrm{T}}\cdot\mathbf{y}$$

z is a complex scaler, which is then fed into the detection threshold process. Over the 16 STAP processes, the values can be integrated for each of the range and steering vector locations.

Shown in Fig. 21.7 is a plot of S_I^{-1}, the inverted covariance matrix. In this case, there is an interfering signal at 60 degrees azimuth angle, and a target of interest at 45 degrees, with range of 1723 m normalized Doppler of 0.11. Note that this is for an airborne radar scanning the ground searching for slow-moving targets, hence the longer range.

Notice the very small values, in the order of −80 dB, present at 60 degrees. The STAP filtering process is detecting the correlation associated with the interfering signal direction at 60 degrees. But inverting the covariance matrix, this jammer will be severely attenuated. Notice also the diagonal clutter line. This is a side-looking airborne radar, so the ground clutter has positive Doppler looking in the forward direction or angle and negative Doppler in the backward direction or angle. This ground clutter is being attenuated at about −30 dB, proportionally less severely than the more powerful interfering signal.

The target is not present in this plot. Recall that the estimated covariance matrix is determined in range bins surrounding but not at the expected range of the target. But in any case, it would not likely be visible anyway. However, the use of STAP with the appropriate target steering vector can make a dramatic difference, as shown in Fig. 8. The top plot shows the high return of the peak ground clutter at range of 1000 m with magnitude of ~0.01, and noise floor of about ~0.0005.

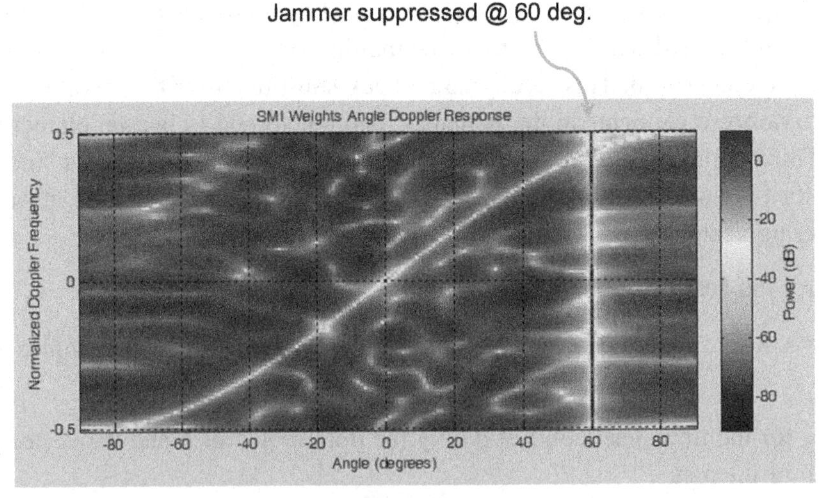

Figure 21.7
Logarithmic plot of inverted covariance matrix.

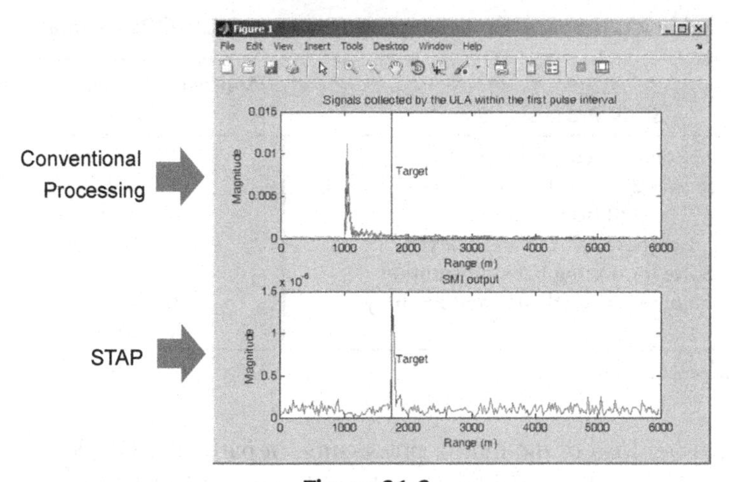

Figure 21.8
Space time adaptive processing (STAP) gain.

With STAP processing, the noise floor is pushed down to $\sim 0.1 \times 10^{-6}$ and the target signal at about 1.5×10^{-6} is now easily detected. It is also clear that floating-point numerical representation and processing will be needed for adequate performance of the STAP algorithm.

The STAP method described is known as the power domain method. It is the called power domain because the covariance matrix estimation results in squaring of the radar data, hence, the algorithm is operating on signal power. This also increases the dynamic range of the algorithm, but this is easily managed as floating point is being used for all processing steps.

21.5 Space Time Adaptive Processing Radar Computational Requirements

It is necessary to consider the processing requirements as shown in Table 21.1. Using the following assumptions:

Pulse repetition frequency (PRF) = 1000 Hz
12 antenna array inputs ($\mathbf{A}_\theta$ vectors are of length 12 or M = 12)
16 pulse processing (Doppler vectors are of length 16 or N = 16)
Minimum required size of $\mathbf{S}_\mathrm{I}$ is 192 × 192, in complex single precision format
Assume 32 likely targets to process (32 target steering vectors)
Use of 200 range bins to estimate $\mathbf{S}_\mathrm{I}$

This actually is a very conservative scenario. The PRF is rather low, and the number of antenna array inputs is very small. Should the number of antenna array inputs increase by

Table 21.1: Space Time Adaptive Processing GFLOPs Estimate

STAP Processing Steps	Approximate Computational Load
Compute covariance matrices	23 GFLOPs
Average over 200 matrices to find estimated $\mathbf{S}_I$	8.0 GFLOPs
QR decomposition	37.7 GFLOPs
Compute $\mathbf{Q}^H \cdot \mathbf{t}^*$ for 32 targets	9.4 GFLOPs
Solve for $\mathbf{u}$ using back substitution	4.7 GFLOPs
Compute $\mathbf{h} = \mathbf{u}/(\mathbf{t}^H \cdot \mathbf{u}^*)$ and $z = \mathbf{h}^T \cdot \mathbf{y}$	0.2 GFLOPs
Total	83 GFLOPs

STAP, space time adaptive processing

12–48, the processing load of the matrix processing, in particular QR decomposition goes up by the third power or 64 times. This would require over 3 TFLOPS of floating-point processing power. Because of this the limitations on STAP are clearly the processing capabilities of the radar system.

The theory of STAP has been known for a long time, but the processing requirements have made it impractical until fairly recently. Many radar applications benefiting from STAP are airborne and often have stringent size, weight, and power (SWaP) constraints. Very few processing architectures can meet the throughput requirements of STAP, and even fewer can simultaneously meet the SWaP constraints. As most radars operate in a continuous manner, receiving an enormous amount of data, this data normally must be processed in real time using powerful compute devices or accelerators. These are commonly graphics processing units or field programmable gate arrays in many radar systems.

CHAPTER 22

Synthetic Array Radar

22.1 Introduction

Synthetic array radar, or "SAR," is normally used to map ground features and terrain. It is also known in literature as synthetic aperture radar. Both names make sense, though we use "Synthetic Array Radar" here. It is used for a wide variety of military and commercial applications. It can be made to map almost arbitrarily fine-resolution ground features or used to more coarsely map larger areas with comparative effort.

This process produces maps, which are often color coded. The color does not represent the actual color of the landscape but is used to indicate the strength of return signal for each resolvable location on the ground. Alternately, the images can be gray scale, with light regions indicating strong return and dark regions indicating little or no return. As different terrain features will reflect radar in differing amounts, features such as buildings, planes, rivers, roads, railroads, and so on can be seen in the SAR images. SAR radar is effective at night and in all weather, so if an effective complement to camera based imaging.

22.2 Synthetic Array Radar Resolution

The key parameter in ground mapping is the resolution. SAR systems can be designed with capabilities to differentiate using dimensions from few centimeters to a hundreds of meters, depending on if the purpose is to map out a military installation, an urban area, or the contours of a mountain range. The range is basically limited by the transmit power of the radar and can operate at resolutions much greater than visual, at long ranges and is unaffected by darkness, haze, or other factors impacting visual detection.

As with video, the quality of images depend on the pixel density (pixel stands for "picture element"). The equivalent of pixel density in radar is a voxel or "volume element." The voxel is defined by the azimuth, elevation, and range. The minimum voxel size is dependent on the radar resolution capabilities. The voxel spacing is basically the distance that two points on the ground can be distinguished from each other. Radar resolution capabilities, in turn, are dependent on range resolution and main lobe beam width capabilities.

The voxel spacing or density should in general be at least 10 times the dimensions of the objects being mapped, to achieve useful images. A 1-m resolution is feasible for detecting buildings that are at least 10 m long and wide.

Digital Signal Processing 101. http://dx.doi.org/10.1016/B978-0-12-811453-7.00022-6

As precision range detection is a fundamental requirement, high pulse repetition frequency (PRF) operation is unsuitable for SAR due to the range ambiguities. Low PRFs are used instead to eliminate range ambiguities over the distances from the aircraft to the ground being mapped. Maximum Doppler rates tend to be low, as the only motion is due to the radar-bearing aircraft motion. Owing to the nature of SAR, the relative motion is often substantially less than the aircraft flight speed. Use of a low PRF, while restricting the usable Doppler range, enhances the precision of Doppler frequency detection within that restricted range. This is an advantage in high-resolution SAR mapping.

22.3 Pulse Compression

Range resolution is dependent on the precision of the receive pulse detection arrival delay. This can be achieved by a very short transmit pulse width, which has the disadvantage of low transmit power level due to the short duration. Or very high levels of pulse compression can be used, which allow relatively long transmit pulses and therefore long integration times at the receiver, with the receiver operating on higher power returns. This raises the signal-to-noise power ratio (SNR) and allows for longer range mapping. A high level of pulse compression can be achieved by using long-matched filters (correlation to the complex conjugate of transmit sequence) and transmit sequences with strong autocorrelation properties. The only consequence is a higher level of computations associated with the long-matched filter. The speed of light, and therefore radar waves, is about 1 m per 3 ns (3×10^{-9}). Since the path is round-trip, the range appears to become half of this. So for about 2 m, or ~6 ft range resolution, requires a 12 ns timing detection precision. To achieve this level of correlation would require a transmit sequence with phase changes of at least 80 MHz rate, resulting at a minimum of the same amount transmit frequency bandwidth within the 10 GHz radar band.

The elevation of the antenna main lobe does not need to be narrowly focused. In an SAR radar system, the antenna is directed to the ground at an angle off to the side, as shown in Fig. 22.1. As the elevation angle decreases, the radar beam will be directed at a steeper angle a ground location closer to the flight path of the aircraft, with a shorter range. The different portions of the beam elevation will therefore map to different ranges, and the return sequence can be directed into different range bins. The precision of the range detection capability translates into the degree of elevation resolution, often utilizing pulse compression.

22.4 Azimuth Resolution

The other requirement for precise ground mapping is a very narrow angular resolution of the main beam in the azimuth. As discussed in a previous chapter, the narrowness of the radar beam depends on the ratio of the antenna size to the wavelength. To achieve a "pencil"-like radar beam will require either a very large antenna or very high frequency (and short wavelength) radar. Both are impractical for airborne radar. The antenna size is

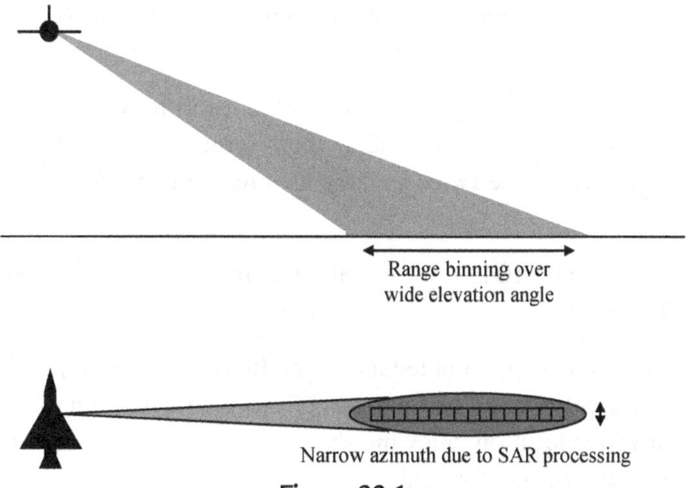

Figure 22.1
Synthetic array radar (SAR) range binning.

necessarily limited by the aircraft size. Extremely high-frequency radars tend to be useful only at very short range, due to both atmospheric absorption and scattering. There is also the practicality of building high power and very high-frequency transmit circuits.

The solution to this is to create an artificially large antenna or synthetic array antenna. The forward motion of the aircraft is used to transmit and receive from many different points along the flight path of the aircraft. By focusing the radar main beam the same area of ground during the aircraft motion, the returns from different angles created by the aircraft motion can be synthesized into a very narrow equivalent azimuth main lobe using signal-processing techniques. The end result is as if an antenna of great length (up to a kilometer) was used. Because this is done using radar returns over several hundred milliseconds, this technique works for stationary targets, so is ideal for ground mapping.

SAR radar typically directs a radar beam at 90 degrees to the plane's flight path. The width of this radar beam does not have to be exceptionally narrow—in fact, the wider beam covers more ground and allows more processing gain in the SAR algorithm. When a large angle main lobe is used, the maximum length of the synthetic antenna is increased. Therefore, small antennas can work well with the SAR technique, as long as the antenna gain is sufficient to meet the SNR requirements for the range involved. The antenna will illuminate a large swath on the ground, typically an oval shape due to the aspect ratio of the beam being aimed outward from the aircraft flight path at downward angle.

To start with, let us assume that we can build an antenna as large as necessary to meet our azimuth resolution requirement. The rule of thumb governing antenna size is

$$d_{azimuth} \approx \lambda R/L$$

where, $d_{azimuth}$ = resolvable distance in the azimuth direction, λ = wavelength of radar, R = range, L = length of the antenna.

(Note, for reasons not explained here, this expression is valid for conventional antennas. A SAR antenna actually has half the resolvable azimuth limit as a real antenna. In other words, a SAR antenna needs to be twice as large as a real antenna for the same resolvable distance).

If we need a 1-m aperture at a 10 km range, with a 3 cm (X band) radar, this requires an antenna length of 300 m.

Imagine we had such an antenna, mounted along the fuselage of an impractically long 300 m long plane. Each pulse could be focused with an azimuth width of 1 m at the 10-km range, with a wide elevation, allowing the radar to scan narrow (1 m) strip of land for each PRF, as the plane travels forward.

This 300-m long antenna could be composed of many radiating elements along this length (for example, 301 separate elements, spaced every meter). The antenna steering is accomplished by setting the phase of each element to ensure that the radar wave transmitted from each element is at a common phase when arriving at the 1-m strip at 10-km distance. The phases will have to be adjusted, or focused, as the distance to the 1-m strip of land will be slightly different for each element, due to the offset relative from the center element.

To aim the antenna beam at a very narrow region, the phase relationship of the different antenna elements must be carefully controlled. In our example, the wavelength is 3 cm. If a radar round-trip path is 1.5 cm longer or shorter than the middle element path, it will be 180 degrees out of phase and add destructively or cancel. Therefore the round-trip path length must be controlled within a few millimeters for each element. The phase error that occurs due to the plane's straight flight path as compared to an arc must be compensated for. This is shown below. The phase correction relative to the middle element of the antenna works is approximately as follows:

$$\Theta_n = \left(2\pi \cdot d_n^2\right) / (\lambda \cdot R)$$

where, Θ_n = phase error of nth antenna element (in radians), d_n = distance between middle element and nth antenna element. (In SAR literature, this term Θ_n is sometimes called the "point target phase history.")

The return echo would be reflected from the ground at all locations and travel back to each element. Owing to the reciprocal path, it would arrive at the same phase offset that was transmitted, and if the same phase compensation is performed on the receive element signal prior to being summed together, the result will be that only the reflections from the

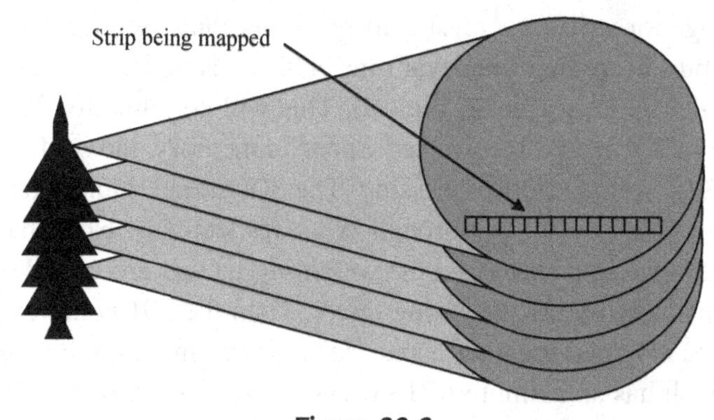

Figure 22.2
Synthetic array radar pulse repition frequency effect.

1-m width azimuth portion will arrive in phase, with all other ground returns being canceled out or at least severely attenuated.

Now suppose the same process is done in sequence, rather than all at once. We start with an element at the one end of the antenna and transmit a pulse and receive the return, using only this element. All the other elements are inactive. Both transmit and receive signals are modified by the phase compensation as before. The return sequence is stored in memory. Then we repeat this process with each separate antenna element in turn, until we have saved all the 301 return sequences. Remember, these return sequences are complex numbers, with a magnitude and phase. Now if we sum all the complex results at the end, we must have the same result as if we did everything in parallel at once. Nothing else has changed—the situation on the ground is assumed static. This is a simplified version of the process the SAR radar performs. Imagine as the plane flies forward, the PRF is such that 301 pulses are transmitted and received along 301 m of flight path, as shown in Fig. 22.2.

The radar could then effectively map the 1-m wide strip at right angles to the flight path. However, while this solves the azimuth resolution problem, this is still not workable because only 1-m wide of strip ground is mapped perpendicular to the plane's flight path, every 301 m.

22.5 Synthetic Array Radar Processing

To go further, we need some conventions. Let us assign an index to each 1-m strip of land, oriented at right angles to the flight path, designated "n." At the PRF when the aircraft is physically aligned with $strip_n$, the real antenna will receive a complex range $sequence_n$. This same index applies to the virtual or synthetic antenna element that is directly perpendicular to the 1-m strip. The next synthetic antenna element forward would be index $n + 1$, continuing on up to $n + 150$. The synthetic antenna element behind would be $n - 1$, extending to $n - 150$. We will have complex weighting factors of proper phase and amplitude for each index, W_{-150} through W_{150}. W_0 is always equal to 1.

To calculate the image for $strip_n$, we must start receiving range sequences at index n−150 and sort into range bins according to arrival time. This is using the single real antenna with wide beam angle (6−12 degrees is typical). This will continue for 300 more PRF intervals, and result in 301 stored receive sequences in memory, indexed from −150 to +150. After PRF_{150}, we can start processing. The 301 stored, binned range sequences will be multiplied or scaled by W_{-150} through W_{150}, respectively. Note that both the values in the binned range sequences and the weighting factors are complex. Each range bin is multiplied separately by the weighting factor. Then the 301 complex range sequence results can then all be summed together across each range bin and will result in a range binned sequence which has an azimuth of 1 m. The range bins correspond to the individual values across the elevation of the 1-m $strip_n$.

$$
\begin{aligned}
& \text{binned range sequence}_{n-150} \cdot W_{-150} \\
+ \quad & \text{binned range sequence}_{n-149} \cdot W_{-149} \\
+ \quad & \text{binned range sequence}_{n-148} \cdot W_{-148} \\
\ldots \quad & \\
+ \quad & \text{binned range sequence}_{n+0} \cdot W_{0} \\
+ \quad & \text{binned range sequence}_{n+1} \cdot W_{+1} \\
\ldots \quad & \\
+ \quad & \text{binned range sequence}_{n+150} \cdot W_{+150} \\
\hline
= \quad & \text{binned range strip}_{n} \text{ (1 meter azimuth)}
\end{aligned}
$$

For $strip_{n+1}$, we wait until PRF_{151}, and the plane has advanced 1 m on its flight path. We can now start processing again. We have now just saved binned range $sequence_{151}$ and can discard binned range $sequence_{-150}$. This updated set of 301 binned range sequences is again multiplied by the weighting factors W_{-150} through W_{150}. In this case, there is an offset as follows:

$$
\begin{aligned}
& \text{binned range sequence}_{n-149} \cdot W_{-150} \\
+ \quad & \text{binned range sequence}_{n-148} \cdot W_{-149} \\
+ \quad & \text{binned range sequence}_{n-147} \cdot W_{-148} \\
\ldots \quad & \\
+ \quad & \text{binned range sequence}_{n+1} \cdot W_{0} \\
+ \quad & \text{binned range sequence}_{n+2} \cdot W_{+1} \\
\ldots \quad & \\
+ \quad & \text{binned range sequence}_{n+151} \cdot W_{+150} \\
\hline
= \quad & \text{binned range strip}_{n+1} \text{ (1 meter azimuth)}
\end{aligned}
$$

In this manner, we can compute each of the strips, one after the other, using a single broad beam antenna but using a long synthetic array to achieve a narrow azimuth. We can make the synthetic antenna arbitrarily long, by using more PRF cycles, more memory, and higher processing rates.

The signal processing achieves the same cancellation of signals coming from azimuths outside out desired 1-m ground strip as an actual 300-m antenna would do. This processing technique is known as line-by-line processing.

22.6 Synthetic Array Radar Doppler Processing

Another alternative method to perform SAR processing is to incorporate Doppler processing. Owing to the use of the efficient fast Fourier transforms (FFT) algorithm, this will lead to a much lower level of computations than line-by-line processing.

Consider the patch of ground being illuminated by the radar pulse directed at right angles to the aircraft flight path. This patch may be 2000 m or more wide (azimuth direction), depending on the range and antenna beam width azimuth. At the midpoint, the Doppler frequency is exactly zero, as the radar is moving parallel to this point and has no relative motion. At the two azimuth end points of the scan area, we will have as follows:

$$\text{positive Doppler frequency} = \sin(\text{azimuth angle})\cdot(\text{aircraft velocity})/(\text{wavelength})$$
$$\text{negative Doppler frequency} = -\sin(\text{azimuth angle})\cdot(\text{aircraft velocity})/(\text{wavelength})$$

As an example, with a range of 10 km and a ground scan area of ± 1000 m, this equates to an angle of ± 5.71 degree. If the aircraft is flying at 250 m/s, this works out to ± 829 Hz in for the 10 GHz radar band. This is shown in Fig. 22.3.

The Doppler frequency variation is not linear across the azimuth, due to the "sin (azimuth angle)" in the equation. At small angles, $\sin(\theta) \approx \theta$ or approximately linear. As the angle increases, the effect becomes more nonlinear, until at 90degrees, the Doppler frequency

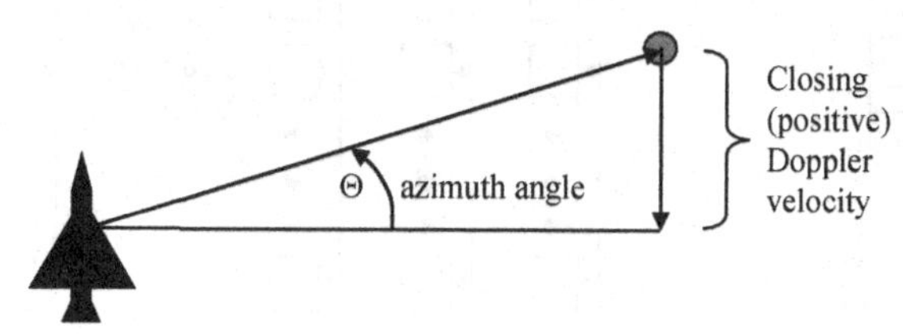

Figure 22.3
Synthetic array radar Doppler effect.

asymptotically approaches the familiar (aircraft velocity/wavelength) or 8333 Hz. However, we want the Doppler frequency response to be completely linear across the azimuth range. This can be compensated by using phase correction multiplier, known as "focusing." The purpose is to make the Doppler frequency variation linear across the azimuth angle, rather than proportional to the sine of the angle. Once the frequency spacing per unit length on the ground is made linear, it allows us to use Doppler filters with equally spaced main lobes along the frequency axis. This filtering is the familiar discrete fourier transform, which can be implemented using the FFT algorithm. This is known as SAR Doppler processing. The advantage of this is that the computational load is made much more manageable than the line-by-line processing technique, by virtue of the FFT algorithm efficiency.

As a side note, this Doppler linearity is an issue only for SAR radar. For conventional radar, the radar is aimed toward the horizon, and there is less variation due to the aspect angles (in this case, $\theta_{elavation}$ is close to 0°, although $\theta_{azimuth}$ can have significant variation), and the sensitivity requirements are much less for SAR.

At each PRF, the return sequence is multiplied by a phase correction (focusing). Each range bin stores a complex value, representing the phase and magnitude of the return at that range. For each range, the values are loaded into all the Doppler frequency filters matching the azimuth angles for each ground element, as shown in Fig. 22.4.

Each pulse has its return processed by azimuth and range, which allows separation over all locations in the radar beam, with resolution determined by the number of range bin and Doppler filter frequency banks. This is repeated at each PRF with the

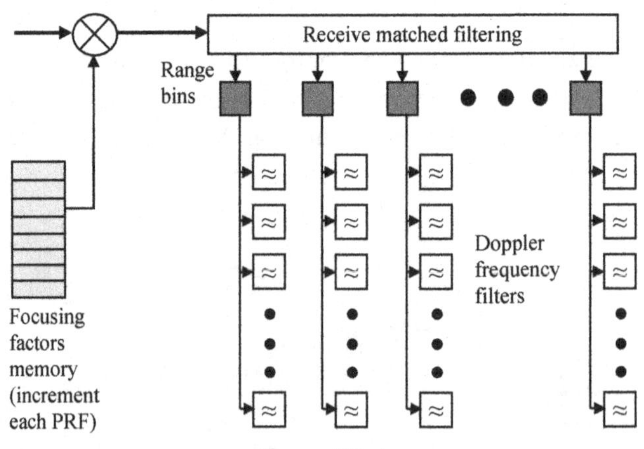

Figure 22.4
Synthetic array radar processing diagram. *PRF*, pulse repetition frequency.

next phase correction value, and the results are accumulated or integrated. Over the set of N PRFs equal to the number of N elements in the synthetic antenna, this is repeated.

After each set of N PRFs, the process repeats. The same point is measured N times, and complex values representing both magnitude and phase are integrated over the measurements for each point. In this architecture, the number of virtual elements in the synthetic array is equal to the number of Doppler filters, which can also be set equal to the number of range bins. This is also the number of times each point is measured, and the results integrated. However, each of the different measurements for a given point is done at a different azimuth angle.

In reality, these two methods provide equivalent results, although the processing steps are different. The first method is conceptually easier for most people to understand. The second method has the advantage of lower computational rate. An intuitive Fig. 22.5 that depicts the two different approaches is shown.

An alternative way to understand this is that line by line, a new narrow beam at right angles to the flight line is synthetically created for each PRF. With Doppler processing, many different azimuth beams are generated by each Doppler frequency bank during each PRF, and the returns from each beam are summed over multiple PRFs. Mathematically, a process called "back-projection" is used to create the SAR image, but this is not covered in this introductory treatment.

22.7 Synthetic Array Radar Impairments

Several factors can degrade SAR performance. One of the most significant is nonlinear flight path of an aircraft. We have seen how sensitive the phase alignments are to proper focusing, in fractions of the radar wavelength. Therefore, deviations in flight path away

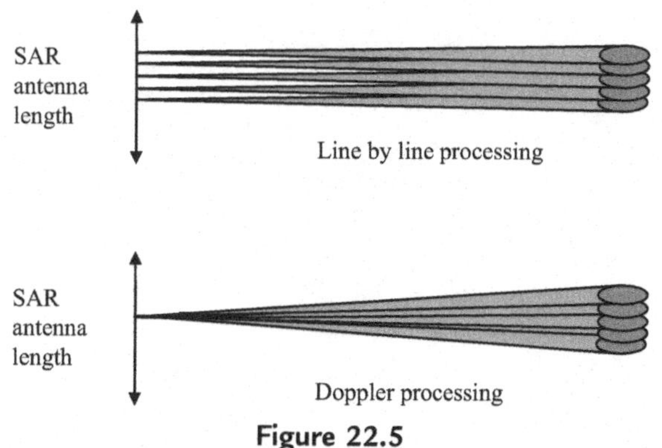

Figure 22.5
Synthetic array radar (SAR) integration over pulses.

from the parallel line of the radar scan path must be determined and accounted for. This motion compensation can be done using inertial navigation equipment and by using GPS location and elevation measurements. Another consideration is side lobe return. When the side lobe return from the ground beneath the plane is integrated over a wide azimuth and elevation angles, this can become significant despite the low antenna gain at the side lobes. The design of the synthetic antenna, just like a real antenna, must take this factor into account. There are methods, similar to windowing in finite impulse response filters, which can reduce side lobes, but at the expense of widening the main lobe and degrading resolution. Another issue is that the central assumption in SAR is that the scanned area is not in motion. If vehicles or other targets on the ground are in motion, they will not be resolved correctly and be distorted in the images. Shadowing is another impairment. This occurs when a tall object shields other object from the radar's illumination, causing a block or blank spot in the range return. This becomes more prevalent when very shallow angles are used—which occurs when the aircraft is at low altitude and scanning at long ranges. At high altitudes, such as satellite-mounted SAR, this is much less of an issue.

CHAPTER 23

Introduction to Video Processing

Video signal processing is used throughout the broadcast industry, the surveillance industry, in many military applications and is the basis of many consumer technologies. Up until the 1990s, nearly all videos were in analog form. In a relatively short time span, nearly all video technologies have become digital. Virtually all videos are now represented in digital form, and the digital signal processing techniques are used in nearly all video signal processing functions.

The picture element or "pixel" is used to represent each location in an image. Each picture element has several components, and each component is usually represented by a 10-bit value.

23.1 Color Spaces

There are several conventions or "color spaces" used to construct pixels. The broadcast industry originally used black-and-white images, so a video signal contained only luminance (or brightness) information. Later, color information was added to provide for color television and movies. This was known as the chrominance information. The color space formats associated with this is known as YCrCb, where Y is the luminance information, Cr and Cb are the chrominance information. Cr tends to contain more reddish hue color information, while Cb tends to contain more bluish hue color information (Fig. 23.1). Each is usually represented as a 10-bit value. One advantage of this system is that image processing and video bandwidth can be reduced by separating the luminance and chrominance. This is because our eyes are much more sensitive to intensity, or

Figure 23.1
Y, Cr, Cb images (which may be difficult to view in black/white book).

Digital Signal Processing 101. http://dx.doi.org/10.1016/B978-0-12-811453-7.00023-8

brightness, than to color. So a higher resolution can be used for luminance, and less resolution for chrominance. There are several formats used.

4: 4: 4 YCrCb each set of four pixels is composed of four Y (luminance) and four Cr and four Cb (chrominance) samples

4: 2: 2 YCrCb each set of four pixels is composed of four Y (luminance) and two Cr and two Cb (chrominance) samples

4: 2: 0 YCrCb each set of four pixels is composed of four Y (luminance) and one Cr and one Cb (chrominance) samples

Most broadcast systems and video signals use the 4:2:2 YCrCb format, where the luminance is sampled at twice the rate of each Cr and Cb chrominance. Each pixel, therefore, requires an average of 20 bits to represent, as compared to 30 bits for 4:4:4 YCrCb.

An alternate system was developed for computer systems and displays. There was no legacy of black and white to maintain compatibility with, and transmission bandwidth was not a concern, as the display is just a short cable connection to the computer. This is known as the RGB format, for red/green/blue (Fig. 23.2). Each pixel is composed of these three primary colors and requires 30 bits to represent. Most all televisions, flat screens, and monitors use RGB video, whereas nearly all broadcast signals use 4:2:2 YCrCb video.

These two color spaces can be mapped to each other as follows:

$$Y = 0.299 \cdot R + 0.587 \cdot G + 0.114 \cdot B$$
$$Cr = 0.498 \cdot R - 0.330 \cdot G + 0.498 \cdot B + 128$$
$$Cb = -0.168 \cdot R - 0.417 \cdot G - 0.081 \cdot B + 128$$

Figure 23.2
R, G, B images (which may be difficult to view in black/white book).

and

$$R = Y + 1.397 \cdot (Cr - 128)$$
$$G = Y - 0.711 \cdot (Cr - 128) - 0.343 \cdot (Cb - 128)$$
$$B = Y + 1.765 \cdot (Cb - 128)$$

There are other color schemes as well, such as CYMK, which is commonly used in printers, but we are not going to cover this further here.

Different resolutions are used. Very common resolution is the NTSC (National Television System Committee), also known as SD or standard definition. This has a pixel resolution of 480 rows and 720 columns. This forms a frame of video. In video jargon, each frame is composed of lines (480 rows) containing 720 pixels. The frame rate is approximately 30 frames (actually 29.97) per second.

23.2 Interlacing

Most NTSC SD broadcast video is interlaced. This was due to early technology where cameras filmed at 30 frames per second (fps), but this was not a sufficient update rate to prevent annoying flicker on television and movie theater screens. The solution was interlaced video where frames are updated at 60 fps, but only half of the lines are updated on each frame. One frame N, the odd lines are updated, and on frame N + 1 the even lines are updated and so forth. This is known as odd and even field updating.

Interlaced video requires half the bandwidth to transmit as noninterlaced or progressive video at the same frame rate, as only one half of each frame is updated at the 60 fps rate.

Modern cameras can record full images at 60 fps, although there are still many low-cost cameras that produce this interlaced video. Most monitors and flat screen televisions usually display full or progressive video frames at 60 fps. When you see a 720p or 1080i designation on a flat screen, the "p" or "i" stand for progressive and interlaced, respectively.

23.3 Deinterlacing

An interlaced video stream is usually converted to progressive for image processing, as well as to for display on nearly all computer monitors. Deinterlacing must be viewed as interpolation, for the result is twice the video bandwidth. There are several methods available for deinterlacing, which can result in different video qualities under different circumstances.

The two basic methods are known as "bob" and "weave." Bob is the simpler of the two. Each frame of interlaced video has only one half the lines. For example, the odd lines (1,3,5, ... 479) would have pixels, and the even lines (2,4,6, ... 480) are blank. On the

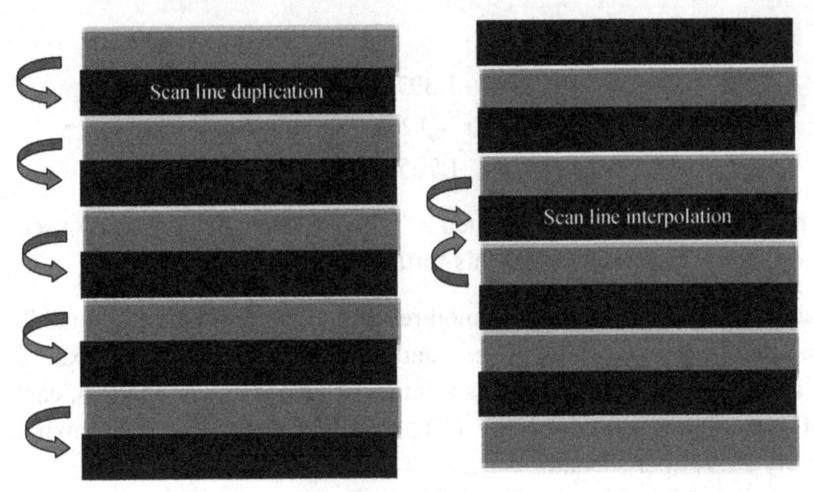

Figure 23.3
Bob verses weave deinterlacing.

following frame, the even lines have pixels, but the odd lines are blank. The simplest bob interlacing is to just copy the pixels from the line above for blank even lines (copy line 1 to line 2), and copy the pixels from the line below for blank odd lines (copy line 2 to line 1). Another method would be to interpolate between the two adjacent lines to fill in a blank line. Both of these methods are shown in Fig. 23.3.

This method can cause blurring of images, because the vertical resolution has been effectively halved.

Weave deinterlacing creates a full frame from the separate interlaced frames with odd and even lines. It then copies this frame twice, to achieve the 60 fps rate. This method tends to work only if there is little change in the odd and even interlaced frames, meaning there is little motion in the video. As the odd and even frame pixels belong to different instances in time (1/60th of a second difference), rapid motion can result in jagged edges in the images rather than smooth lines. This is shown in Fig. 23.4.

Both of these methods have drawbacks. A better method, which requires more sophisticated video processing, is to use motion adaptive deinterlacing. Where there is motion on the image, the bob technique works better, and slight blurring is not easily seen. In still areas of the image, the weave method will result in crisper images. A motion adaptive deinterlacer scans the whole image and detects areas of motion, by comparing to previous frames. It will use the bob method in these areas of the frame and use the weave method on the remaining areas of the frame. In this way, interlaced video can be converted to progressive with little loss of quality.

Figure 23.4
Deinterlacing effects.

23.4 Image Resolution and Bandwidth

Early televisions used a cathode ray gun inside a fluorescent tube. The gun traversed the screen horizontally from left to right for each line. There was a brief horizontal blanking period, while the gun swung back to the left side of the screen, as indicated by the horizontal sync (HSYNC) signal. After 480 lines, the gun would be at the bottom right corner of the screen. It would swing back to the top left corner to begin a new frame, as indicated by the vertical sync (VSYNC) signal. This time period was the vertical blanking time.

Owing to these blanking periods, the frame size is larger than the image size. For example, in SD definition, the actual image size viewed is 480 × 720 pixels. When blanking times are included, this is as if a 525 × 858 pixel image was sent, with the pixels blanked out, or zeroed, during the extra horizontal and vertical space. This legacy of allowing time for the cathode ray gun to return to the beginning of the video line or frame is still present in video standards today. In digital video and displays, these empty pixels can be filled with what is called ancillary data. This could be a display text at the bottom of the screen, the audio information, data on the program currently being viewed, and so forth. The extra blanking times must be taken into account when determining video signal bandwidths.

Higher definition or resolution, or HD, video formats are now common. HD can refer to 720p, 1080i, or 1080p image resolutions. Popular video resolutions are shown in Table 23.1. Many people are unaware, but often the HD 3 720p video. The quality difference is small, and it requires half the transmission bandwidth to transmit. Most HD flat screens can display 1080p resolution, but this resolution video is normally available only through DVRs or from other in home media sources.

Recently, ultrahigh resolution 4K video support is being made to the consumer market. This increases the frame resolution to 2160 × 3840 or four times greater than 1080p.

Table 23.1: Common Video Resolutions

Image Size	Frame Size	Color Plane Format at 60 frames per second	Bit/s Transfer Rate
1080p × 1920	1125 × 2200	4:2:2 YCrCb	2200 × 1125 × 20 × 60 = 2.97 Gbps
1080i × 1920	1125 × 2200	4:2:2 YCrCb	2200 × 1125 × 20 × 60 × 0.5 = 1.485 Gbps
720p × 1280	750 × 1650	4:2:2 YCrCb	1650 × 750 × 20 × 60 = 1.485 Gbps
480i × 720	525 × 858	4:2:2 YCrCb	858 × 525 × 20 × 60 × 0.5 = 270 Mbps

23.5 Chroma Scaling

Chroma scaling is used to convert between the different YCrCb formats, which have various resolution of color content. The most popular format is 4:2:2 YCrCb. In this format, each Y pixel has alternately a Cr or a Cb pixel associated with it but not both. To convert to the RGB format, 4:4:4 YCrCb representation is needed, where each Y pixel has both a Cr and Cb pixel associated with it. This requires interpolation of the chroma pixels. In practice this is often done by simple linear interpolation or by nearest neighbor interpolation. It can also be combined with the mapping to the RGB color space. Going the other direction is even simpler, as the excess chroma pixels can be simply not computed during the RGB → 4:2:2 YCrCb conversion.

23.6 Image Scaling and Cropping

Image scaling is required to map to either a different resolution, or to a different aspect ratio (row/column ratio). This requires upscaling (interpolation) over two dimensions to go to a higher resolution, or downscaling (decimation) over two dimensions to go to a lower resolution. Several methods of increasing complexity and quality are available such as:

- Nearest neighbor (copy adjacent pixel);
- Bilinear (use 2 × 2 of array of four pixels to compute new pixel);
- Bicubic (use 4 × 4 of array of 16 pixels to compute new pixel); and
- Polyphase (larger array of N × M pixels to compute new pixel).

When filtering vertically over several lines of video, the memory requirements increase, as multiple lines of video must be stored to perform any computations across vertical pixels. The effects of increasing filtering when downscaling can be easily seen using a circular video pattern. Downscaling, like decimation, will cause aliasing if high frequencies are not suitable filtered, as shown in Fig. 23.5.

Upscaling is far less sensitive as downscaling. Bicubic (using a 4 × 4 array of pixels) is sufficient. The effect of performing upscaling by using a smaller array or nearest neighbor is limited to slight blurriness, which is far less objectionable than aliasing.

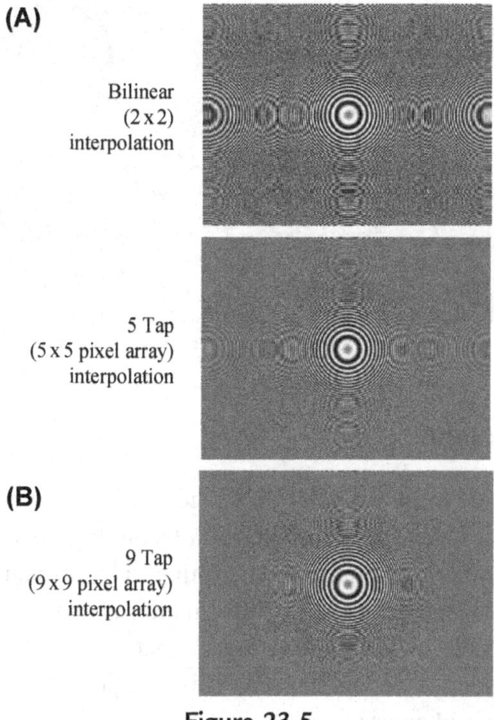

Figure 23.5
Image downscaling effects.

Cropping is simply eliminating pixels, to allow an image to fit within the frame size. It does not introduce any visual artifacts.

23.7 Alpha Blending and Compositing

Alpha blending is the merging of multiple images. One image can be placed over the top of one image, as shown below. This is known as compositing, a common process used to implement "picture in picture" (Fig. 23.6) functionality.

The more general case is a blending or weighting of the pixels in each image. This is controlled by a factor, alpha (α). This is done on a pixel by pixel basis, for each color as shown.

$$\text{New pixel}_{red} = \alpha \bullet \text{pixel}_{red} \text{ from image2} + (1 - \alpha) \bullet \text{pixel}_{red} \text{ from image1}$$
$$\text{New pixel}_{green} = \alpha \cdot \text{pixel}_{green} \text{ from image2} + (1 - \alpha) \cdot \text{pixel}_{green} \text{ from image1}$$
$$\text{New pixel}_{blue} = \alpha \cdot \text{pixel}_{blue} \text{ from image2} + (1 - \alpha) \cdot \text{pixel}_{blue} \text{ from image1}$$

Figure 23.6
Picture in picture (PiP).

23.8 Video Compression

Video data is very large, due to two spacial dimensions, high resolution, and requires 60 fps. To store or transmit video, data compression technology is used. This is essential to allow service such as video on demand or streaming video to handheld wireless devices. Video compression is a lossy process—some information is lost. The goal is to achieve as much compression as possible, while minimizing the data loss and restoring same perceptual quality when video is decompressed. This is especially important in fast motion video, such as during televised sports.

Video compression ratios depend on both the compression technology or standard used, and the video content itself. The newer video compression algorithms can deliver better quality, but at a price of very high computational requirements. Video compression processing is almost always done in hardware due to the computational rate, either in field programmable gate arrays (FPGAs) or application-specific integrated circuits (ASICs).

The most popular video compression algorithms are part of the MPEG4 standard. MPEG4 evolved from earlier H.263. At the time of this writing, MPEG4 part 10, also known as MPEG4 AVC or H.264 is mature and has been widely adopted by industry.

As you might guess, video compression is a very complex topic. It involves analysis across both spatial and temporal dimensions and uses complex algorithms. The two images in Fig. 23.7 depict comparative quality of an early version of MPEG4 and quality of the later MPEG4-10, in a video with a high degree of motion. An introduction to this topic will be covered in a subsequent chapter.

23.9 Digital Video Interfaces

There are several common video interfaces, which are used both in broadcast industry and among consumer products. These are briefly described below:

Figure 23.7
Video compression quality comparison.

SDI: This is a broadcast industry standard (Fig. 23.8), used to interconnect various professional equipment in broadcast studios and mobile video processing centers (like those big truck trailers seen at major sports events). SDI stands for "serial data interface," which is not very descriptive. It is an analog signal, modulated with digital information. This is usually connected using a coaxial cable. It is able to carry all of the data rates listed in Table 23.1 and dynamically switch between them. Most FPGAs and broadcast ASICs can interface directly with SDI signals.

DVI: Digital visual interface (DVI) is a connection type commonly used to connect computer monitors. It is a multipin connector carrying separated RGB digitized video information at the desired frame resolution (Fig. 23.9).

Figure 23.8
Serial data interface coax connector.

Figure 23.9
Digital visual interface monitor connector.

HDMI: High definition multimedia interface (HDMI) is also commonly used on computer monitors and on big screen to connect home theater equipment such as flat panels, computers, and DVDs together. Also carrying video and audio information in digital form, HDMI has backward electrical compatibility to DVI but utilizes a more compact connector. Later versions of HDMI support higher video frame sizes, rates, and higher bits per pixel (Fig. 23.10).

DisplayPort: The latest state of the art video interface is DisplayPort. This digital interface uses a packetized protocol to transmit video and audio information. DisplayPort can have 1, 2, 3, or 4 serial differential interfaces to support various data rates and very high

Figure 23.10
High definition multimedia interface connector.

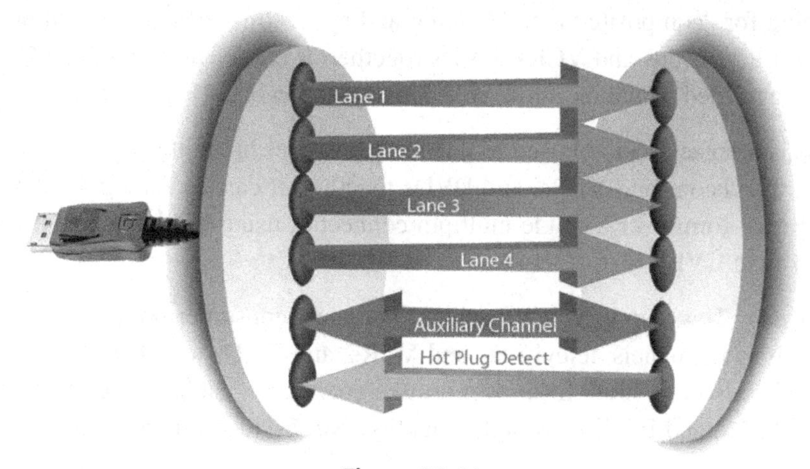

Figure 23.11
Display port interface and connector.

resolutions. Each serial interface supports about 5 Gbps data rate. It also uses 8/10-bit encoding, which allows the clocking to be embedded with the data. It also has a compact form factor connector, similar in size to HDMI (Fig. 23.11).

23.10 Legacy Analog Video Interfaces

VGA: Video graphics array connector is used to connect legacy computer monitors. Still common today, it is the familiar multipin "sub-D" connector located on the back or side of many laptop or desktop computers, used to connect to monitors or for laptops to connect to projectors for whole room viewing. This carries separated RGB analog video information at the desired frame resolution (Fig. 23.12).

Figure 23.12
Video graphics array monitor connector.

CVBS: Standing for "composite video blanking and sync," this is the basic yellow cable used to connect televisions and VCRs, DVDs together. It carries an SD 4:2:2 YCrCb combined analog video signal on a low-cost coax "patch cable" (Fig. 23.13).

S-Video: This is a legacy method used to connect consumer home theater equipment such as flat panels, televisions and VCRs, and DVDs together. It carries analog 4:2:2 YCrCb signals in separate form over a single multipin connector, using a shielded cable. It is of higher quality than CVBS (Fig. 23.14).

Component Video: This also is legacy method to connect consumer home theater equipment such as flat panels, televisions and VCRs, and DVDs together. It carries analog 4:2:2 YCrCb signals in separate form over a three coax patch cables. Often the connectors are labeled as Y, P_B, and P_R. It is of higher quality than S-video due to separate cables (Fig. 23.15).

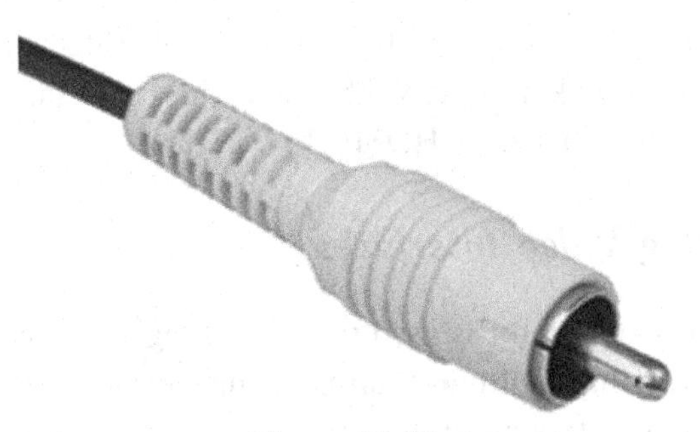

Figure 23.13
Composite video cable.

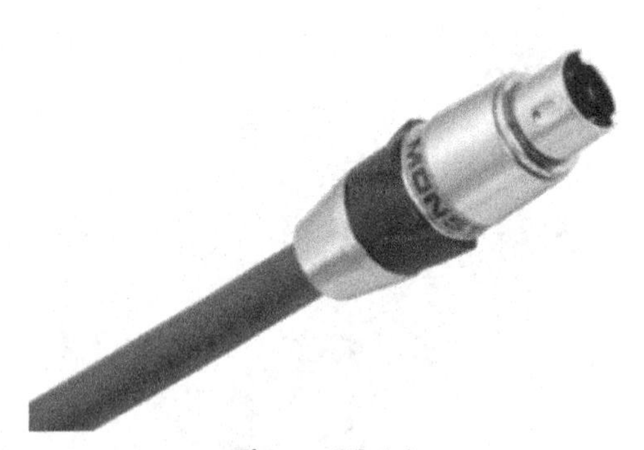

Figure 23.14
S-video cable.

Figure 23.15
Component video cables.

CHAPTER 24

DCT, Entropy, Predictive Coding, and Quantization

In this chapter, we will discuss some of the basic concepts used in data compression, including video and image compression. These concepts may seem unrelated, but will come together in the following chapter. Up until now, we have considered only uncompressed video formats, such as RGB or YCrCb, where each pixel is individually represented (although this is not strictly true for 4:2:2 or 4:2:0 forms of YCrCb). However, much greater levels of compression are possible with little loss of video quality. Reducing the data needed to represent an individual image or a sequence of video frames is very important when considering how much storage is needed on a camera flash chip or computer hard disk, or the bandwidth needed to transport cable or satellite television, or stream video to a computer or handheld wireless device.

24.1 Discrete Cosine Transform

The discrete cosine transform, or DCT is normally used to process two dimensional data, such as an image. Unlike the DFT or FFT that operates on one dimensional signals, which may have real and quadrature components., the DCT is usually used as an image presented by a rectangular array of pixels, which is a real signal only. When we discuss frequency, it will be how rapidly the sample values change. With the DCT, we will be sampling spatially across the image in either the vertical or horizontal direction.

The DCT is usually applied across an N by N array of pixel data. For example, if we take a region composed or 8 by 8 pixels, or 64 pixels total, we can transform this into a set of 64 DCT coefficients, which is the spatial frequency representation of the 8×8 region of the image. This is very similar to what we say in the DCT. However, instead of expressing the signal as a combination the complex exponentials of various frequencies, we will be expressing the image data as a combination of cosines of various frequencies, in both vertical and horizontal dimensions.

Now recall in the discussion on DFT, that the DFT representation is for a periodic signal or one that is assumed to be periodic. Now imagine connecting a series of identical signals together, end to end. Where the end of the sequence connects to the beginning of the next, there will be a discontinuity, or a step function. This will represent high frequency.

Digital Signal Processing 101. http://dx.doi.org/10.1016/B978-0-12-811453-7.00024-X

For the DCT, we make an assumption that the signal is folded over on itself. So an 8 long signal depicted becomes 16 long when appended as flipped. This 16 long signal is then symmetric about the midpoint. This is the same property of cosine waves. A cosine is symmetric about the midpoint, which is at π (since the period is from 0 to 2π). This property is preserved for higher frequency cosines, that only 8 of the 16 samples are needed as shown by the figures below, showing the sampled cosine waves. The waveforms in Fig. 24.1 are eight samples long, and if folded over to create 16 long sampled waveforms which will be symmetric, start and end with the same value, and has "*u*" cycles across the 16 samples.

To continue requires some terminology. The value of a pixel at row x and column y is designated as f(x,y), as shown in Fig. 24.2. We will compute the DCT coefficients, *F*(*u*,*v*) using equations that will correlate the pixels to the vertical and horizontal cosine frequencies. In the equations, "*u*" and "*v*" correspond to both the indices in the DCT array, and the cosine frequencies as shown in Fig. 24.1.

The relationship is given in the DCT equation, shown below for the 8 × 8 size.

$$F(u,v) = \tfrac{1}{4}C_uC_v \sum_{x=0 \text{ to } 7} \sum_{y=0 \text{ to } 7} f(x,y) \cdot \cos((2x+1)u\pi/16) \cdot \cos((2y+1)v\pi/16)$$

where

$C_u = \sqrt{2}/2$ when $u = 0$, $C_u = 1$ when $u = 1\ldots7$
$C_v = \sqrt{2}/2$ when $v = 0$, $C_u = 1$ when $v = 1\ldots7$
f(x,y) = the pixel value at that location

This represents 64 different equations, for each combination of *u*, *v*. For example:

$$F(0,0) = 1/8 \cdot \sum_{x=0 \text{ to } 7} \sum_{y=0 \text{ to } 7} f(x,y)$$

Or simply the summation of all 64 pixel values divided by 8. *F*(0,0) is the DC level of the pixel block.

$$F(4,2) = 1/4 \cdot \sum_{x=0 \text{ to } 7} \sum_{y=0 \text{ to } 7} f(x,y) \cdot \cos((2x+1)4\pi/16) \cdot \cos((2y+1)2\pi/16)$$

The nested summations indicate for each of the 64 DCT coefficients, we need to perform 64 multiply and adds. This requires 64 × 64 = 4096 calculations, which is very processing intensive.

The DCT is a reversible transform (provided enough numerical precision is used), and the pixels can be recovered from the DCT coefficients as shown below.

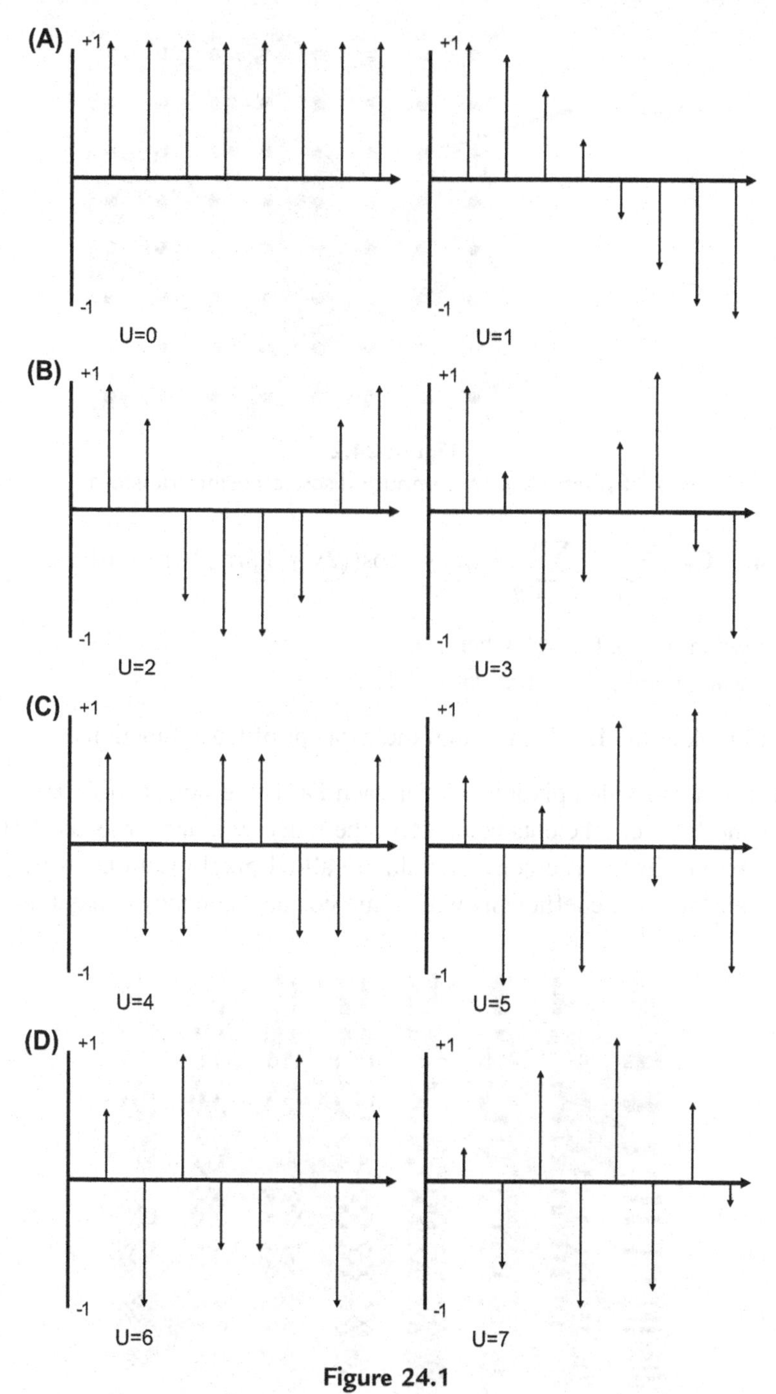

Figure 24.1
Sample cosine frequencies used in discrete cosine transform.

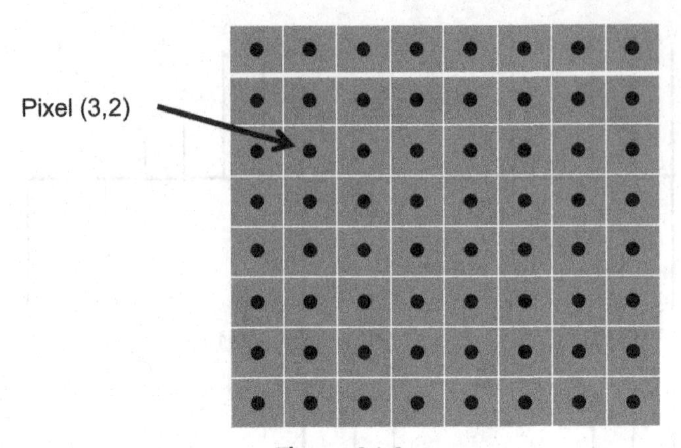

Figure 24.2
Block of pixels used to compute discrete cosine transform.

$$f(x, y) = \frac{1}{4} C_u C_v \sum_{u=0 \text{ to } 7} \sum_{v=0 \text{ to } 7} F(u, v) \cdot \cos((2x + 1)u\pi/16) \cdot \cos((2y + 1)/v\pi/16)$$

$C_u = \sqrt{2}/2$ when $u = 0$, $C_u = 1$ when $u = 1 \ldots 7$
$C_v = \sqrt{2}/2$ when $v = 0$, $C_u = 1$ when $v = 1 \ldots 7$

Another way to look at the DCT is through the concept of basis functions.

These tiles represent the video pixel block for each DCT coefficient. If $F(0,0)$ is nonzero, and the rest of the DCT coefficients equal zero, the video will appear as the {0,0} tile in Fig. 24.3. This particular tile is a constant value in all 64 pixel locations, which is what is expected since all the DCT coefficients with some cosine frequency content are zero.

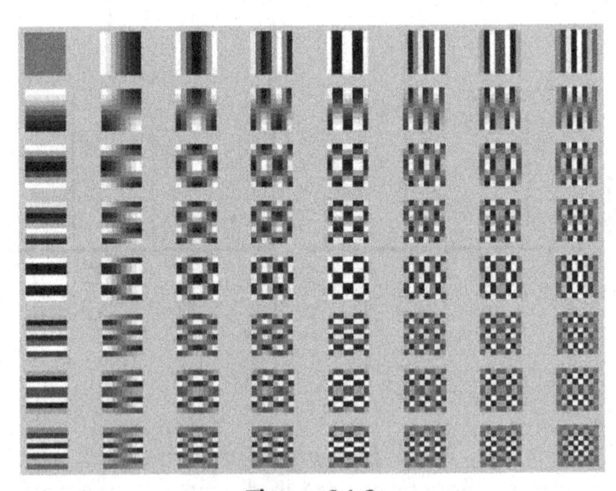

Figure 24.3
Discrete cosine transform basis functions.

If $F(7,7)$ is nonzero, and the rest of the DCT coefficients equal zero, the video will appear as the {7,7} tile in Fig. 24.3, which shows high frequency content in both vertical and horizontal direction. The idea is that any block of 8×8 pixels, no matter what the image, can be represented as the weighted sum of these 64 tiles in Fig. 24.3.

The DCT coefficients, and the video tiles they represent, form a set of basis functions. From linear algebra, any set of function values f(x,y) can be represented as a linear combination of the basis functions.

The whole purpose of this is to provide an alternate representation of any set of pixels, using the DCT basis functions. By itself, this exchanges one set of 64 pixel values with a set of 64 DCT values. However, it turns out frequently in many image blocks, many of the DCT values are near zero, or very small and can be presented with few bits. This can allow the pixel block to be presented more efficiently with fewer values. However, this representation is approximate, because when chose to minimize the bits representing various DCT coefficients, we are quantizing. This is a loss of information, meaning the pixel block cannot be restored perfectly.

24.2 Entropy

We will start with the concept of entropy. Some readers may recall from studies in thermodynamics or physics that entropy is a measure of the disorderliness of a system. Further, the second law of thermodynamics states that in a closed system, entropy can only increase and never decrease. In the study of compression, and also a related field of err correction, entropy can be thought of as the measure of unpredictability. This can be applied to a set of digital data.

The less predictable a set of digital data is, the more information it carries. Here is a simple example. Assume that a bit can be equally likely to be either 0 or 1. By definition, this will be 1 bit of data information. Now assume that this bit is known to be a 1 with 100% certainty. This will carry no information, because the outcome is predetermined. This relationship can be generalized by:

$$\text{Info of outcome} = \log_2 (1/\text{probability of outcome}) = -\log_2 (\text{probability of outcome})$$

Let us look at another example. Suppose there is a four outcome event, with equal probability of outcome_1, outcome_2, outcome_3, or outcome_4.

Outcome 1: Probability = 0.25, encode as 00
Outcome 2: Probability = 0.25, encode as 01
Outcome 3: Probability = 0.25, encode as 10
Outcome 4: Probability = 0.25, encode as 11

The entropy can be defined as the sum of the probabilities of each outcome multiplied by the information conveyed by that outcome.

$$\text{Entropy} = \text{prob}(\text{outcome}_1) \cdot \text{info}(\text{outcome}_1) + \text{prob}(\text{outcome}_2) \cdot \text{info}(\text{outcome}_2) + \ldots \text{prob}(\text{outcome}_n) \cdot \text{info}(\text{outcome}_n)$$

In our simple example,

$$\begin{aligned}\text{Entropy} &= 0.25 \cdot \log_2(1/0.25) + 0.25 \cdot \log_2(1/0.25) + 0.25 \cdot \log_2(1/0.25) \\ &\quad + 0.25 \cdot \log_2(1/0.25) \\ &= 2 \text{ bits}\end{aligned}$$

This is intuitive—2 bits is normally what would be used to convey one of four possible outcomes.

In general, the entropy is the highest when the outcomes are equally probable, and therefore totally random. When this is not the case, and the outcomes are not random, the entropy is lower, and may be possible to take advantage of this and reduce the number of bits to represent the data sequence.

Now what if the probabilities are not equal, for example:

Outcome 1: Probability = 0.5, encode as 00
Outcome 2: Probability = 0.25, encode as 01
Outcome 3: Probability = 0.125, encode as 10
Outcome 4: Probability = 0.125, encode as 11

$$\begin{aligned}\text{Entropy} &= 0.5 \cdot \log_2(1/0.5) + 0.25 \cdot \log_2(1/0.25) + 0.125 \cdot \log_2(1/0.125) \\ &\quad + 0.125 \cdot \log_2(1/0.125) \\ &= 1.75 \text{ bits}\end{aligned}$$

24.3 Huffman Coding

Since the entropy in the previous example is less than 2 bits, then in theory, we should be able to convey this information in less than 2 bits. What if we encoded these events differently as shown below:

Outcome 1: Probability = 0.5, encode as 0
Outcome 2: Probability = 0.25, encode as 10
Outcome 3: Probability = 0.125, encode as 110
Outcome 4: Probability = 0.125, encode as 111

One-half the time, we would present the data with 1 bit, one-fourth of the time with 2 bits, and one-fourth of the time with 3 bits.

$$\text{Average number of bits} = 0.5 \cdot 1 + 0.25 \cdot 2 + 0.125 \cdot 3 + 0.\ 125 \cdot 3 = 1.75$$

So this outcome can be represented with less than 2 bits, in this case 1.75 bits. The nice thing about this encoding is that the bits can be put together into a continuous bit stream and unambiguously decoded.

010110111110100....

The sequence above can only represent one possible outcome sequence.

outcome$_1$, outcome$_2$, outcome$_3$, outcome$_4$, outcome$_3$, outcome$_2$, outcome$_1$...

An algorithm known as Huffman coding works in this manner, by assigning the shortest codewords (the bit-word that each outcome is mapped to, or encoded) to the events of highest probability. For example, Huffman codes are used in JPEG-based image compression. Actually, this concept was originally used in Morse code for telegraphs over 150 years ago, where each letter of the alphabet is mapped to a series of short dots and long dashes.

Common letters.

E	•
I	••
T	-

Uncommon letters.

X	-••-
Y	-•--
Z	--••

In Morse code, the higher probability letters are encoded in a few dots or a single dash. The less likely probability letters use several dashes and dots. This minimizes, on the average, the amount of time required by the operator to send a telegraph message, and the number of dots and dashes transmitted.

24.4 Markov Source

Further opportunities for optimization arise, when the probability of each successive outcome or symbol is dependent on previous outcomes. An obvious example is that if the nth letter is a "q", you can be pretty sure the next letter will by a "u". Another example is that if the nth letter is a "t", this raises the probability that the following letter will be an "h". Data sources that have this kind of dependency relationship between successive symbols are known as Markov sources. This can lead to more sophisticated encoding schemes. Common groups of letters with high probabilities can be mapped to specific codewords. Examples are the letter pairs "st" and "tr", or "the".

This falls out when we map the probabilities of a given letter based on the few letters preceding. In essence, we are making predictions based on the probability of certain letter groups appearing in the any given construct of the English language. Of course, different languages, even if using the same alphabet, will have a different set of multiletter probabilities, and therefore different codeword mappings. An easy example in the United States is the sequence of letters: HAWAI_. Out of 26 possibilities, it is very likely the next letter is an "I".

In a first order Markov source, a given symbol's probability is dependent on the previous symbol. In a second order Markov source, a given symbol's probability is dependent on the previous two symbols, and so forth. The average entropy of a symbol tends to decrease as the order of the Markov source increases. The complexity of the system also increases as the order of the Markov source increases.

A first order binary Markov source can described in the diagram given in Fig. 24.4. The transition probabilities depend only on the current state.

In this case, the entropy can be found as the weighted sum of the conditional entropies corresponding to the transitional probabilities of the state diagram. The probability of a zero given the previous state is a 1 is described as $P(0|1)$.

The probability of each state can be found by solving the probability equations:

$$P(0) = P(0) \cdot P(0|0) + P(1) \cdot P(0|1) = P(0) \cdot 0.8 + P(1) \cdot 0.4$$

Both methods a 0 can be output (meaning to arrive at state = 0 circle), either starting from state 0 or 1

$$P(1) = P(1) \cdot P(1|1) + P(0) \cdot P(1|0) = P(1) \cdot 0.6 + P(0) \cdot 0.2$$

Both methods a 1 can be output (meaning to arrive at state = 0 circle), either starting from state 0 or 1.

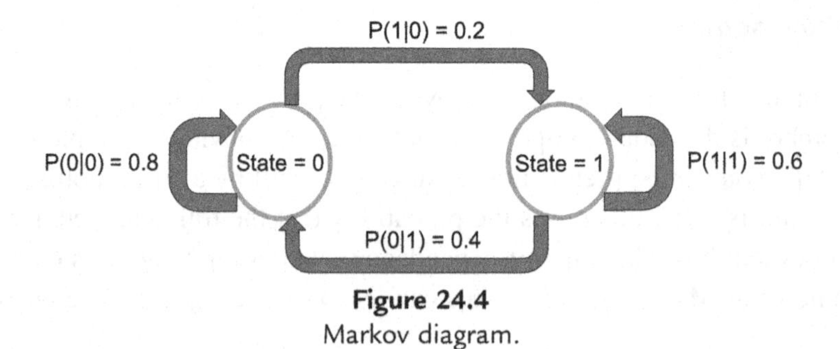

Figure 24.4
Markov diagram.

By definition $P(0) + P(1) = 1$.

From this, we can solve for P(0) and P(1)

$$P(0) = 2/3$$
$$P(1) = 1/3$$

Now we can solve for the entropy associated with each state circle.

For state 0 and state 1:

$$\text{Entropy}_0 = 0.8 \cdot \log_2(1/0.8) + 0.2 \cdot \log_2(1/0.2) = 0.258 + 0.464 = 0.722 \text{ bits}$$
$$\text{Entropy}_1 = 0.6 \cdot \log_2(1/0.6) + 0.4 \cdot \log_2(1/0.4) = 0.442 + 0.529 = 0.971 \text{ bits}$$

The entropy of the Markov source or system is then given by:

$$P(0) \cdot \text{Entropy}_0 + P(1) \cdot \text{Entropy}_1 = 1/3 \cdot 0.722 + 2/3 \cdot 0.971 = 0.888 \text{ bits}$$

24.5 Predictive Coding

Finding the sequence "HAWAII" is a form of predictive coding. The probabilities of any Markov source, such as language, can be mapped in multiletter sequences with an associated probability. These in turn, can be encoded using Huffman coding methods, to produce a more efficient representation (fewer number of bits) of the outcome sequence.

This same idea can be applied to images. As we have previously seen, video images are built line by line, from top to bottom, and pixel by pixel, from left to right. Therefore, for a given pixel, the pixels above and to the left are available to help predict the next pixel. For our purposes here, let us assume an RGB video frame, with 8 bits per pixel and color. Each color can have a value from 0 to 255.

The value unknown pixel, for each color, can be predicted from the pixels immediately left, immediately above, and diagonally above. Three simple predictors are given below, in Fig. 24.5.

Now the entire frame could be iteratively predicted from just three initial pixels, but this is not likely to be a very good prediction. The usefulness of these predictors becomes apparent when used in conjunction with differential encoding.

24.6 Differential Encoding

Suppose instead we take the actual pixel value, and subtract the predicted pixel value. This is differential encoding, and this difference is used to represent the pixel value for that

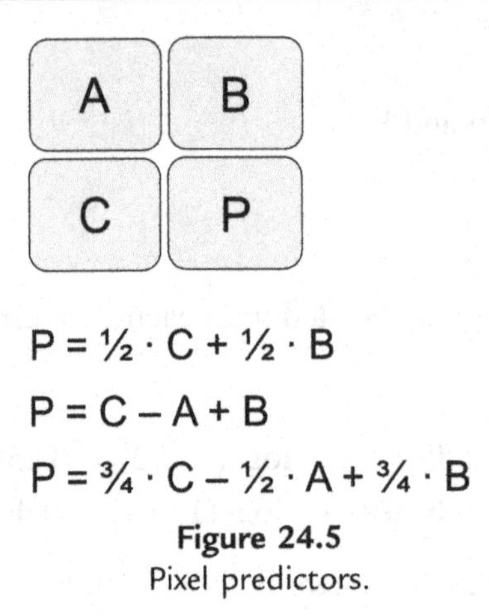

Figure 24.5
Pixel predictors.

location. For example, to store a video frame, we would perform the differential encoding process, and store the results.

To restore the original video data for display, simply compute the predicted pixel value (using the previous pixel data) and then add to the stored differential encoded value. This is the original pixel data. (The three pixels on the upper left corner of the frame are stored in their original representation and are used to create the initial predicted pixel data during the restoration process).

All of this sounds unnecessarily complicated, so what is the purpose here? The reason to do this is that the differential encoder outputs are most likely to be much smaller values than the original pixel data, due to the correlation between nearby pixels. Since statistically, the differentially encoded data is just representing the errors of the predictor, this signal is likely to have the great majority of values concentrated around zero, or to have a very compact histogram. In contrast, the original pixel values are likely to span the entire color value space, and therefore are a high entropy data source with equal or uniform probability distribution.

What is happening with differential encoding is that we are actually sending less information than if the video frame was simply sent pixel by pixel across the whole frame. The entropy of the data stream has been reduced through the use of differential encoding, as the correlation between adjacent pixels has been largely eliminated. Since we have no information as to the type of video frames will be processed, the initial pixel values outcomes are assumed to be equally distributed, meaning each of the 256 possible values is equally with probability of 1/256, and entropy of 8 bits per pixel. The possible outcomes of the differential encoder tend be much more probable for small values (due to the correlation to nearby pixels) and much less likely for larger values. As we saw in our simple example above, when the probabilities are not evenly distributed, the entropy is lower. Lower entropy means less information. Therefore, on average, significantly less bits is required to represent the image, and the only cost is increased complexity due to the predictor and differential encoding computations.

Let us assume that averaged over a complete video frame, the probabilities work out as such:

Pixel color value Probability = 1/256 for values in range of 0–255

And that after differential encoder, the probability distribution comes out as:

Differential color Probability = 1/16 for value equal to 0
Differential color Probability = 1/25 for values in range of −8 to −1, 1 to 8
Differential color Probability = 1/400 for values in range of −32 to −9, 9 to 32
Differential color Probability = 1/5575 for values in range of −255 to −33, 33 to 255

Recall the entropy is defined as:

$$\text{Entropy} = \text{prob}\,(\text{outcome}_1)\cdot\text{info}\,(\text{outcome}_1) + \text{prob}\,(\text{outcome}_2)\cdot\text{info}\,(\text{outcome}_2) + \ldots\ \text{prob}\,(\text{outcome}_n)\cdot\text{info}\,(\text{outcome}_n)$$

with info of outcome = $\log_2$ (1/probability of that outcome)

In the first case, we have 256 outcomes, each of probability 1/256.

$$\text{Entropy} = 256\cdot(1/256)\cdot\log_2(1/(1/256)) = 8\ \text{bits}$$

In the second case, we have 511 possible outcomes, with one of four probabilities. We will see that this has less entropy and can be represented in less than 8 bits of information.

$$\begin{aligned}\text{Entropy} &= 1\cdot(1/16)\cdot\log_2\cdot(1/(1/16)) + 16\cdot(1/25)\cdot\log_2(1/(1/25))\\ &\quad + 48\cdot(1/400)\cdot\log_2(1/(1/400)) + 446\cdot(1/5575)\cdot\log_2(1/(1/5575))\\ &= 0.250 + 2.972 + 1.037 + 0.996 = 5.255\ \text{bits}\end{aligned}$$

24.7 Lossless Compression

The entropy has been significantly reduced through the use of the differential coder. Although this step function probability distribution is obviously contrived, typical entropy values for various differential encoders across actual video frame data tend to be in the range of 4–5 ½ bits. With use of Huffman coding or similar mapping of the values to bit codewords, the number of bits used to transmit each color plane pixel can be reduced to ~5 bits compared to the original 8 bits. This has been achieved without any loss of video information, meaning the reconstructed video is identical to the original. This is known as lossless compression, as there is no loss of information in the compression (encoding) and decompression (decoding) processing.

Much higher degrees of compression are possible if we are willing to accept some level of video information loss, which can result in video quality degradation. This is known as lossy compression. Note that with lossy compression, each time the video is compressed and decompressed, some amount of information is lost, and there will be a resultant impact on video quality. The trick is to achieve high levels of compression, without noticeable video degradation.

One of the issues causes information loss in quantization, which we will examine next.

24.8 Quantization

Many of the techniques used in compression such as the DCT, Huffman coding, and predictive coding are fully reversible, with no loss in video quality. Quantization is often the principal area of compression where information is irretrievably lost, and the decoded video will suffer quality degradation. With care, this degradation can be made reasonably imperceptible to the viewer.

Quantization occurs when a signal with many or infinite number of values must be mapped into a finite set of values. In digital signal processing, signals are presented in binary numbers, with 2^n possible values mapping into an n-bit representation.

For example, suppose we want to present the range -1 to $+1$ (well, almost $+1$) using an 8-bit fixed point number. With 2^n or 256 possible values to map to across this range, the step size is 1/128, which works out to 0.0078125. Let us say the signal has an actual value of 0.5030. How closely can this value be presented? What if the signal is 1/10 the level of first sample, or 0.0503. And again, consider a signal with value 1/10 the level as the second sample, at 0.00503. Below is a table showing the closest representation just above and below each of these signal levels, and the error that will result in the 8-bit representation of the actual signal to sampled signal value.

Signal Level	Closest 8 Bit Representation	Hexadecimal Value	Actual Error	Error as a Percent of Signal Level (%)
0.50300	0.5000000	0x40	0.00300	0.596
0.50300	0.5078125	0x41	−0.0048128	0.957
0.05030	0.0468750	0x06	0.003425	6.809
0.05030	0.0546875	0x07	−0.0043875	8.722
0.00503	0.000000	0x00	0.00503	100
0.00503	0.0078125	0x01	−0.0027825	55.32

The actual error level remains more or less in the same range over the different signal ranges. This error level will fluctuate, depending on the exact signal value, but with our

8-bit signed example will always be less than 1/128, or 0.0087125. This fluctuating error signal will be seen as a form of noise or unwanted signal in the digital video processing. It can be modeled as an injection of noise when simulation an algorithm with unlimited numerical precision. It is called quantization noise.

When the signal level is fairly large for the allowable range (0.503 is close to one-half the maximum value) the percentage error is small—less than 1%. As the signal level gets smaller, the error percentage gets larger, as the table indicates.

What is happening is that the quantization noise is always present and is, on average, the same level (any noiselike signal will rise and fall randomly, so we usually concern ourselves with the average level). But as the input signal decreases in level, the quantization noise becomes more significant in a relative sense. Eventually, for very small input signal levels, the quantization noise can become so significant that it degrades the quality of whatever signal processing is to be performed. Think of it as like static on a car radio. As you get further from the radio station, the radio signal gets weaker, and eventually the static noise makes it difficult or unpleasant to listen to, even if you increase the volume.

So what can we do if our signal is sometimes strong (0.503, for example), and other times weak (0.00503, for example)? Another way of saying this is that the signal has a large dynamic range. The dynamic range describes the ratio between the largest and smallest value of the signal, in this case 100.

Suppose we exchange our 8-bit representation with 12-bit representation? Then our maximum range is still from −1 to +1, but our step size is now 1/2048, which works out to 0.000488. Let us make a 12-bit table similar to the 8-bit example.

Signal Level	Closest 12 Bit Representation	Hexadecimal Value	Actual Error	Error as a Percent of Signal Level (%)
0.50300	0.502930	0x406	0.000070	0.0139
0.50300	0.503418	0x407	−0.000418	0.0831
0.05030	0.050293	0x067	0.000007	0.0140
0.05030	0.050781	0x068	−0.000481	0.9568
0.00503	0.004883	0x00A	0.000147	2.922
0.00503	0.005371	0x00B	−0.000341	6.779

This is a significant difference. The actual error is always less than our step size, 1/2048. But the error as a percent of signal level is dramatically improved. This is what we usually care about in signal processing. Because of the much smaller step size of the 12-bit representation, the quantization noise is much less, allowing even small signals to be represented with very reasonable precision. Another way of describing this is to introduce the concept of signal-to-noise power ratio, or SNR. This describes the power of the largest signal compared to the background noise. This can be very easily seen on a frequency domain or spectral plot of a signal. There can be many sources of noise, but for now, we are only considering the quantization noise introduced by the digital representation of the video pixel values.

What we have just described is the uniform quantizer, where all the step sizes are equal. However, there are alternate quantizing mappings, where the step size varies across the signal amplitude. For example, a quantizer could be designed to give the same SNR across the signal range. This would require a small step size when the signal amplitude is small, and a larger step size as the signal increases in value. The idea is to provide a near constant quantization error as a percentage of the signal value. This type of quantizing is performed on the voice signals in the US telephone system, known as μ-law encoding.

Another possible quantization scheme could be to use a small step size for regions of the signal where there is a high probability of signal amplitude occurring, and larger step for regions where the signal has less likelihood of occurring.

However, in video signal processing, uniform quantizing is by far the most common. The reason is that often we are not encoding simply amplitude, representing small or large signals. In many cases, this could be color information, where the values map to various color intensities, rather than signal amplitude.

As far as trying to use likelihood of different values to optimize the quantizer, this assumes that this probability distribution of the video data is known. Alternatively, the uniform quantizer can be followed by some type of Huffman coding or differential encoding, which will optimize the signal representation for the minimum average number of bits.

Vector quantization is also commonly used. In the preceding discussion, a single set of values or signal is being quantized with a given bit representation. However, a collection of related signals can be quantized to a single representation using a given number of bits.

For example, assume the pixel is represented in RGB format, with 8 bits used for each color. This means that the color red can be represented in 2^8 or 256 different intensities.

Each pixel uses a total of 24 bits, for a total of 2^{24}, or about 16 million possible values. Intuitively, this seems excessive, can the human eye really distinguish that many colors? Instead, a vector quantizer might map this into a color table of 256 total colors, presenting

256 combinations of red, green, and blue combinations. This mapping results in requiring only 8 bits to present each pixel.

This seems like this might be reasonable, but the complexity is in trying to map the 16 million possible inputs to the allowed 256 representations, or color codewords. If done using a look up table, as memory of 16 million bytes would be required for this quantization, with each memory location containing one of the 256 color codewords. This is excessive, so some sort of mapping algorithm or computation is required to map these possible 16 million color combinations to the closest color codeword. As we are starting to see, most methods to compress video, or other data for that matter, come at the expense of increased complexity and increased computational rates.

24.9 Decibels

SNR is usually expressed in decibels (denoted dB), using a logarithmic scale. The SNR of a digital represented signal can be determined by the following equation:

$$\mathrm{SNR}_{\mathrm{quantization}}(\mathrm{dB}) = 6.02 * (\text{Number of bits}) + 1.76$$

Basically, for each additional bit of the signal representation, 6 dB of SNR is gained. Eight bit representation is capable of representing a signal with an SNR of about 48 dB, a 12 bits can do better at 72 dB, and 16 bits will give up to 96 dB. This only accounts for the effect of quantization noise; in practice there are other effects that could also will degrade SNR in a system.

There is another important point on decibels. These are very commonly used in many areas of digital signal processing subsystems. A decibel is simply a signal power ratio, similar to percentage. But because of the extremely high ratios commonly used (a billion is not uncommon), it is convenient to express this logarithmically. The logarithmic expression also allow chains of circuits or signal processing operations each with its own ratio (say of output power to input power) to simply be added up to find the final ratio.

Where people commonly get confused is in differentiating between signal levels or amplitude (voltage if an analog circuit) and signal power. Power measurements are virtual in the digital world, but can be directly measured in analog circuits in which video systems interface with, such as the RF amplifiers and analog signal levels for circuits in head-end cable systems, or video satellite transmission.

There are two definitions of dB commonly used.

$$\mathrm{dB}_{\mathrm{voltage}} = \mathrm{dB}_{\mathrm{digital\ value}} = 20 \cdot \log(\text{voltage signal } 1/\text{voltage signal } 2)$$

$$\mathrm{dB}_{\mathrm{power}} = 10 \cdot \log(\text{power signal } 1/\text{power signal } 2)$$

The designations of "signal 1" and "signal 2" depend on the situation. For example, with an RF power amplifier, the dB of gain will be the 10 log (output power/input power). For digital, the dB of SNR will be the 20 log (maximum input signal/quantization noise signal level).

The use of dB can refer to many different ratios in video system designs. But it is easy to get confused whether to use to multiplicative factor of 10 or 20, without understanding the reasoning behind this.

Voltage squared is proportional to power. If a given voltage is doubled in a circuit, it requires four times as much power. This goes back to a basic Ohm's law equation.

$$\text{Power} = \text{Voltage}^2 / \text{Resistance}$$

In many analog circuits, signal power is used, because that is what the lab instruments work with, and while different systems may use different resistance levels, power is universal (however, 50 Ω is the most common standard in most analog systems).

The important point is that since voltage is squared, this effect needs to be taken into account in the computation of logarithmic decibel relation. Remember, $\log x^y = y \log x$. Hence, the multiply factor of "2" is required for voltage ratios, changing the "10" to a "20".

In the digital world, the concept of resistance and power do not exist. A given signal has specific amplitude, expressed in a digital numerical system (such as signed fractional or integer, for example).

Understanding dB increases using the two measurement methods is important. Let us look at doubling of the amplitude ratio and doubling of the power ratio.

$$6.02\ \text{dB}_{\text{voltage}} = 6.02\ \text{dB}_{\text{digital value}} = 20 \cdot \log(2/1)$$
$$3.01\ \text{dB}_{\text{power}} = 10 \cdot \log(2/1)$$

This is why shifting a digital signal left 1 bit (multiply by 2) will cause a 6 dB signal power increase, and why so often the term 6 dB/bit is used in conjunction with ADCs, DACs, or digital systems in general.

By the same reasoning, doubling in power to an RF engineer means a 3 dB increase. This will also impact the entire system. Coding gain, as used with error-correcting code methods, is based on power. All signals at antenna interfaces are defined in terms of power, and the decibels used will be power ratios.

In both systems, ratio of equal power or voltage is 0 dB. For example, a unity gain amplifier has a gain of 0 dB.

$$0\ \text{dB}_{\text{power}} = 10 \cdot \log(1/1)$$

A loss would be expressed as a negative dB. For example a circuit whose output is equal to ½ the input power.

$$-3.01\ \text{dB}_{\text{power}} = 10 \cdot \log(1/2)$$

CHAPTER 25

Image and Video Compression Fundamentals

Now that we have the basics of entropy, predictive coding, DCT, and quantization, we are ready to discuss image compression. Image compression, in contrast to video compression, deals with a still image, rather than a continuous sequence of images, which makes up a video stream.

JPEG is often ubiquitous with image compression. JPEG stands for Joint Photographic Experts Group, a committee that has published international standards on image compression. JPEG is an extensive portfolio of both lossy and lossless image compression standards and options. In this section, we will focus on baseline JPEG.

25.1 Baseline JPEG

With baseline JPEG, each color plane is compressed independently. A monochrome image would have 8 bits per pixel. Generally, lossy compression can achieve less than 1 bit on average per pixel with high quality.

For RGB images, each of the three color planes is treated independently. With YCrCb representation, Y, Cr, and Cb are treated independently. For 4:2:2 or 4:2:0 YCrCb, the Cr and Cb are undersampled, and these undersampled color planes will be compressed. For example, standard definition images are 720 (width) by 480 (height) pixels. For 4:2:2 YCrCb representation, the Cr and Cb planes will be 360 by 480 pixels. Therefore, a higher degree of compression can be achieved using JPEG on 4:2:2 or 4:2:0 YCrCb images. Intuitively, this makes sense, as more bits are used to represent the luminance to which the human eye is more sensitive and less to the chrominance to which the human eye is less sensitive.

25.2 DC Scaling

Each color plane of the image is divided up into 8×8 pixel blocks. Each 8-bit pixel can have a value ranging from 0 to 255. The next step is to subtract 128 from all 64 pixel values, so the new range is -128 to $+127$. The 8×8 DCT is next applied to this set of 64 pixels. The DCT output is the frequency domain representation of the image block.

Digital Signal Processing 101. http://dx.doi.org/10.1016/B978-0-12-811453-7.00025-1

The upper left DCT output is the DC value, or average of all the 64 pixels. Since we subtracted 128 prior to the DCT processing, the DC value can range from −1024 to 1016, which can be represented by an 11-bit signed number. Without the 128 offset, the DC coefficient would range from 0 to 2040, while other 63 of the DCT coefficients would be signed (due to the cosine range). The subtraction of 128 from the pixel block has no effect on the 63 AC coefficients (an equivalent method could be to perform subtraction of −1024 of DC coefficient after the DCT).

25.3 Quantization Tables

The quantization table used has a great influence on the quality of JPEG compression. It also influences the degree of compression achieved. These tables are often developed empirically (by trial and error) to give the greatest number of bits to the DCT values, which are most noticeable and have the most impact to the human vision system.

The quantization table is applied to the output of the DCT, which is an 8 × 8 array. The upper left coefficient is the DC coefficient, and the remaining are the 63 AC coefficients, of increasing horizontal and vertical frequencies as one moves rightward and downward. As the human eye is more sensitive to lower frequencies, less quantization and more bits are used for the upper and leftmost DCT coefficients which contain the lower frequencies in vertical and horizontal directions.

Example baseline tables are provided in the JPEG standard, as shown below.

Luminance Quantization Table of $Q_{j,k}$ values

16	11	10	16	24	40	51	61
12	12	14	19	26	58	60	55
14	13	16	24	40	57	69	56
14	17	22	29	51	87	80	62
18	22	37	56	68	109	103	77
24	35	55	64	81	104	113	92
49	64	78	87	103	121	120	101
72	92	95	98	112	100	103	99

Chrominance Quantization Table of $Q_{j,k}$ values

17	18	24	47	99	99	99	99
18	21	26	66	99	99	99	99
24	26	56	99	99	99	99	99
47	66	99	99	99	99	99	99
99	99	99	99	99	99	99	99
99	99	99	99	99	99	99	99
99	99	99	99	99	99	99	99
99	99	99	99	99	99	99	99

Many other quantization tables claiming greater optimization to the human visual system have been developed for various JPEG versions.

The quantized output array is formed as follows:

$$B_{j,k} = \text{rounded}\left(A_{j,k}/Q_{j,k}\right) \quad \text{for } j = \{0\ldots7\}, k = \{0\ldots7\}$$

where $A_{j,k}$ is the DCT output array value, $Q_{j,k}$ is the quantization table value.

A couple of examples would be:

$$\text{Luminance DCT output value of } \left(A_{0,0}\right) = 426.27$$
$$B_{0,0} = \text{ round } \left(A_{0,0}/Q_{0,0}\right) = \text{round } (426.27/16) = 27$$
$$\text{Chrominance DCT output value of} \left(A_{6,2}\right) = -40.10$$
$$B_{6,2} = \text{round } \left(A_{6,2}/Q_{6,2}\right) = \text{round } (-40.10/99) = 0$$

When the quantization values $Q_{j,k}$ are large, this results in few possible values of the output $B_{j,k}$. For example, with a quantization value of 99, the rounded output values can only be -1, 0, or $+1$. In many cases, especially when j or k is 3 or larger, the $B_{j,k}$ will be rounded to zero, indicating little high frequency in the image region.

This is lossy compression. Data are lost in quantization and cannot be recovered. The principle is to reduce the data by discarding only data that have little noticable impact on the image quality.

25.4 Entropy Coding

The next step is to sequence the quantized array values $B_{j,k}$ as in the order shown in Fig. 25.1. The first value $B_{0,0}$ is the quantized DC coefficient. All the subsequent values are AC values.

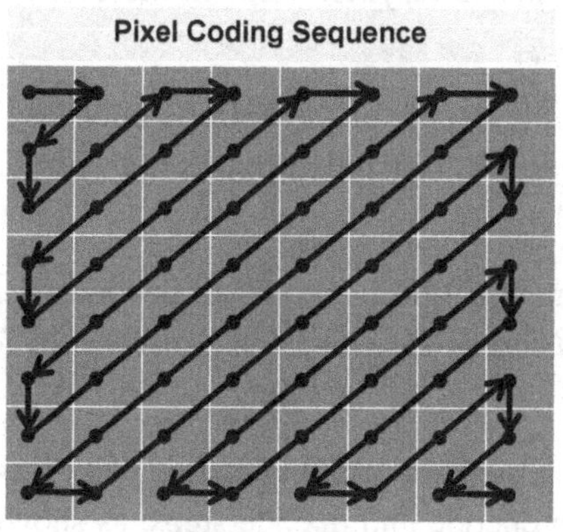

Figure 25.1
Sequencing of quanitzed pixel coding.

The entropy encoding scheme is fairly complex. The AC coefficients are coded differently than that of the DC coefficient. The output of the quantizer often contains many zeros, so special symbols are provided. One is an EOB of end of block symbol, used when the remaining values from the quantizer are all zero. This allows the encoding to be terminated when the rest of the quantized values are zero. The coded symbols also include ability to specify the zero run length following a nonzero symbol, again to help efficiently take advantage of the zeros present within the quantizer output. This is known as run length encoding.

The DC coefficients are differentially coded across the image blocks. There is no relationship between the DC and AC coefficients. However, DC coefficients in different blocks are likely to be correlated, as adjacent 8 × 8 image blocks are likely to have a similar DC or average luminance and chrominance across nearby image blocks. So the only difference is coded for the next DC coefficient, relative to the previous DC coefficient.

Four Huffman code tables are provided in the baseline JPEG standard:

- DC coefficient, luminance,
- AC coefficients, luminance,
- DC coefficient, chrominance,
- AC coefficients, chrominance.

These tables give encoding for both individual values, and values plus a given number of zeros. Following the properties of Huffman coding, the tables are constructed so that the most statistically common input values are coded using the fewest number of bits. The Huffman symbols are then concatenated into a bitstream, which forms the compressed image file. The use of variable length coding makes recovery difficult if any data corruption occurs. Therefore, special symbols or markers are inserted periodically to allow the decoder to resynchronize in the event of any bit errors in the JPEG file.

The JPEG standard specifies the details of the entropy encoding followed by Huffman coding. It is quite detailed and is not included in this text. For nonbaseline JPEG, alternate coding schemes may be used.

For those planning to implement a JPEG encoder or decoder, the following book is recommended: *JPEG Digital Image Compression Standard*, by William Pennebaker and Joan Mitchell.

We have described the various steps in JPEG encoding. The Baseline JPEG process can be summarized by the following encode and decode steps, as shown in Fig. 25.2.

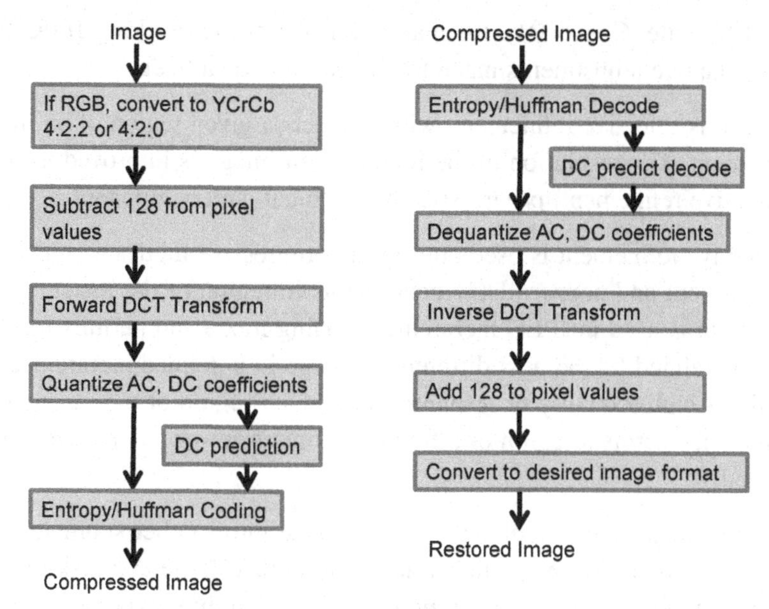

Figure 25.2
JPEG encode and decode steps.

25.5 JPEG Extensions

The JPEG standard provides several extensions, some of which are summarized below.

Huffman coding is popular and has no intellectual property restrictions. However, some variants of JPEG use an alternate coding method known as arithmetic coding. Arithmetic coding is more efficient and will adapt to changes in the statistical estimates of the input data stream, and it is subject to patent limitations.

Variable quantization is an enhancement to the quantization procedure of DCT output. This enhancement can be used with the DCTs in JPEG except for the baseline JPEG. The quantization values can be redefined prior to the start of an image scan but must not be changed once they are within a scan.

With variable quantization, the quantization values are scaled at the start of each 8×8 block, but matching the scale factors used to the AC coefficients stored in the compressed data. Quantization values may then be located and changed as needed. This provides for variable quantization based on the characteristics of an image. The variable quantizer continually adjusts during decoding to provide higher quality at the expense of increasing the size of the JPEG file. Conversely, the maximum size of the resulting JPEG file can be set by constant adaptive adjustments made by the variable quantizer.

Another extension is selective refinement, which selects a given region of an image for further enhancement. The resolution of the region of an image is improved using three methods of selective refinement: progressive, hierarchical, and component.

Progressive selective refinement is used only in the progressive modes to add more bit resolution of near zero and nonzero DCT coefficients in region of the image. Hierarchical selective refinement is used in JPEG hierarchical coding mode and permits for a region of an image to be refined by the next differential image in a defined hierarchical sequence. It allows higher quality or resolution in a given region of the image. Component selective refinement permits a region of a frame to contain fewer colors than are originally defined.

Image tiling is an enhancement to divide a single image into smaller subimages, which allows for smaller memory buffers, quicker access in both volatile and disk memory, and the storing and compression of very large images. There are three types of tiling: simple, pyramidal, and composite.

Simple tiling divides an image in multiple fixed-size tiles. All simple tiles are coded from top to bottom, left to right, and are adjacent. The tiles are all the same size, and encoded using the same procedure.

Pyramidal tiling also partitions the image into multiple tiles, but each tile can have different levels of resolution. This is known as the JPEG Tiled Image Pyramid (JTIP) model, resulting in a multiresolution pyramidal JPEG image. The JTIP image has successive layers of the same image but using different resolutions. The top of the pyramid has an image that is 1/16th of the defined screen size. It is called the vignette that can be for quick displays of image contents. The next image is one-fourth of the screen and is called the imagette. This is often used to display multiple images simultaneously. Then comes is a lower-resolution, full-screen image. After that are higher-resolution images. The last is the original image. Each of the pyramidal images can be JPEG encoded, either separately or together in the same data stream. If done separately, then it can allow for faster access of the selected image quality.

Multiple-resolution versions of images can also be stored and displayed using composite tiling known as a mosaic. The difference from pyramidal tiling is that composite tiling permits the tiles to overlap, be different sized, and be encoded using different quantization scaling. Each tile is encoded independently, so they can be easily combined.

Other JPEG extensions are detailed in the JPEG standards.

25.6 Video Compression Basics

MPEG is often considered ubiquitous with image compression. MPEG actually stands for Moving Pictures Experts Group, a committee which publishes international standards on video compression. MPEG is a portfolio of video compression standards, which will be discussed further in the next chapter.

Image compression theory and implementation focus on taking the advantage of the spatial redundancy present in the image. Video is composed of a series of images, usually referred to as frames, and so can be compressed by compressing the individual frames as discussed in the last chapter. However, there are temporal (or across time) redundancies present across video frames. This means that the frame immediately following has a lot in common with the current and previous frames. In most videos, there will be a significant amount of repetition in the sequences of frames. This property can be used to reduce the amount of data used to represent and store a video sequence. To take advantage of the temporal redundancy, this commonality must be determined across the frames.

This is known as predictive coding and is effective to reduce the amount of data that must be stored or streamed for a given video sequence. Usually, only parts of the image changes from frame to frame, which permits prediction from previous frames. Motion compensation is used in the predictive process. If an image sequence contains moving objects, then the motion of these objects within the scene can be measured, and the information used to predict the location of the object in frames later in the sequence.

Unfortunately, this is not as simple as just comparing regions in one frame to another. If the background is constant, and objects move in the foreground, there will be significant areas of the frames that will not change. But if the camera is panning across a scene, then there will be areas of subsequent frames that will be the same, but will be shifted in location from frame to frame. One way to measure this is to sum up all the absolute differences (without regard to sign), pixel by pixel, between two frames. Then the frame can be offset by one or more pixels and the same comparison run. After many such comparisons, the sum of differences results can be compared, and the minimum result corresponds to the best match. This will provide for a method to determine the location offset of the match between frames. This is known as the *minimum absolute differences* (MAD) method, or sometime referred to as *sum of absolute differences* (SAD). An alternative is the *minimum mean square error* (MMSE) method that measures the sum of the squared pixel differences. This method can be useful because it accentuates large differences, due to the squaring, but the trade-off is it requires multiplies in addition to subtractions.

25.7 Block Size

A practical trade-off is determining the block size to run the comparison over. When the block size is too small, there is no benefit in trying to reuse pixel data from a previous frame rather than just use the current frame data. If the block is too large, there will be many more differences likely and is difficult to get a close match. One obvious block size to run comparisons across would be the 8×8 block size used for DCT. Experimentation has shown that a 16×16 pixel area works well, and this is commonly used. This size is referred to as a "macroblock."

The computational effort to run these comparisons can be enormous. For each macroblock, 256 pixels must be compared against a 16×16 area of the previous frame. If we assume the macroblock data could shift by up to, say, 256 pixels horizontally or 128 pixels vertically (a portion of an HD 1080×1920) from one frame to another, there will be 256 possible shifts to each side and 128 possible shifts up or down. This is a total of $512 \times 256 = 131{,}072$ permutations to check, each requiring 256 difference computations. This is over 33 million per macroblock, with $1080/16 \times 1920/16 = 8100$ macroblocks per HD image or frame (Fig. 25.3).

25.8 Motion Estimation

However, there are methods to reduce this brute force method. A larger block size can be used initially, over larger portions of the image, to estimate the motion trend between frames. For example, enlarging the block size to include macrocells on all sides, for a 48×48 pixel region, can be used for motion estimation. Once the motion estimation is completed, this can be used to dramatically narrow down the number of permutations to perform the MAD or MMSE over. This motion estimation process is normally performed using luminary (Y) data. Also, motion estimation is performed locally for each region. The case of the camera panning will move large portions of an image with a common motion vector, but in cases with moving objects within the video sequence, different objects or regions will have different motion vectors.

Note that these motion estimation issues are present only on the encoder side. The decoder side is much simpler, as it must recreate the image frames based on data supplies in compressed form. This asymmetry is common in audio, image, video, or general data compression algorithms. The encoder must search for the redundancy present in the input data using some iterative method, whereas the decoder does not require this functionality (Fig. 25.4).

Once a good match is found, the macroblocks in the following frame data can be minimized during compression by referencing the macroblocks in previous frames.

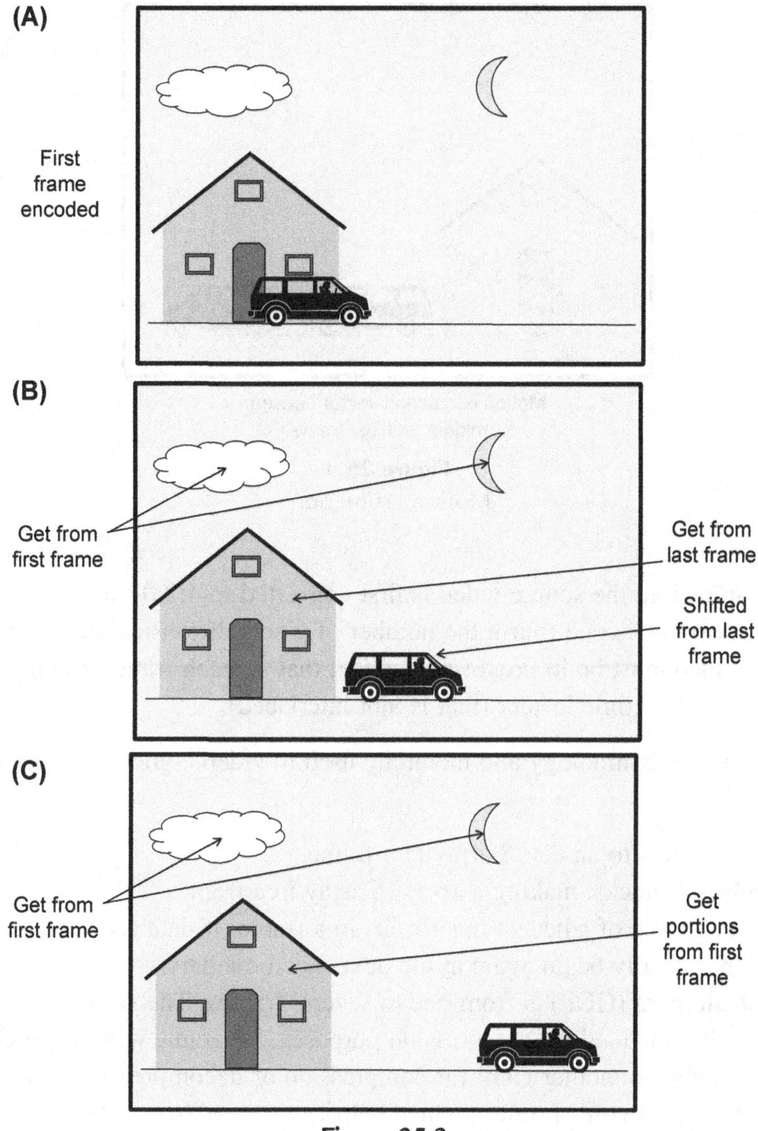

Figure 25.3
Macroblock matching across nearby frames.

However, even a "good" match will have some error. That is referred to a residuals or residual artifacts. These artifacts can be determined by differences in the four 8×8 DCTs in the macroblock. These differences can be coded and compressed as part of the frame data. Of course, if the residual is too large, coding the residual might require more data compared to just compressing the image data without any reference to the previous frame.

Motion estimation vector between middle and last frame

Figure 25.4
Motion estimation.

For MPEG compression, the source video is first converted to 4:2:0 format, so the chrominance data frame is one-fourth the number of pixels, ½ vertical and ½ horizontal resolution. The video must be in progressive mode; that is, each frame is composed of pixels all from the same time instant (that is, not interlaced).

At this point, a bit of terminology and hierarchy used in video compression needs to be introduced.

- A **pixel block** refers to an 8×8 array in a frame.
- A **macrocell** is 4 blocks, making a 16×16 array in a frame.
- A **slice** is a sequence of adjacent macrocells in a frame. If data are corrupted, the decoding can typically begin again at the next slice boundary.
- A **group of pictures** (GOP) is from one to several frames. The significance of the GOP is that it is self-contained for compression purposes. No frame within one GOP uses data from a frame in another GOP for compression or decompression. Therefore, each GOP must begin with an I frame (defined below).
- A video is made up of a sequence of GOPs.

Most video compression algorithms have three types of frames:

- **I frames**—These are frames which are compressed using only information in the current frame. The video compression term for this is intracoded, meaning the coding uses information within the frame only. A GOP always begins with an I frame, and no previous frame information is required to compress or decompress an I frame.
- **P frames**—These are predicted frames. P frames are compressed using image data from an I or P frame (may not be the immediate preceding frame) and comparing to the

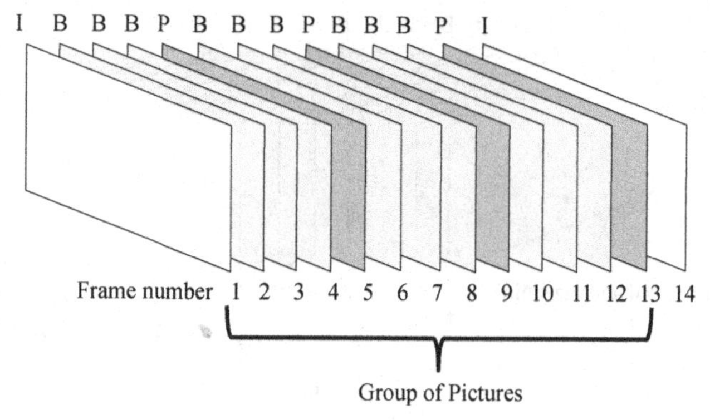

Figure 25.5
Group of pictures example.

current P frame. Restoring or decompressing the frame requires compressed data from a previous I or P frame plus residual data and motion estimation corresponding to the current P frame are used. The video compression term for this is intercoded, meaning the coding uses information across multiple video frames.

- **B frames**—These are bidirectional frames. B frames are compressed using image data from proceeding and following I or P frames. This is compared to the current B from a data to form the motion estimation and residuals. Restoring or decompressing the B frame requires compressed data from proceeding and following I or P frames plus residual data and motion estimation corresponding to the current B frame are used. B frames are intercoded (Fig. 25.5).

Also, by finishing the GOP with a P frame, the last few B frames can be decoded without needing information from the I frame of the following GOP.

25.9 Frame Processing Order

The order of frame encode processing is not sequential. Forward and backward prediction mean that each frame has a specific dependency that mandates processing order and requires buffering of the video frames to allow out-of-sequential order processing. This also introduces multiframe latency in the encoding and decoding process.

At the start of the GOP, the I frame is processed first. Then the next P frame is processed, as it needs current frame plus information from the previous I or P frame. Then the B frames in between are processed, as in addition to the current frame, information from both previous and postframes are used. Then the next P frame, and after that the intervening B frames, as shown in the processing order given in Fig. 25.6.

Note that since this GOP begins with an I frame and finishes with a P frame, it is completely self-contained, which is advantageous when there is data corruption during

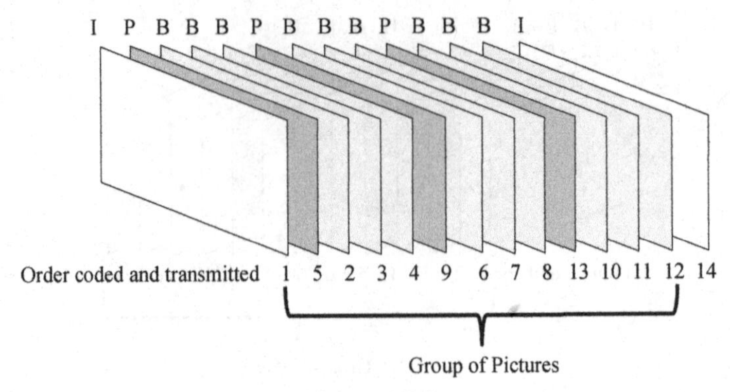

Figure 25.6
Video frame sequencing example.

playback or decoding. A given corrupted frame can only impact one GOP. The following GOP, since it begins with an I frame, is independent of problems in a previous GOP.

25.10 Compressing I Frames

I frames are compressed in very similar fashion using the JPEG techniques covered in the last chapter. The DCT is used to transform pixel blocks into the frequency domain, after which quantization and entropy coding is performed. I frames do not use information in any other video frame, therefore an I frame is compressed and decompressed independently from other frames. Both the luminance and chrominance are compressed, separately. Since 4:2:0 format is used, the chrominance will have only ¼ as many macrocells as the luminance. The quantization table used is given below. Notice how again larger amounts of quantization are used for higher horizontal and vertical frequencies, as the human vision is less sensitive to high frequency.

Luminance and Chrominance Quantization Table

8	16	19	22	26	27	29	34
16	16	22	24	27	29	34	37
19	22	26	27	29	34	34	38
22	22	26	27	29	34	37	40
22	26	27	29	32	35	40	48
26	27	29	32	35	40	48	58
26	27	29	34	38	46	56	69
27	29	35	38	46	56	69	83

Further scaling is provided by means of a quantization scale factor, which will be discussed further in the rate control description.

The DC coefficients are coded differentially, using the difference from the previous frame, which takes into account the high degree of the average or DC level of adjacent blocks.

Entropy encoding similar to JPEG is used. In fact, if all the frames are treated as I frames, this is pretty much the equivalent of JPEG compressing each frame of image independently.

25.11 Compressing P Frames

With P frames, a decision needs to be made for each macroblock, based on the motion estimation results. If the search for a match with another macroblock does not yield a good match in the previous I or P frame, then the macroblock must be coded as I frames are coded; that is, no temporal redundancy can be taken advantage of. If on the other hand, a good match is found in the previous I or P frame, then the current macroblock can be represented by a motion vector to the matching location in the previous frame, and by computing the residual, quantizing, and encoding (intercoded). The residual uses a uniform quantizer, and the DC component is treated just as the rest of the AC coefficients.

Residual Quantization Table

16	16	16	16	16	16	16	16
16	16	16	16	16	16	16	16
16	16	16	16	16	16	16	16
16	16	16	16	16	16	16	16
16	16	16	16	16	16	16	16
16	16	16	16	16	16	16	16
16	16	16	16	16	16	16	16
16	16	16	16	16	16	16	16

Some encoders also compare the number of bits used to encode the motion vector and residual, to ensure that there is a savings by using the predictive representation. Otherwise, the macroblock can be intracoded.

25.12 Compressing B Frames

B frames try to find a match using both preceding and following I or P frames for each macroblock. The encoder searches for a motion vector resulting in a good match over both the previous and following I or P frame and, if found, uses that frame to intercode the macroblock. If unsuccessful, than the macroblock must be intracoded. Another option is to use the motion vector but with both preceding and following frames simultaneously. When computing the residual, the B macroframe pixels are subtracted to the average of the macroblocks in the two I or P frames. This is done for each pixel, to compute the 256 residuals values, which are then quantized and encoded.

Many encoders then compare the number of bits used to encode the motion vector and residual using the preceding, following, or average of both to see which provides the most bit savings with predictive representation. Otherwise, the macroblock is intracoded.

25.13 Rate Control and Buffering

Video rate control an important part of image compression. The decoder has a video buffer of fixed size, and the encoding process must ensure that this buffer never underruns or overruns the buffer size, as either of these events can cause very noticeable discontinuities to the viewer.

The video compression and decompression process requires processing of the frames in a nonsequential order. Normally, this is done in "real-time". Consider these three scenarios:

1. Video storage—Video frames arriving at 30 frames per minute must be compressed (encoded) and stored to a file.
2. Video readback—A compressed video file is decompressed (decoded), producing 30 video frames per minute
3. Streaming video—A video source at 30 frames per minute must be compressed (encoded) to transmit over a bandwidth limited channel (able to support maximum number of bits/second throughput), the decompressed (decoded) video is to be displayed at 30 frames per minute at the destination.

All of these scenarios will require buffering or temporary storage of video files during the encoding and decoding process. As mentioned above, this is due to the out-of-order processing of the video frames. The larger the GOP, containing longer sequences of I, P, and B frames, the more buffering is potentially needed.

The number of bits to encode a given frame depends on the video content, complexity and the type of video frame. A video frame with lots of constant background (the sky, a concrete wall) will take few bits to encode, whereas a complex static scene like a nature film will require much more bits. Fast moving, blurring scenes with a lot of camera panning will be somewhere in between as the quantizing of the high frequency DCT coefficients will tend to keep the number of bits moderate.

I frames take the most bits to represent, as there is no temporal redundancies to take advantage. Then comes P, and the fewest are B frames, since B frames can leverage motion vectors and predictions from both previous and following frames. P and B frames will require much less bits than I frames, if there is little temporal difference between successive frames, or conversely, may not be able to save any bits through motion estimation and vectors if the successive frames exhibit little temporal correlation.

Despite this, the average rate of the compressed video stream often needs to be held constant. The transmission channel carrying the compressed video signal may have a fixed bit rate, and keeping this bit rate low is the reason for compression in the first place. This does not mean the bits for each frame, say at a 30 Hz rate, will be equal, but that the average bit rate over a reasonable number of frames may need to be constant. This requires a buffer, to absorb higher numbers of bits from some frames, and provide enough bits to transmit for those frames encoded with few bits.

25.14 Quantization Scale Factor

Since the video content cannot be predicted in advance, and this content will cause variation amount of bits required to encode the video sequence, provision is made to dynamically force a reduction in the number of bits per frame. A scale factor is applied to the quantization process of the AC coefficients of the DCT, which is referred to as Mquant.

The AC coefficients are first multiplied by 8, then divided by the value in the quantization table, and then divided again by Mquant. Mquant can vary from 1 to 31, with a default value of 8. For the default level, Mquant just cancels the initial multiplication of 8.

One extreme is an Mquant of 1. In this case, all AC coefficients are multiplied by 8, then divided by the quantization table value. The results will tend to be larger, nonzero numbers, which will preserve more frequency information at the expense of using more bits to encode the slice. This results in higher quality frames.

The other extreme is an Mquant of 31. In this case, all AC coefficients are multiplied by 8, then divided by the quantization table value, then divided again by 31. The results will tend to be small, mostly zero numbers, which will remove most spatial frequency information and reduce bits to encode the slice. This results in lower quality frames. Mquant provides a means to trade quality verses compression rate, or number of bits to represent the video frame. Mquant is normally updated at the slice boundary and is sent to the decoder as part of the header information.

This process is complicated by the fact that human visual system is sensitive to video quality, especially for scenes with little temporal or motion activity. Preserving a reasonable consistent quality level needs to be considered as the Mquant scale factor is varied (Fig. 25.7).

Here the state of the decoder input buffer is shown. The buffer fills at a constant rate (positive slope) but empties discontinuously as various frame size I, P, and B data are read for each frame decoding process. The amount of data for each of these frames will vary according to video content and the quantization scale factors the encode process has chosen.

To ensure the encoder process never causes the decode buffer to overflow or underflow, it is modeled using a video buffer verifier (VBV) in the encoder. The input buffer in the decoder as conservatively sized, due to consumer cost sensitivity. The encoder will use the VBV model to mirror the actual video decoder state, and the state of the VBV can be used to drive the rate control algorithm, which in turn dynamically adjusts the quantization

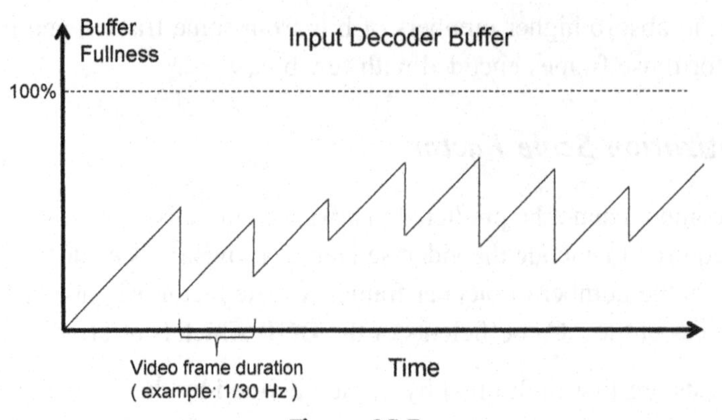

Figure 25.7
Video frame buffer behavior.

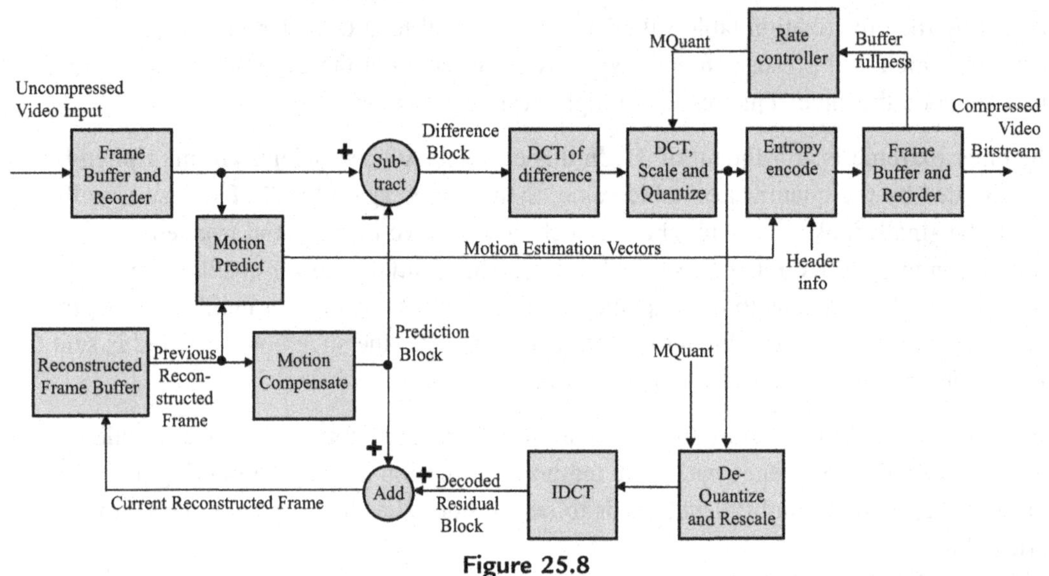

Figure 25.8
Video compression encoder.

scale factor used in the encode process and sent to the decoder. This process closes the feedback loop in the encoder to ensure the decoder buffer does not over or under flow (Fig. 25.8).

The video encoder is shown in a simplified block diagram. The following steps take place in the video encoder:

1. Input video frames are buffered and then ordered. Each video frame is processed macroblock by macroblock.

2. For P or B frames, the video frame is compared to an encoded reference frame (another I or P frame). The motion estimation function searches and looks for matches between macroblocks of the current and previously encoded frames. The spatial offset between the macroblock position in the two frames is the motion vector associated with macroblock.
3. The motion vector points to be best matched macroblock in the previous frame, which is called a motion compensated prediction macroblock. It is subtracted from the current frame macroblock to form the residual or difference.
4. The difference is transformed using the DCT and quantized. The quantization also uses a scaling factor to regulate the average number of bits in compressed video frames.
5. The quantizer output, motion vector, and header information are entropy (variable length) coded, resulting in the compressed video bitstream.
6. In a feedback path, the quantized macroblocks are rescaled and transformed using the IDCT to generate the same difference or residual as the decoder. It has the same artifacts as the decoder due to the quantization processing, which is irreversible.
7. The quantized difference is added to the motion compensated prediction macroblock (see Step 2 above). This is used to form the reconstructed frame, which can be used as the reference frame for encoding the next frame. Recall the decoder will have access only to reconstructed video frames, not the actual original video frames, to use a reference frames (Fig. 25.9).

The video encoder is shown in a simplified block diagram. The following steps take place in the video encoder:

1. Input compressed stream entropy decoded. This extracts header information, coefficients, and motions vectors.

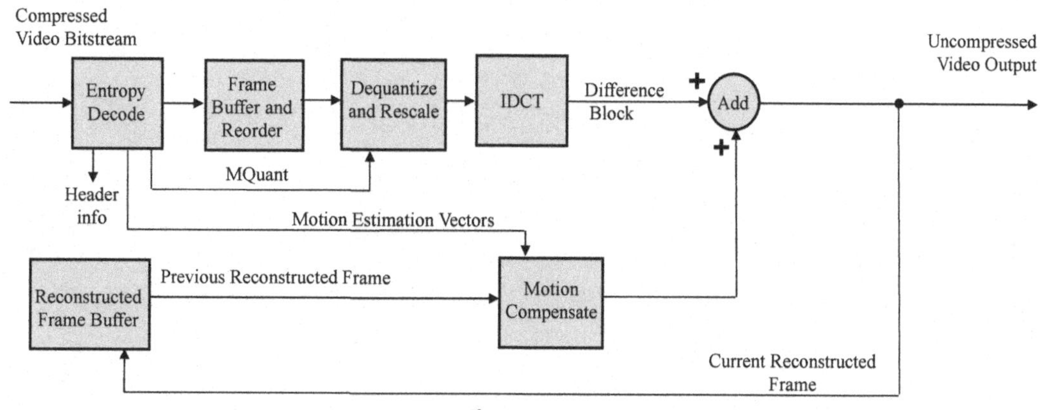

Figure 25.9
Video compression decoder.

2. The data are ordered into video frames of different types (I, P,B), buffered and reordered.
3. For each frame, at the macroblock level, coefficients are rescaled and transformed using IDCT, to produce the difference or residual block.
4. The decoded motion vector is used to extract the macroblock data from a previous decoded video frame. This becomes the motion compensated prediction block for the current macroblock.
5. The difference block and the motion compensated prediction block are summed together to produce the reconstructed macroblock. Macroblock by macroblock, the video frame is reconstructed.
6. The reconstructed video frame is buffered and can be used to form prediction blocks for subsequent frames in the GOP. The buffered frames are output from the decoder in sequential order for viewing.

The decoder is basically a subset of the encoder functions, and it fortunately requires much less computations. This supports the often asymmetric nature of video compression. A video source can be compressed once, using higher performance broadcast equipment, and then distributed to many users, who can decode or decompress using much less costly consumer type equipment.

CHAPTER 26

Introduction to Machine Learning*

Most complex machines or software utilize algorithms, which is a methodology defined to solve a specific problem. In some cases, the algorithm has an adaptive capability, allowing the algorithm to adapt for varying conditions. In contrast, machine learning does not have specific algorithm. Rather, it has a basic structure, to process data and produce outputs. The structure typically contains thousands or millions of weights, or coefficients. These weights are determined by passing data, lots of data, through the structure and updating the weights during a feedback process where the results are usually known in advance, to produce the desired output. This weight updating is called training, and large amounts of data, often labeled with the "right" answer, is used. Labeled data training is called supervised training, where the correct outcome is labeled and used as feedback. When the data has no labeling, the training is unsupervised. In either case, the machine learning is trained by using enormous amounts of data and trying many scenarios and incrementally updating the weights to produce the desired results. This eventually allows correct results to be produced at the great majority of the time. It also means that a human designer or engineer is not required to painstakingly develop and tune an algorithm for a specific task. Enormous amounts of compute power are able to replace "intelligence" and as powerful compute devices as well as large databases become evermore available; machine learning will be more prevalent in many aspects of everyday life. Some everyday examples of machine learning are as follows: internet search engine behavior, email spam identification, credit card fraud alerts, tax return anomaly detections, online product shopping suggestions, speech recognition, facial recognition, automotive driving assist functionality, and much more.

26.1 Convolutional Neural Networks

This introductory chapter will focus on a type of machine learning called "convolutional neural networks" or CNNs, which is commonly used to process image or video data. Convolution is the same process used in finite impulse response (FIR) filters—running a series of data past a set of coefficients. At each step, the sum of all the products between the data and coefficients is computed. In the case of image data, the filtering is performed across the spatial, not time dimension. With image data, this will be two spatial dimensions. The filter coefficients can be described in a two dimensional function, F(x,y).

* With contributions from Utku Aydonat, Gordon Chiu, Shane O'Connell, Davor Capalija and Andrew Ling.

Digital Signal Processing 101. http://dx.doi.org/10.1016/B978-0-12-811453-7.00026-3

$$P_{new}[x][y] = \sum_{x'=-i}^{i} \sum_{y'=-i}^{i} P_{old}[x + x'][y + y'] \times F[x'][y']$$

A classic problem is image recognition—identifying the object(s) in the image which may be varying in size, color, light affects, pose, aspect ratio, may be partially obscured or surrounded by clutter, and many other issues. The motivation for neural network processing is to try imitating the type of processing used in the nature to process vision information coming from the sensors in the eye. This is also sometimes called deep learning, and it attempts to model high-level abstractions and derive features without having to be explicitly coded into the system by the designer.

Neural networks are a "graph" of many layers and of different types. The image data passes through each layer in turn. It uses many layers (also known as deep graphs) with both nonlinear and linear processing layers to model various data features, at both fine-grained and coarse level. The graph is designed so the early layers process nearby data, identifying small-scale features—edges, shapes, and stripes. As features are identified, later layers will process data across large swaths of the image. The layers successively extract higher level of features.

Each layer can have M input 2-D planes of feature data (is really just an image at the first layer), and N output 2-D planes of feature data, which will depend on the particulars of each layer. Often the 2-D planes of feature data are called channels. For example, the initial input to graph is normally three channels, corresponding to the three RGB color planes of the same image. But subsequent layers will be much deeper, as there are multiple filters, each separately applied to the RBG separate color planes.

There are different types of layers. The common layer types are as follows:

- Convolutional
- Nonlinear function (ReLU)
- Normalization
- Max-Pooling
- Fully Connected

26.2 Convolution Layer

Convolution is FIR filtering. In this case, the filtering is spatial, rather than temporal. A small patch of the image (say 7 × 7 pixels) is processed using a spatial filter, with separate coefficients or parameters for each pixel (in this case, 49). Each pixel in the output feature plane is the accumulation of the 49 products of each pixel value multiplied by its parameter. Then the 7 × 7 pixel map is stepped 1 pixel to the right, and the process is repeated. At the right edge of the feature plane, the 7 × 7 patch is moved back to the left edge, and down 1 pixel, and repeated, in a raster scan type of order. The output feature plane has the same number of pixels as the input feature plane. (Some zero padding may be necessary at the input data edges to deal with edge effects.) In the initial input layer, this is repeated three times (not shown in Fig. 26.1) as the input image is typically RGB and has three color planes, each to be processed.

Moreover, there are typically many filters (as defined by their coefficients) used for each convolutional layer. In the early graphs, filters are used to identify edges, corners, or other fine-grained features, at various orientations. These filters are dependent on the type of objects being searched for by the graph. Near the input of the graph, there tends to be fewer numbers of filters, but use of larger size patches. Deeper into the graph, the number of filters increases, but the patch sizes tend to decrease. Both number of filters and patch size will determine the computational effort of each convolutional layer. The convolutional layers tend to mimic neural processing across a small number of inputs that are close together and are used to detect localized features throughout the initial image.

The depth of the convolution output will depend on the number of filters, each of which produces separate output feature planes or channels.

The convolution can also be performed with different strides. Stride is amount of pixels that the filter patch will slide for each convolution. The previous description was for a

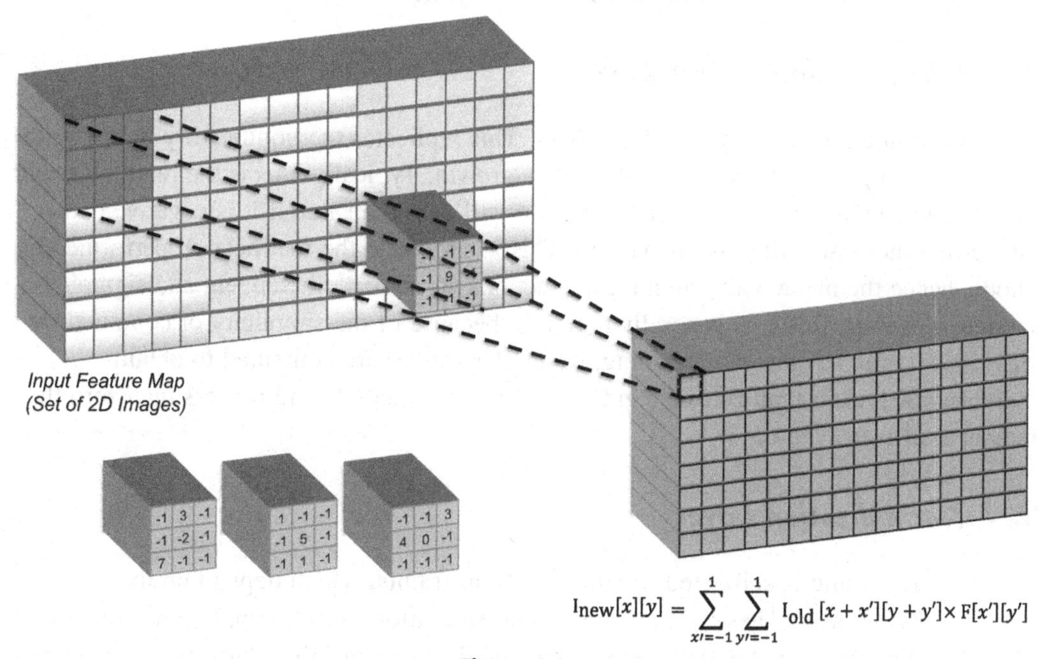

Figure 26.1
Convolutional or partially connected processing.

stride of one, with the 3×3 filter patch shown in Fig. 26.1 being moved by 1 pixel to the right for the next convolution. Using stride of one will result in the output patch size being equal to the input patch size, meaning there is a convolution computed for each pixel of the input image. In some layers, a different stride will be used. For example, using a stride of two in both horizontal and vertical directions means that for the next convolution operation, the filter patch will slide over by 2 pixels, not 1. This results in downsampling of the input channel during filtering, and in this case, results in an output channel with half of the pixels in both horizontal and vertical dimensions, or reduction in pixels by a factor of 4.

For example, one of the popular CNN graphs, AlexNet, uses a stride of four for the first convolution layer of the input. This reduces the image size (number of pixels) by a factor of 16 for subsequent layers. While the input image size is reduced (horizontal and vertical directions), the depth will increase. Starting with just three layers of the RGB color plane, after application of multiple filters to each layer, the depth will increase dramatically. For example, there may be quite a few filters to detect edges and corners in all possible directions. The low-level features such as edges and corners can be used to subsequent layers to detect shapes and objects.

Some graphs will also perform a secondary filter operation across the depth dimension, at each pixel location. The initial filter results across the image can themselves be filtered, to provide various levels of weighting to each of those filters.

26.3 Rectified Linear Unit Layer

A nonlinear function is always used in CNNs. This replicates the nonlinearity in biological image processing, as well as helps reduce data growth. Typically, this is the rectified linear unit, abbreviated as ReLU. The ReLU function will zero the value if it is negative (less than zero), otherwise will pass unchanged. This is similar to the rectification performed by a diode, hence the name. Other nonlinear functions are hyperbolic tangent and sigmoid function. However, ReLU is generally preferred because of the simplicity of computation. ReLU also has an advantage in training, where derivatives are computed to obtain gradients. For greater than zero, the gradient of ReLU equals 1, and for less than zero the gradient is of course zero.

26.4 Normalization Layer

The CNN processing is subjected to numerical constraints, which depend on the numerical format used. However, at some point, saturation must be applied to prevent overflow. This process is typically similar to CFAR (constant false alarm rate) processing in radar detection. The convolution results are divided by a value presenting some sort of mean of the neighboring values. Often these nearby values are squared, summed, and then a square root is taken over the result. A general equation is given for normalization, where a^j and b^i are the input and output maps, respectively, of the normalization layer.

$$b^i_{x,y} = a^i_{x,y} \Big/ \left(k + \alpha \sum_{j=\max(0,i-n/2)}^{\min(N-1,i+n/2)} \left(a^i_{x,y}\right)^2 \right)^{\beta}$$

26.5 Max-Pooling Layer

This is a method to reduce feature plane size, downsampling, or decimating the channel. However, rather using a stride greater than one for convolution, a nonlinear method is used. Normally, this is just the maximum operation. This takes the input channel and partitions into a set of nonoverlapping rectangles of pixels. Then the pixel with the maximum value is chosen to represent all the pixels for that partition in the corresponding output image. The idea is that, once feature has a strong correlation to the convolutional filter, the precise location is less important compared with the relative location to other features. The pooling operation provides a form of translation invariance, when an object or feature has movement between image frames. Max pooling also reduces the input channel size, which reduces the number of weights and computation in subsequent layers. Typically, the graph is organized into subgraphs, which contain convolutional, ReLU, normalization, and max pooling. The fully connected layers are used in the later stages of the graph. The pooling layer operates independently on every depth slice of the input and downsizes it spatially (See Fig. 26.2).

26.6 Fully Connected Layer

Fully connected layers are applied after all the lower level and local features have been detected via the convolutional and max-pooling layers. In a fully connected layer, every

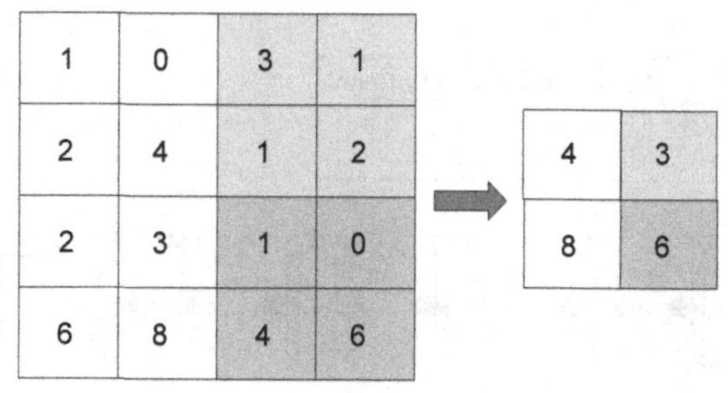

Figure 26.2
Max pooling with 2 × 2 partitions.

pixel in the input-feature plane can have an effect on every pixel in the output-feature plane. A fully connected layer, for a given input-feature plane size, has the highest computational requirement. All activations from the previous layers are weighted and allowed for parameters in the fully connected layers to encompass the full range of the feature plane. The processing is performed using matrix multiplication. Normally, there is also a bias offset of each of the matrix multiplication dot products (See Fig. 26.3).

CNN graphs have no state information. The feature plane flows through, with no intermediate storage, until the result is obtained. With optimized parameters, the graph will respond, or activate, to the desired features and objects. The resulting output will indicate how strong of a match—whether looking for dogs, for a particular human face, or character recognition, or many other classes of objects.

The initial network topology to gain wide popularity was AlexNet. Subsequently, many other networks were developed, usually with higher number of layers and more computational complexity. Some examples are GoogLeNet, ResNet, FractalNet, SqueezeNet, and DriveNet. Many researchers are actively working to develop new graph networks, which give a combination of better accuracy for a reasonable amount of computational throughput, for a given application.

The computational load for an AlexNet graph to process a single RGB image of 224 × 224 pixels is over a GFLOP and requires over 60 million parameters. The amount

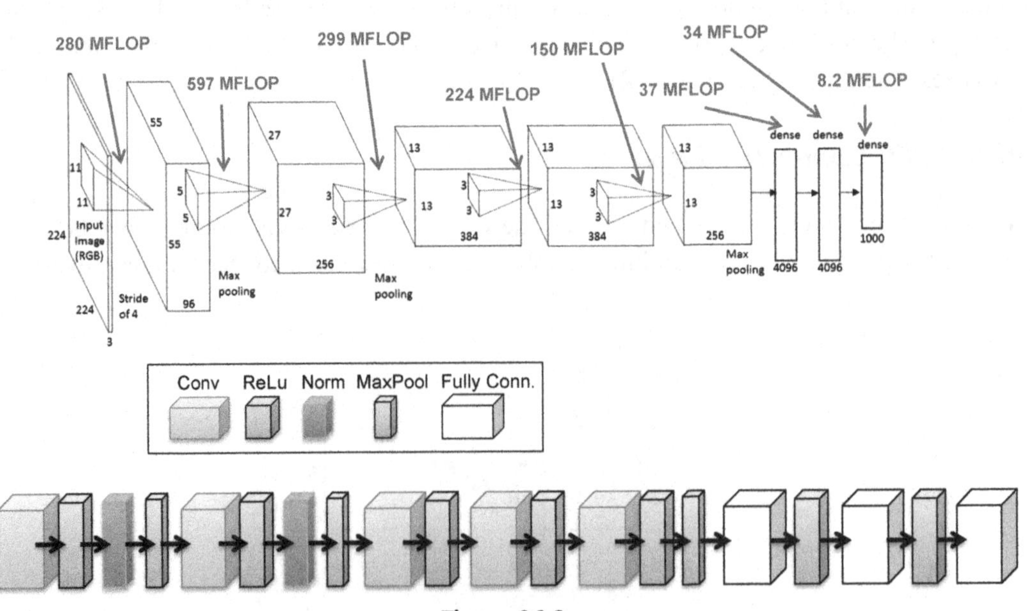

Figure 26.3
AlexNet graph.

of signal processing required for CNNs is astronomical, and there is great effort in the semiconductor industry to develop increasingly more powerful and efficient signal processing architectures to support machine learning.

26.7 Training Computational Neural Networks

A CNN graph contains many millions of coefficients, or parameters, which encapsulate its behavior, or ability to distinguish, often with great accuracy, between specific objects of interest in various poses, surrounded by a variety of backgrounds and objects which are not of interest. Given the sheer quantity of the parameters, they cannot be computed directly. The only methodology that could cope with such a huge number of unknown parameters is an iterative approach, and by using vast amounts of data.

Training is generally performed offline and requires much greater amounts of computation. The key requirement for effective training is to have enormous amounts of data—hundreds of thousands or millions of images of interest for the task the CNN is being trained for (Fig. 26.4). Companies such as Google, Facebook, Netflix have access to very large databases to utilize in training. This also applies for non-image data. For example, Amazon possesses enormous databases of customer buying patterns and preferences, which can be leveraged to better market products to these customers. Automotive manufacturers and their key supplier companies are acquiring large databases of real-world driving experiences through vehicles equipped with cameras for driving assist which have networking capabilities. These databases can be invaluable in developing completely autonomous vehicles.

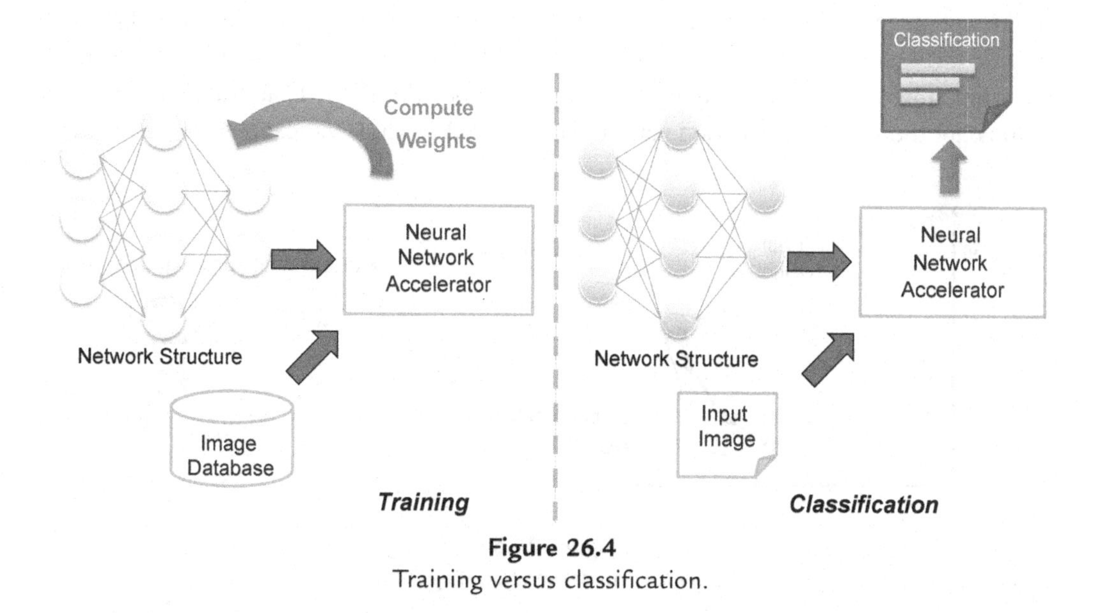

Figure 26.4
Training versus classification.

Once a network is trained, it can be retrained to detect or classify reasonably similar objects with much less effort than the original training. For example, a CNN might be trained to detect and classify different breeds of dogs. If the task was changed to horses instead, the CNN would only need to be partially retrained. The very fine-grained filters in the early stage of the CNN graph could likely remain the same. However, the fully connected layers toward the end of the CNN graph would be retrained, as these are the layers that put together the complete picture and classify the entire object of interest.

Once a CNN has been trained, it can process input data such as images or video streams in a real-time fashion, given sufficient computational capability of the hardware.

Most CNNs use supervised training. This means that labeled data is available to train the network. This can provide a cost function, differentiating correct classification outputs and incorrect outputs. The general procedure is to create a cost function between the correct outcome and the actual outcome and try to minimize this cost by updating weights and parameters in the direction of steepest descent. Given J(w) describes surface and w is the current location, the way to get to the minimum level (lowest cost or error) is to determine the negative gradient of J, which is the steepest descent and to take a small step in that direction. Over many iterations, even with the presence of noise, eventually the lowest level (or reasonable close) will normally be found (See Fig. 26.5).

CNNs are trained using backpropagation. This means that the last layers are trained first, and gradually the training moves backward until the first layer is trained. Since a graph is a composition of layer functions, the gradient of a graph is the product of layer gradients.

Classification data flow is forward, from the input to the output. Training is backward, from the last stage backward to the first. This works because the gradient follows the chain rule through the graph. The key functions are being able to find the gradient of the error (a function of the actual output vs. the known labeled result). That gradient is what is used to adapt the weights, or parameters in each layer. Fig. 26.6 indicates how the result

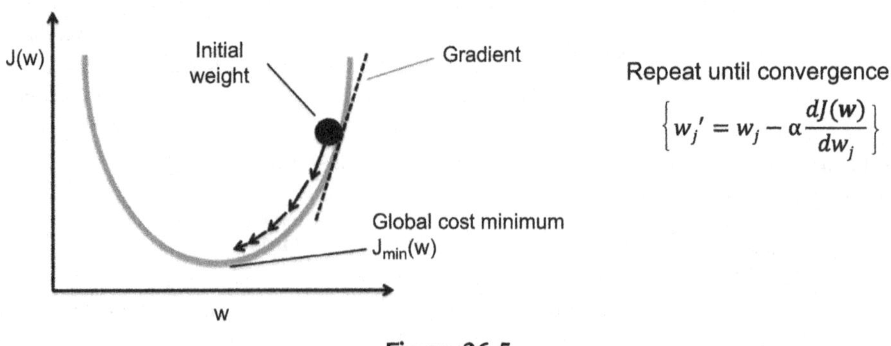

Figure 26.5
Gradient descent concept.

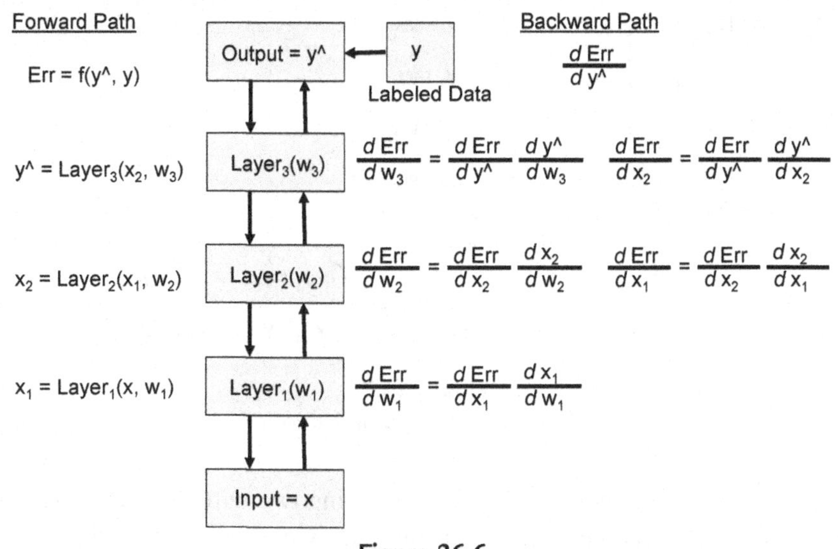

Figure 26.6
Gradient descent chain rule.

from the later layer is needed to calculate the gradient for the earlier layer in the graph. Training the graph using gradient descent involves computing the gradients with respect to the inputs and weights. The computation of the gradient requires convolution of the layer input feature plane with the backpropagated errors, using the same algorithm used for forward propagation. The calculations are given here, which is beyond the scope of an introduction.

26.8 Winograd Transform

The Winograd transform, named after Shmuel Winograd, is a technique to reduce the computational effort for convolution and can be used when the patch being convolved is small. Specifically, it reduces the amount of multipliers, even though the number of additions may increase significantly. For fixed point processing, often used in inference, the hardware cost and power consumption of multipliers is far greater than adders.

This will involve matrix multiplication, and use a new matrix operation shown in Fig. 26.7. Rather than traditional matrix multiply, this operation is an element by element multiplication, with no sum of products.

The Winograd transform **W**(y, b) produces a vector of y outputs and using a vector of b coefficients. For the two-dimensional case, the transform **W**(y * z, b * c) produces an array of y * z outputs by using an array of b * c coefficients.

$$\begin{pmatrix} V_1 \\ V_2 \\ V_3 \\ V_4 \end{pmatrix} \odot \begin{pmatrix} W_1 \\ W_2 \\ W_3 \\ W_4 \end{pmatrix} = \begin{pmatrix} V_1 * W_1 \\ V_2 * W_2 \\ V_3 * W_3 \\ V_4 * W_4 \end{pmatrix}$$

$$\begin{pmatrix} X_{1,1} & X_{1,2} \\ X_{2,1} & X_{2,2} \end{pmatrix} \odot \begin{pmatrix} Y_{1,1} & Y_{1,2} \\ Y_{2,1} & Y_{2,2} \end{pmatrix} = \begin{pmatrix} X_{1,1} * Y_{1,1} & X_{1,2} * Y_{1,2} \\ X_{2,1} * Y_{2,1} & X_{2,2} * Y_{2.2} \end{pmatrix}$$

Figure 26.7
Matrix element multiplication.

First, consider the two one-dimensional case, computing two outputs, each having three coefficients: **W**(2,3). This can be computed using matrix multiplication, shown in Fig. 26.8 and requires six multiply operations.

Alternately, the computation can be broken up into subcomponents $\mathbf{a_i}$. Notice that by performing the computation in this fashion, only four multiply operations, although a much higher adder count, is required.

This operation can be expressed in matrix form as shown in Fig. 26.9. Note that $\mathbf{G} * \mathbf{B}$ is a column vector that can be computed once for any set of coefficients. Also, that $\mathbf{D^T} * \mathbf{X}$, another column vector, does not require multiplications; rather it is composed of sums of the inputs. The only multiplies are between the two column vectors, to perform the element by element multiplication, and the number of multiplies equals the length of the column vector, which is 4 in this example. Multiplication of these results with $\mathbf{A^T}$ also

$$\mathbf{W}(2,3) = \begin{pmatrix} x_0 & x_1 & x_2 \\ x_1 & x_2 & x_3 \end{pmatrix} \begin{pmatrix} b_0 \\ b_1 \\ b_2 \end{pmatrix} = \begin{pmatrix} x_0 * b_0 + x_1 * b_1 + x_2 * b_2 \\ x_1 * b_0 + x_2 * b_1 + x_3 * b_2 \end{pmatrix} = \begin{pmatrix} y_0 \\ y_1 \end{pmatrix}$$

$$y_0 = a_1 + a_2 + a_3 \qquad \text{where} \qquad a_1 = (x_0 - x_2) * b_0$$

$$y_1 = a_2 - a_3 - a_4 \qquad a_2 = (x_1 + x_2) * (b_0 + b_1 + b_2) / 2$$

$$a_3 = (x_2 - x_1) * (b_0 - b_1 + b_2) / 2$$

$$a_4 = (x_1 - x_3) * b_2$$

Figure 26.8
Winograd operation for W(2,3).

$$W(2,3) = Y = A^T * [(G * B) \odot (D^T * X)]$$

$$A^T = \begin{pmatrix} 1 & 1 & 1 & 0 \\ 0 & 1 & -1 & -1 \end{pmatrix} \qquad B = \begin{pmatrix} b_0 \\ b_1 \\ b_2 \end{pmatrix}$$

$$G = \begin{pmatrix} 1 & 0 & 0 \\ ½ & ½ & ½ \\ ½ & -½ & ½ \\ 0 & 0 & 1 \end{pmatrix}$$

$$D^T = \begin{pmatrix} 1 & 0 & -1 & 0 \\ 0 & 1 & 1 & 0 \\ 0 & -1 & 1 & 0 \\ 0 & 1 & 0 & -1 \end{pmatrix} \qquad X = \begin{pmatrix} x_0 \\ x_1 \\ x_2 \\ x_3 \end{pmatrix}$$

Figure 26.9
Winograd matrix expression W(2,3).

does not require multiplication operations; rather it is composed of sums of the multiplication products.

With CNNs, the Winograd is normally used in two-dimensional cases. The equations are modified to perform the convolution in two dimensions. The result **Y** is a 2 × 2 matrix, containing four outputs.

$$\mathbf{W}(2 \times 2, 3 \times 3) = \mathbf{Y} = \mathbf{A}^{\mathrm{T}} * \left[\left(\mathbf{G} * \mathbf{B} * \mathbf{G}^{\mathrm{T}}\right) \odot \left(\mathbf{D}^{\mathrm{T}} * \mathbf{X} * \mathbf{D}\right)\right] * \mathbf{A}$$

The **A**, **G**, and **D** matrices are the same as in the one-dimensional W(2,3) example. The **B** matrix is a 3 × 3 matrix of the two-dimensional filter coefficients, and **X** is a 4 × 4 matrix of the input data tile.

The 4 × 4 matrix of ($\mathbf{G} * \mathbf{B} * \mathbf{G}^{\mathrm{T}}$) can be precomputed once. The 4 × 4 matrix of ($\mathbf{D}^{\mathrm{T}} * \mathbf{X} * \mathbf{D}$) can be computed with just adder operations. The element by element multiplication of these two 4 × 4 matrices required 16 multiplies. The last step of multiplying $\mathbf{A}^{\mathrm{T}}$ again requires no multiply operations, as $\mathbf{A}^{\mathrm{T}}$ contains only unity values.

Using straightforward convolution requires 3 × 3 = 9 multiplies for each output. Computing four outputs required 36 multiply operations, more than twice the 16 of the Winograd approach.

This can be extended to compute larger output patches, in this case an output tile of 4 × 4 rather than 2 × 2. This will require a larger input patch of 6 × 6, when using a 3 × 3-size filter.

The matrices are given in Fig. 26.10. In this case, the Winograd uses 6 × 6 or 36 multiplies. Using convolution, the number of multiplies is 9 × 16 = 144 (9 coefficients, 16 outputs). The Winograd reduces the number of multiplies by a factor of 4.

$$W(4x4,3x3) = Y = A^T * [(G * B * G^T) \odot (D^T * X * D)] * A$$

$$A^T = \begin{pmatrix} 1 & 1 & 1 & 1 & 1 & 0 \\ 0 & 1 & -1 & -2 & -2 & 0 \\ 0 & 1 & 1 & 4 & 4 & 0 \\ 0 & 1 & -1 & 8 & -8 & 1 \end{pmatrix}$$

$$G = \begin{pmatrix} 1/4 & 0 & 0 \\ -1/6 & -1/6 & -1/6 \\ -1/6 & 1/6 & -1/6 \\ 1/24 & 1/12 & 1/6 \\ 1/24 & -1/12 & 1/6 \\ 0 & 0 & 1 \end{pmatrix} \qquad B = \begin{pmatrix} b_{0,0} & b_{0,1} & b_{0,2} \\ b_{1,0} & b_{1,1} & b_{1,2} \\ b_{2,0} & b_{2,1} & b_{2,2} \end{pmatrix}$$

$$D^T = \begin{pmatrix} 4 & 0 & -5 & 0 & 1 & 0 \\ 0 & -4 & -4 & 1 & 1 & 0 \\ 0 & 4 & -4 & -1 & 1 & 0 \\ 0 & -2 & -1 & 2 & 1 & 0 \\ 0 & 2 & -1 & -2 & 1 & 0 \\ 0 & 4 & 0 & -5 & 0 & 1 \end{pmatrix} \qquad X = \begin{pmatrix} x_{0,0} & x_{0,1} & x_{0,2} & x_{0,3} & x_{0,4} & x_{0,5} \\ x_{1,0} & x_{1,1} & x_{1,2} & x_{1,3} & x_{1,4} & x_{1,5} \\ x_{2,0} & x_{2,1} & x_{2,2} & x_{2,3} & x_{2,4} & x_{2,5} \\ x_{3,0} & x_{3,1} & x_{3,2} & x_{3,3} & x_{3,4} & x_{3,5} \\ x_{4,0} & x_{4,1} & x_{4,2} & x_{4,3} & x_{4,4} & x_{4,5} \\ x_{5,0} & x_{5,1} & x_{5,2} & x_{5,3} & x_{5,4} & x_{5,5} \end{pmatrix}$$

Figure 26.10
Winograd matrix expression $W(4 \times 4,3 \times 3)$.

The number of multipliers to perform $\mathbf{W}(y * z, b * c)$ using convolution is $y * z * b * c$. The number of multipliers required using Winograd function is $(y + b - 1) * (z + c - 1)$.

26.9 Convolutional Neural Network Numerical Precision Requirements

Training algorithms are evolving rapidly and can be mathematically very complex. Many optimizations are determined empirically. There are also functions to try to zero some portion of parameters, to ensure noise does not creep up and compete with the actual activations being generated by the data. In addition, numerical effects can destabilize the system, when gradients become very small.

With the immense computational requirements for both CNN-based classification and training, engineers are always seeking the most efficient implementation methods. Early CNN training and classification used single-precision floating point, which requires a 32-bit word representation. Experimentation showed that classification did not require this degree of precision and dynamic range. Classification is also needed in environments where meeting cost, power consumption, and computational processing requirements are challenging. This could be for autonomous vehicles, for self-piloting drones, for law enforcement or soldier real-time camera analysis, robotic applications, augmented reality, smart phone apps, and many more situations.

It has been demonstrated that 8-bit integer is sufficient to provide the same classification quality as single-precision floating point. Even lower precision integer can be used, especially if one is willing to tolerate some degradation in CNN results (higher percentage of “wrong decisions”). Some researchers have even claimed that as low as 1-bit precision is sufficiently accurate for certain applications.

Use of 8-bit integer will provide a fourfold effective memory bandwidth increase, and in architectures optimized for integer, a fourfold increase in computational rates. Using 8-bit representation for parameters also reduces on-chip memory requirements. Some GPUs (graphics processing units) have been optimized to support 8-bit integer processing, and specialized semiconductor products are designed to natively support 8-bit integer processing. FPGAs have long been efficient at integer processing, and newer FPGAs are further optimized for 8-bit integer operations. Further research has shown that use of binary weights (+1, −1) or ternary weights (+1, 0, −1) is sufficient precision for some applications. This has the advantage of not needing multipliers—the data is added or subtracted depending on the weight values.

Training of CNNs is normally done in a data center environment, where access to many millions of training images is available. Due to the gradient computations involved in training algorithms, a greater degree of dynamic range is required. Here, it has been found that half-precision floating point, which requires a 16-bit word representation, is generally sufficient. Some accumulation operations may, however, benefit from the greater precision and dynamic range of single-precision floating point. Switching to half precision effectively doubles the memory bandwidth and computational processing rates. GPUs, which are popular accelerators for training in the data center, have been optimized to support half-precision floating point with twice the capacity of single precision. FPGAs are also used as machine-learning accelerators in data centers for machine-learning training and can support reduced-precision floating point operations. Another use case is reinforced training—where the network is pre-trained offline, but the weight values in the final fully connected layers are updated, or retrained in real time while performing inference processing on real data. This can allow for a network to be customized for its own local data, and produce better results in some cases.

CHAPTER 27

Implementation Using Digital Signal Processors

A digital signal processor is a special form of microprocessor that is optimized to perform digital signal processing (DSP) operations. The first DSP processors became available in 1982. They are designed to perform DSP functions in an efficient manner, using conventional software programming design and verification techniques. Owing to the special features and parallelism added for DSP operations, conventional programming languages, such as "C", often do not have the syntax to express the needed operations, although advanced compilers developed by the DSP processor manufacturers do try to interpret the programmer's intent and map the DSP processor features in the most effective manner. For this reason, a vendor-specific language, known as assembly, is instead used to code the most DSP-intensive portions of the software. Assembly language is a proprietary set of instructions for a given processor. This is not as big a restriction as it may sound, because the portion of the software code implemented in assembly is often 5%−10% or less of the entire code base, and the DSP vendors themselves often provide the reference code for many common DSP algorithmic implementations. But writing in assembly code does require knowledge of the DSP processor hardware architecture and how to optimally write code that takes advantage of the DSP processor architecture features.

27.1 Digital Signal Processing Processor Architectural Enhancements

DSP processors enhancements, as compared to microprocessors, are generally divided into three categories:

- Enable maximum data bandwidth in and out of the DSP core engine;
- Efficiently implement the mathematical operations commonly required in DSP; and
- Addition of multiple cores or hardware coprocessors to meet demands of high computational rate DSP applications.

27.1.1 Data I/O Bandwidth

All processors can be limited by data bandwidth in and out of the processor core or engine. However, this is especially a concern for DSP processors. This is due to the nature of the processing tasks. Many microcontrollers are primarily used to make decisions based

Digital Signal Processing 101. http://dx.doi.org/10.1016/B978-0-12-811453-7.00027-5

on various inputs, and to control various tasks in response. A flow chart or state diagram is often used to describe the required behavior. In contrast, a DSP processor is typically focused on processing or performing specific mathematical operations on continuous streaming input data, often in an assembly line fashion. Where decisions are made, it is typically based on some mathematical characteristic of the input data, which the DSP processor has determined after on-the-fly analysis of the data.

Data is transferred on buses, which can be 16, 32, 64, or even larger bit widths. Data buses are used to connect the core to various memories and to various peripheral units. The memories might be on-chip random access memory, or off-chip flash or dynamic random access memory. Peripheral units could be serial ports, parallel ports, USP ports, Ethernet multiply accumulates (MACs), and so on. The DSP core is primarily interfacing with on-chip memory. On-chip memory is usually very fast and low latency. Data usually can be read or written in a single processor clock cycle. This is also known as Level 1 or L1 memory. Larger multicycle access memories, which can also be on chip, is known as L2 memory. Much larger off-chip memory, which will have even longer access times, is often referred to as L3 memory. Caches can also be used but typically only for the instructions, as DSP data is often read or written once from memory.

DSP processors at a minimum need three input data buses. On any given DSP processor clock cycle, the core must be able to read the next instruction word and two data words. This is sometimes referred to as the "Harvard" architecture. The two data words are typically both the digitized signal being processed and a fixed coefficient. This would certainly apply to finite impulse response (FIR) filters, as well as fast Fourier transforms (FFTs) where the coefficient is the complex exponential or twiddle factors. Often a fourth bus is employed, to write back to the on-chip memory. Data buses can sometimes have additional flexibility, such as for a 32-bit bus to be able to read/write 16-bit complex symbol pairs in a single cycle, by concatenating the quadrature component.

In addition, DSP processors often need to read/write data in a specific order to optimize the processing steps. Recall in our discussion on FFTs that the FFT data is either read in or written out in "bit reversed" format. Or imagine performing digital filtering, where after processing one sample through all the filter coefficients, the next sample processing is required to read the same set of coefficients again. Or data may need to be read/written in various strides (reading every N^{th} sample) for decimation filters. To facilitate these functions, hardware units known as data address generators (DAGs) are used. A DSP processor will typically have multiple DAGs (four or more) to avoid the need to reconfigure DAGs during a given algorithm. A DAG can be programmed to support a circular buffer, for example. In this case, the DAG might read in a set of coefficients sequentially, and at the end, wrap around to start anew at the top of the coefficient memory location. This is known as circular addressing. Another typical function might be to read/write data with a programmable stride. Often, the I and Q values might be stored in an interleaved fashion. If a function needs to access the real and imaginary data separately, two DAGs might be programmed to access data at every other memory location, with the quadrature data DAG starting an offset relative to the real data DAG.

For other data transfers, such as from one type of memory to memory, or from a peripheral unit to memory, special hardware units known as direct memory access controllers or DMA controllers are used. The advantage is to allow the processor to preprogram these DMA controllers to move a block of data from one place to another, relieving the processor core from spending time and clock cycles in performing these functions itself. This is especially important in DSP processors, as data is usually processed in chunks or blocks, and must be available to the DSP core when ready for processing. Typically, the DSP core will process one block of data while the DMA engine is moving the next block of data to a designated location in L1 memory.

A simple example might be an input signal coming through an analog-to-digital convertor (ADC). The ADC is sampling a signal and converting into a digitized form at a specific sampling rate. The ADC data could be transferred to the DSP through several possible serial or parallel interface methods. The DMA engine might be configured to move these samples sequentially into a specific processor memory space, and interrupt the processor when a specific number of samples have been received. This eliminates the need for the DSP core to perform individual read operations and be interrupted only when a complete block of data is ready for processing. Like the DAGs, the DMA engines in DSP processors are often capable of supporting different strides and offsets. DMA engines may have the flexibility to support multidimensional data structures, which might be composed of frames, which in turn are composed of slots, which are composed of symbols, which are composed of real and quadrature data words.

27.1.2 Core Processing

Owing to the high processing rates required and the need to complete processing in a deterministic time interval (known as real-time processing), DSP processors usually execute all instructions in a single clock cycle. This is similar to RISC processors, but in the case of DSP processors, the instructions can be quite complex. In fact, an architecture known as very long instruction word has become popular with several DSP processor vendors, as it allows for complex instructions to be easily defined and executed in a single cycle, using very parallel core hardware architectures. It can also be adapted to work with as a variable length instruction word, depending on each instruction's complexity. The most common instructions would tend to be the shortest, reducing code memory size.

The most fundamental DSP operation is to multiply two operands and add the product to an accumulator. This is known as an MAC. Naturally, the DSP core contains one or more multiply accumulator circuits. In fact, this is so important that DSP processors are often rated in either MIPs or MMACs (millions of multiply accumulates per second). The most common precision size is a 16 × 16-bit multiplier, feeding a 40-bit accumulator. This means that there are eight extra bits in the accumulator, and 2^8 or 256 products can be summed into the accumulator before any accumulator overflow could occur. Some DSP processors also have provision for 32 × 32 bit multiplication, often by combining several 16-bit multipliers. Architectural support for single cycle complex multiplication using multiple multipliers in the core is also advantageous.

Larger multipliers can be built from smaller multipliers, allowing a flexible multiplier size. As shown below, as single 32 × 32 multiplier can be built from four 16 × 16 multipliers, using several adders.

$$\text{Inputs:}\quad A[31:0]\quad B[31:0]$$
$$\text{Output:}\quad R[63:0]$$

The four 16 × 16 multipliers will produce the following products:

$$A[15:0] * B[15:0] = P1[31:0]$$
$$A[31:16] * B[15:0] = P1[47:16] \text{ must shift left by 16}$$
$$A[15:0] * B[31:15] = P1[47:16] \text{ must shift left by 16}$$
$$A[31:16] * B[31:16] = P1[63:32] \text{ must shift left by 32}$$

The sum of these four products forms the final 32 × 32 result R[63:0].

A 16 × 16 complex multiplier is also implemented using four 16 × 16 multipliers.

$$\begin{aligned}&(A + jB)\cdot(C + jD)\\&= A\cdot C + jB\cdot C + A\cdot jD + jB\cdot jD\\&= AC + jBC + jAD - BD\\&= (AC - BD) + j(BC + AD)\end{aligned}$$

The real part of the complex product is the difference between the two multiplier products. The imaginary part of the complex product is the sum of the two multiplier products. Notice that four multiply operations are required.

Barrel shifting, rounding and saturation circuits, and other shift instructions are also important for DSP processing. A simple example could be implementation of FIR filters. The input data is multiplied by all the coefficients, each output requiring N cycles for an N-tap FIR filter. As the result is accumulated over the N cycles, the result will grow. Assuming a 16-bit DSP and a 40-bit accumulator, the filter output will be 40 bits. The accumulator can often be saved as the LSW (least significant word, 16 bits), MSW (most significant word, 16 bits), and overflow (8 bits). Often, only the MSW result will be saved back to memory as 16-bit result. This requires shifting the result in need to align with MSW boundaries, rounding the result using the LSW content, and performing saturation in the event the result exceeds this representation. All of these functions should be able to be performed in a single cycle as the 16-bit result is saved back to memory, and so provisions for this must be provided in the vendor-specific assembly instructions.

Other necessary functions are support for Boolean operations such as "AND," "OR," and "EXOR." These are normally performed across a 16- or 32-bit word. In addition, bit field packing, concatenating, rotation, and extraction are often highly useful for various error correction and cryptography applications and included in a DSP processor instruction set.

DSP processors spend most of their cycles in tight loops. Therefore, they have provisions for what is called "zero overhead looping." In a conventional processor, the programmer adds code to decrement a counter and test the result at the end of the loop to determine whether to exit or jump to top of the code loop again. In DSP processors, this is built into hardware, so no extra instruction cycles need to be used in this testing or jumping back to top.

Owing to the need to respond quickly in external inputs, DSP processors have resources to facilitate low latency interrupt service routines (ISRs). Features such as interrupt shadow registers, vectored ISRs rather than use of single global ISR locations, nested interrupt capabilities, and other features allow for a DSP processor to react quickly, prioritize, and perform necessary processing from a number of different interrupt sources, including the DMA controller engines.

DSP processors may also have application-specialized instructions. For example, vector instructions might allow simultaneous processing of real and quadrature portions of a symbol. Often, there is an "ADD, COMPARE, and SELECT" instruction to allow efficient implementation of the Viterbi algorithm.

27.1.3 Multiple Cores or Hardware Coprocessors

DSP vendors have also responded the high-computational demands of long term evolution (LTE) wireless baseband processing in particular by building hardware circuits to perform specialized LTE algorithmic tasks, thereby off-loading the DSP cores. Otherwise, the DSP cores would be unable to keep up with the required LTE baseband processing. Owing to the large LTE basestation market for DSP processors, and the well-defined requirements in the LTE standard, it has been economically feasible to optimize DSP processor products for this application. Functions such as turbo error correction and orthogonal frequency division multiple access FFT processing can be implemented in hardware coprocessors, with the data flow being managed by one or more of the multiple DSP cores contained in a single device.

27.2 Scalability

In many applications, the trend is for increasing DSP computational requirements. For example, use of high definition in video processing, 4G LTE systems, advanced radar and sonar, and some new error correction algorithms all require increasingly high rates of DSP.

In the 1990s, DSPs evolved by increasing clock rates, increasing amounts of L1 memory, and some architectural improvements. In the next decade, further improvements such as more on-board multiply accumulate circuits per core, better C compilers, addition of Ethernet MAC and other high-bandwidth interfaces were included. However, silicon process technology improvements were unable to deliver the same improvements in clock circuit speeds. Clock rates have topped out at about 1.2 GHz in the highest performance DSPs. To deliver more processing power, DSP processor vendors have responded by adding more DSP cores per chip, currently up to six independent cores.

Having more processor cores is effective when many different and independent DSP tasks can be partitioned and run simultaneously. It is a poor solution when the tasks are interrelated, or individual tasks have very high computational requirements. It is also not especially scalable. As the number of cores increases, the difficulty in partitioning the tasks across the cores and managing the intercore communications becomes increasingly difficult. Several DSP processor "startup" or venture-funded companies have developed products with dozens or hundreds of cores, but these products were not significantly adopted by industry. In a more recent trend, some DSP processors have evolved into vector processors, where several hundred compute units (multipliers, accumulators, shifters ect) operate in parallel using very wide buses. These types of processors can be used for vision processing applications, or other highly parallelized signal processing use cases.

27.3 Floating Point

Most DSP processors are fixed point processors. This is due to the larger floating point circuit requirements, higher power consumption and lower performance compared to fixed point. For the vast majority of DSP applications, fixed point arithmetic is suitable. However, this does require care on part of designer to ensure dynamic range of the signal is mapped into the limited fixed point precision.

Several DSP processors do offer floating point DSPs products, initially at much lower performance levels than the fixed point products. One popular application for floating point DSP processors is high fidelity audio processing. This is due to relatively low processing rates, high dynamic range requirements, and extensive use of infinite impulse response filters, which can have stability issues when implemented in fixed point.

Some more recent DSP processors offer the ability to do either 16-bit fixed point or single precision floating point at fairly high rates. The peak floating point performance of DSPs can exceed 100 GFLOPS.

However, high-performance floating point is more commonly implemented on Pentium-type processors, graphics processors, field programmable gate arrays, or some specialty floating point processors. Performance levels in the multiple TFLOPS are available in some devices. Usage tends to be on military applications, such as radar back-end processing, or high performance computing or machine learning. Some examples requiring high rates of floating point processing are financial modeling, options pricing, climate simulations, seismic research, and genomics.

27.4 Design Methodology

The design methodology used with DSP processors is very similar to that used on other types of processors. The software-based approach offers the optimal flexibility to build and debug complex algorithms. Owing to the serial nature of the software flow, the implementation and debugging is simplified, all variables in memory are accessible and the flow tends to be more "natural" to the designer's thought process.

In summary, the DSP processor is a specialized processing engine being reconfigured each clock cycle for many different functions, mostly DSP related, others more control or protocol oriented. Resources such as processor core registers, internal, and external memory, DMA engines, and I/O peripherals are shared by all tasks, often referred to as "threads." This creates ample opportunities for the design or modification of one task to interact with another, often in unexpected or nonobvious ways. In addition, most DSP algorithms must run in "real time," so unanticipated delays of latencies can cause system failures. Some of the challenges of DSP programming include the following:

- Mixture of "C" or high-level language subroutines with assembly language subroutines;
- Possible pipeline restrictions of some assembly instructions;
- Nonuniform assumptions regarding processor resources by multiple engineers simultaneously developing and integrating disparate functions;
- Ensure interrupts completely restore processor state on completion;
- Blocking of critical interrupt by another interrupt or by an uninterruptible process;
- Undetected corruption or noninitialization of pointers;
- Must properly initialize and disable circular buffering addressing modes;
- Preventing memory leaks, the gradual consumption of available volatile memory due to failure of a thread to release all memory when finished;
- Dependency of DSP routines on specific memory arrangements of variables;
- Unexpected memory rearrangement by optimizing linkers and compilers;

- Use of special "DSP mode" instruction options in core;
- Conflicts or unexpected latencies of data transfers peripherals and memory, when using DMA controllers;
- Corrupted stack or semaphores;
- Subroutine execution times dependent on input data or configuration; and
- Pipeline restrictions of some assembly instructions.

27.5 Managing Resources

Interaction between different tasks or threads can cause intermittent and sometimes hard to detect problems in all processor architectures. Microprocessor, DSP, and operating system (OS) vendors have attempted to address these problems with different levels of protection or isolation of one task or "thread" from each other. An OS can be used to manage access processor resources, such as allowable execution time, memory, or common peripheral resources. However, there tends to be an inherent compromise between processing efficiency and the level of protection offered by the OS. In DSPs, where processing efficiency and deterministic latency are often critical, the result is usually minimal or no level of RTOS (real-time OS) isolation between tasks. Each task or thread often requires unrestricted access to many processor resources to run.

Fig. 27.1 helps illustrate how complex a DSP processor system is. All of these functions exist to service the DSP processor core, which can only execute one instruction at a time. But generally, only a subset of the hardware in a DSP processor is needed at any given time, as the hardware must be designed to support every instruction. Generally, DSP processors

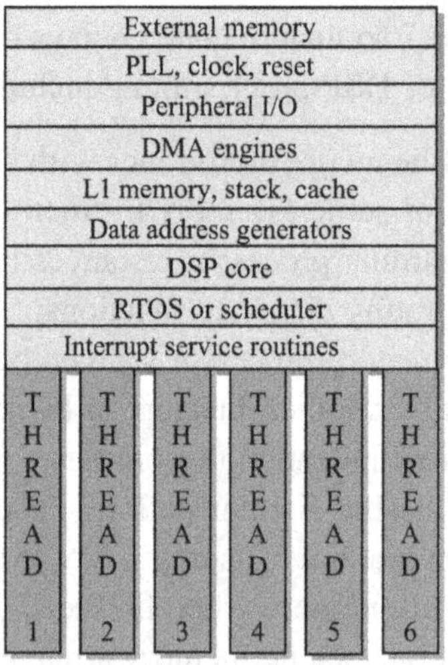

Figure 27.1
Task sharing of hardware resources.

have relatively few computation elements (multipliers, adders, accumulators, shifters ect) which is supported by this complex infrastructure. This is the inherent inefficiency in any processor architecture, compared to a custom hardware implementation. The penalty of the flexibility of the DSP processor is this hardware inefficiency.

27.6 Ecosystem

A significant factor in choosing a silicon platform for DSP applications is the available tool and IP (intellectual property). Different DSP processor vendors have various amounts of DSP software IP available. Smaller, independent DSP IP companies also license software modules for many functions on popular DSP families. This can be a major factor in the choice of DSP processor. For example, someone wishing to build media gateways will need implementing voice compression and decompression for a variety of industry standard voice codes, or vocoders, as well as echo cancellation capability. These algorithms are available for several DSP processor families, eliminating the need for a proprietary development. Similar IP is available for image compression used in video surveillance industry, motor control algorithms used in industrial applications, code for facsimile and modem standards, and many other applications.

In addition, major DSP processor manufacturers have sophisticated and robust development tools, which include compilers, assemblers, linkers, and debuggers. They also often supply RTOS, Ethernet transmission control protocol/internet protocol stacks and many other commonly used device drivers and IP stacks. In addition, many third party companies also supply RTOS products and various IP stacks for specific DSP processor architectures.

CHAPTER 28

Implementation Using FPGAs

Field programmable gate arrays (FPGAs) can be used to perform digital signal processing (DSP). From their origins as custom logic and interface functions, FPGAs have grown to include nearly every function that can be implemented digitally and are able to interface to nearly any other circuit at nearly any data rate desired. From a DSP point of view, FPGAs are often used to interface analog-to-digital converters (ADCs) and digital-to-analog converters (DACs), as well as backplanes. Wherever a DSP processor is used, the FPGA often will be used in the system for interfacing as well. The two dominant FPGA vendors are Intel (formerly Altera) and Xilinx.

FPGA fabric is composed of configurable logic elements and programmable routing, which allows any digital function to be built, including multipliers, adders, accumulators, shifters, registers, and any other functions that might be found in a DSP processor. Distributed memory in different size arrays or blocks is also integrated within FPGAs. In addition, to optimize FPGAs for DSP applications, special DSP blocks are available in most FPGA devices. Logic intensive functions such as multipliers can be hardened, which besides leaving more programmable logic available for other functions, allows the multipliers to run much faster and consume less power. There are other circuits that can be hardened in addition to the multipliers. Typically, these circuits are collected into DSP blocks, which are architected to implement common DSP operations efficiently. This will be discussed in more detail below.

FPGA devices can be very small, from a few thousand logic elements, or very large, with millions of logic elements. The number of hardened multipliers can vary from a few dozen to many thousands in a single device. Because of the sheer number of available circuits, FPGAs are much more powerful DSP platforms than any DSP processor. A high-end multicore DSP might be able to perform perhaps 100 GMACs (giga multiply accumulates) per second, whereas a high-end FPGA could perform in excess of 10,000 GMACs per second. This is an increase of about 100-fold, or two orders of magnitude.

With this disparity in processing power, one might wonder why DSP processors are still used. First, DSP processors were first available in 1982, whereas FPGAs suitable for DSP applications were not available until the mid-1990s. Many applications have a legacy of using DSP processors. More importantly, the design methodology of the two technologies is completely different. This is very important and helps explain why either a DSP

Digital Signal Processing 101. http://dx.doi.org/10.1016/B978-0-12-811453-7.00028-7

processor(s) or an FPGA is more suitable for a given application. And a third alternative has also started to become popular, the GP-GPU (general purpose graphics processing unit), discussed in the next chapter.

28.1 FPGA Design Methodology

FPGAs are fundamentally hardware devices even though they are programmable. Unlike processors, the operation of the device is not controlled by scheduling the operations of a processing engine but by configuring the hardware itself to perform the necessary operations for a particular design.

A DSP processor core is a reconfigurable engine, which on a cycle-by-cycle basis is configured to support a specific operation. Aside from the parallelism built into the core (for example, parallel circuits to fetch the next instruction plus two data operands, while multiplying and accumulating previous operands), each algorithm is fundamentally implemented in a serial fashion. Processing rates are primarily determined by the DSP processor clock rate. The software designer of a processor-based system is limited by the available instructions of a given processor, which in turn depends on the processor architecture. Operations not supported in the processor hardware can still be implemented in software but usually inefficiently. For example, a 16-bit division operation in most DSP processors is implemented iteratively over at least 16 clock cycles, as there is no single cycle division instruction.

In contrast, using an FPGA, the hardware is normally configured to support a predetermined DSP datapath function. The entire application is normally implemented by building a separate hardware circuit for each operation and passing the data through in a process not unlike the assembly line used in factories. Just like on an assembly line, the distribution of operations to separate circuits results in a dramatic speed up in throughput, as many operations are performed in parallel, in a pipelined or step-by-step fashion. The DSP processor analogy would be a single worker to build the entire product, step by step. Throughput is increased by adding more workers (DSP processor cores).

Owing to the distributed nature of memory in an FPGA, the needed data can be always quickly accessed from memory by each different circuit. The composite memory I/O bandwidth in the distributed FPGA memory far exceeds that of a DSP processor bandwidth, despite the multiple memory buses of the DSP processor.

An FPGA circuit can be made adaptable, often by using registers to set the operational mode. For example, an fast Fourier transform (FFT) hardware circuit can be built with a register configurable number of points (although often limited to 2^N, where N is an integer), or a finite impulse response (FIR) filter can have coefficients stored in memory block or registers, which can be updated during operation. FPGAs can also execute DSP designs with high degrees of parallelism, with a throughput of many samples per clock cycle in the DSP datapath, whether FFTs, FIR filters, or more complex matrix processing. However, an FPGA simply does not have the same programming flexibility as a DSP processor. The processor can execute any valid instruction on the next cycle, whereas the flexibility needed in an FPGA hardware datapath must be anticipated and provisioned for by the designer.

DSP processors, while having much smaller memory bandwidth, do have the ability to access any location across the entire memory space from cycle to cycle. FPGAs, while having the much higher memory bandwidth due to inherent parallelism, can only access the data available in each local memory block, which is normally determined by the dataflow through the datapath as anticipated by the designer. Again, the FPGA achieves a massive increase in aggregate memory bandwidth, but at the expense of run time flexibility.

28.2 DSP Processor or FPGA Choice

Given these differences, FPGAs and DSP processors tend to be used in applications where their respective merits can provide the optimum implementation platform. DSP processors tend to be more suitable when the application is very complex algorithmically or requires many different configurations, and the processing rates are low enough to allow for DSP processor implementation. Complex algorithms are often easier to implement and debug in a software-based flow. Some applications may require very data-dependent processing algorithms, where different DSP operations may be needed, as determined by the input data. In an FPGA, this would require several alternative hardware circuits, which can be complex and inefficient to implement in hardware. In some organizations, a large DSP processor legacy code base used in product development and an experienced DSP processor programming team on staff can make use of an FPGA unattractive, despite its greater processing capabilities.

FPGA usage is prevalent when high rates of data throughput and processing are required. In terms of GMAC/s, a single FPGA can replace a whole board full of DSPs. Many times, the application processing rate dictates the implementation method. For example, high-definition image compression algorithms such as H.265 require use of a hardware-based solution such as an FPGA, whereas standard (low) definition image compression may be able to be processed in real time by suitable DSP processor. Radar systems often have extremely high rates of data throughput and may use multiple FPGAs chained together.

Even in many higher performance DSP systems, it is common to find both FPGAs and DSP processors in use. In such systems, FPGAs are generally used for preprocessing, filtering, and the main DSP datapath operations. DSP processors may be used for back-end processing, once the data rate has been reduced, or for more complex portions of an algorithm. An example is an adaptive filter. The FPGA may perform the actual filtering, but the DSP processor could be used to process any feedback information used by an adaptive algorithm to update the filter coefficients.

Small FPGAs can be useful coprocessors for DSP processors. By off-loading standard but high millions of multiply accumulates per second functions such as digital filters and FFTs, the DSP processor millions of instructions per seconds (MIPs) are freed up for more value-added functions. This is also often very feasible to implement, as in many DSP processor-based systems, an FPGA is already present in many cases to interface between the data convertors and the DSP processor, or between the backplane and the DSP processor.

28.3 Design Methodology Considerations

FPGAs are inherently more difficult to design with than a DSP processor. This is due to the high degree of choice in an FPGA device. The designer is able to create any hardware circuit desired, any size data bus, configure the dataflow as needed, plus synthesize internal microprocessors out of logic and support nearly all possible serial and parallel external interfaces. This results in many available degrees of design freedom. The standard design entry method is known as "HDL" or hardware description language. Two variants are commonly used, Verilog and VHDL. One drawback of HDL design is the long compile times. On processors, new software updates can be compiled in a matter of seconds. With FPGAs, compile times can take hours or even a day. This is due to nearly unlimited degrees of freedom the compiler can explore to find a near-optimum solution. The structures described in the HDL code must be synthesized, a process not unlike that of compiling on a processor. The HDL code is broken down and simplified into many small logic functions. It then must be mapped to FPGA hardware resources, and all interconnections be made using the FPGA routing resources. The FPGA vendor provided tools, known as place and route, performs this function while simultaneously ensuring that the connection and logic delays will still allow the design to operate at the clock at the rate specified by the designer. Verification is also much more arduous that on DSP processors, due to the need to verify not only logical operation, but the timing of all circuits and routed connections. Complex test benches that simulate as many possible states of the design in operation as possible are used for verification.

Surprisingly, FPGA designs can be made even more robust than DSP processor code implementations, despite the increased design effort. This is fundamentally due to the independence of the different tasks because of the inherent parallelism of the FPGA. Each task has separate hardware, including memory structures. This tends to limit the number of

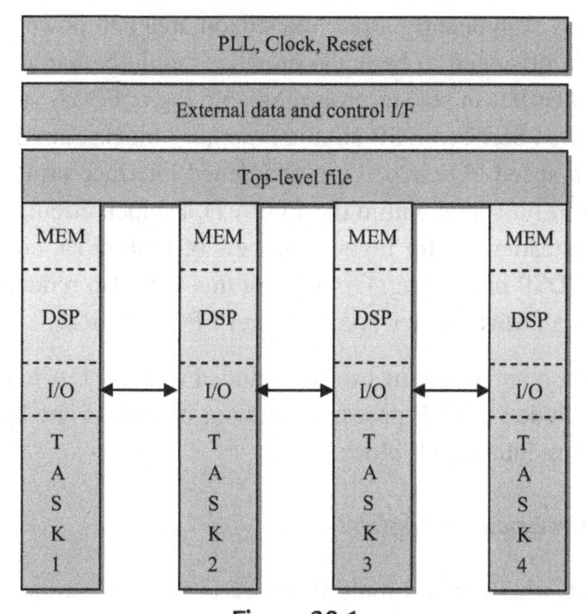

Figure 28.1
Separation of task resources.

unexpected interactions that can occur when all functions share the same hardware and memory. The separation in an FPGA is shown in Fig. 28.1, which is in contrast to the structure used in processors as shown in the previous chapter.

Additionally, FPGA verification methodologies are patterned after application-specific integrated circuit (ASIC) design flows. With ASICs, millions of dollars and dozens of man-years are often spent on each chip, so verification is a major function in the design flow. Bug fixes such as a simple recompile and software download are not possible. This has led to a huge investment in design verification tools by the electronic design automation industry, which can also be used to verify FPGA designs.

28.4 Dedicated Digital Signal Processing Circuit Blocks in FPGAs

Initially, it may appear that the hardened DSP block circuits in FPGAs should be designed to have the same capabilities as DSP processor cores. After all, most of the DSP algorithms to be implemented are the same. However, there are two major differences. First, in addition to DSP blocks, the FPGA also contains large amounts of programmable logic, which can also be used to implement DSP functions. So, unlike a DSP processor, not all functional capabilities need be implemented in the FPGA's DSP block circuits. Second, the large FPGAs may contain several thousand DSP block circuits. Therefore, the size of these blocks may significantly affect the silicon area and power consumption of the FPGA. Intelligent trade-offs need to be made on what should be hardened in the DSP block and what should be left in soft or programmable logic. FPGA vendors need to partition their silicon area between DSP circuits, memory blocks, configurable logic, routing resources, high-speed I/O circuits, and hardened interface protocol circuits. By eliminating seldom used functions within the FPGA DSP block circuit, this can reduce DSP block area and instead allow for higher numbers of DSP block circuits for the silicon area that is devoted to DSP processing. Optimizing this will also reduce power consumption, which is a critical factor in ever larger FPGA devices.

There are a number of potential features to be included in an FPGA hard DSP circuit. These are normally included in DSP processor and are discussed one by one, with a view on the merit of including into a DSP block circuit.

28.4.1 Adjustable Precision Multipliers

In most FPGA products, the standard multiplier size is 18 × 18, rather than the 16 × 16 size used in DSP processors. This is sufficient for the majority of applications. Some FPGAs, like some DSP processors, can use four 18 × 18 multipliers to implement a single 36 × 36 multiplier or else a complex 18 × 18 multiplier within a DSP block.

There is also a growing need for higher precision multipliers in many applications, at least in some parts of the datapath. Most data convertors are 12–16 bits, and 18-bit multipliers are sufficient. Yet in some applications, the processing gain in decimation filters, or by algorithmically combining several correlated receiver data (MIMO) can lead to an increased datapath precision. In FFTs, data precision naturally increases on only one side of the multiplier, as the datapath precision grows with each successive butterfly stage, but the complex exponential coefficients remain of fixed precision. Also, some DSP algorithms and applications just naturally require higher precision to meet their performance requirements. This can be accommodated by using higher precision multipliers. However, use of a 25 × 25-bit multiplier requires twice the multiplier area of an 18 × 18 multiplier. This is a large penalty on applications requiring only 18-bit multipliers or less. And use of 36 × 36 multipliers built from 18 × 18 multipliers is also very expensive, particularly if the needed precision is only a few bits more than 18. Fortunately, FPGA multiplier circuits can be designed with more flexibility than DSP processors, as they are not intrinsically tied to the data bus widths and instructions of a DSP processor. Using 9 × 9 multipliers as building blocks, FPGA DSP blocks could support 9 × 9, 18 × 18, 27 × 27, 36 × 36, and 54 × 54 multiplier sizes with a roughly proportionate increase in DSP block resources as multiplier precision increases. In addition, recent developments in machine learning are driving the need for smaller multipliers, such as 8 × 8 or 9 × 9.

Some applications can benefit by an asymmetric multiplier size, such as 18 × 27 bits. In particular, FFTs have a natural increase in data width during the progression through the radix stages of the FFT, whereas the twiddle precision requirement remains constant. The FPGA vendors take different approaches to supporting these various conflicting requirements to provide the highest compute density with the best power efficiency and as much flexibility as possible.

One approach to try to serve all applications is to have the DSP circuits configurable to allow either a high count of lower precision multipliers or a lower count of high precision multipliers, which would allow system designers to design with the precision they need, rather than try to tailor to the limitations of the hardware device. Using this approach, only the applications actually using this additional precision need to allocate more multiplier resources.

28.4.2 Accumulator

Accumulators are integral to many DSP operations. They are necessary when using a single multiplier to calculate in series a number of multiply and add operations. The accumulator circuit can also be reused in another mode, the distributed adder circuit, described next. Accumulators are essentially large adders with feedback on one operand. They can be implemented in logic but then may not run as fast as the hard multiplier. Since they are used in a high percentage of DSP applications, this is a good capability to include in the DSP block of an FPGA. The size of the accumulator depends on the common multiplier precisions used. Normally, at least eight extra bits should be used above the product size of the multiplier, to allow at least 256 product accumulations. An 18 × 18 multiplier size needs 44 or more bit accumulator, a 27 × 27 multiplier size needs 62 or more bits accumulator, and a 36 × 36 multiplier size needs 80 or more bits. However, the larger accumulator size is a penalty on all DSP blocks, even when smaller multiplier sizes are used, although a much smaller penalty than an oversized multiplier. A good compromise would be able to accommodate at least 18 × 18, 18 × 27, and 27 × 27 multiplier sizes, as these are likely to be frequently used. This would lead to an accumulator size of about 60−64 bits.

28.4.3 Postadder (Subtractor) and Distributed Adder

Postadders are used both to construct larger adders from smaller adders, to perform the sum/subtraction in complex multiplications, and to perform sums of products used in FIR filters. The postadder should also be able to perform subtraction if desired. This is an excellent function to include in a DSP block. An example of postadder used in an FIR filter is depicted in Fig. 28.2.

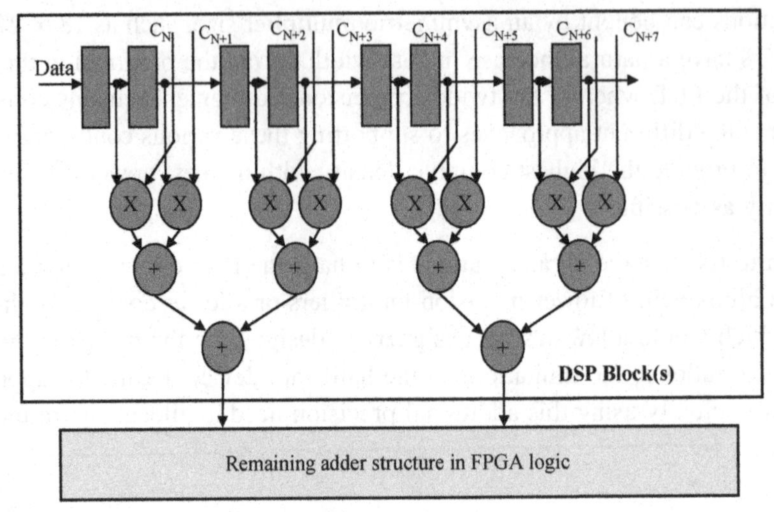

Figure 28.2
Traditional finite impulse response filter implementation.

An alternate FIR filter architecture is known as the systolic architecture. It uses a distributed output adder, and therefore, no matter how large the FIR filter, no programmable logic-based adder circuits are required. This can allow for more efficient and higher Fmax (clock frequency) FIR filter implementations. The only penalty is an increase in filter latency, and a slightly more complex sequencing of input data. This is a small price to pay for the benefits. With the inclusion of an accumulator or postadder, there is little extra cost to add this useful feature to an FPGA DSP block. A vertical cascade path is needed between DSP blocks, which are normally placed in columns in an FPGA. This fixed path is of insignificant cost and could also be used to build large multiplier circuits that span several DSP blocks, such as 54×54 or complex 27×27. For these reasons, the cascade path is an excellent function to include in a DSP block.

An example of postadder used in a systolic FIR filter is depicted in Fig. 28.3. The postadder is connected as a distributed adder cascaded from block to block. Note the additional registers in the input data chain to compensate for the postadder register delay stages. This performs the same algorithmic function as the diagram above, although with greater latency (clock delay).

28.4.4 Preadder (Subtractor)

Preadders are primarily used in hardware circuits and not commonly found in DSP processors. The main application is for symmetric FIR filters. As the filter data is shifted across the coefficient set, two data samples can be multiplied by a common coefficient due to the symmetry. The preadder adds the two samples *prior* to multiplication, which allows

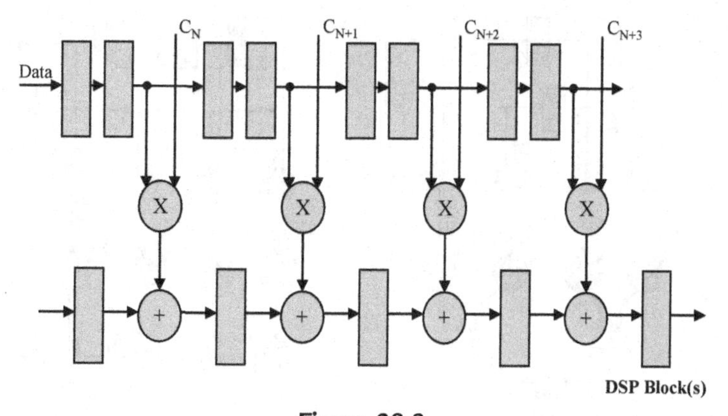

Figure 28.3
Systolic finite impulse response filter implementation.

the use of one multiplier for every two taps, rather than two multipliers. This preadder function can be implemented in FPGA logic, but as most FIR filters are symmetric and FIR filters are the most common DSP application, this is a reasonable feature to include in a DSP block. Configured as a subtractor, this block can also be used to perform "sum of absolute differences," a less commonly used function. Also, although not discussed here, in some cases a preadder can also be used to more efficiently implement complex multiplication architectures.

The preadder is difficult to implement with DSP processors, because normally a single multiplier is used, with the input data being in a circular buffer. The input data is therefore not organized in a way to easily take advantage of a preadder.

Two symmetric, eight tap filters are diagramed in Fig. 28.4. The filter taps are [C_0, C_1, C_2, C_3, C_3, C_2, C_1, C_0]. A preadder is used, on the left in a conventional FIR filter and on the right in a systolic FIR filter implementation. The extra input data registers in the systolic diagram used to align dataflow are depicted in red. The input data needs to "wrap around" to utilize the preadder, which makes it difficult for implementation in DSP processors. In both conventional and systolic FIR filter architectures, it can reduce multiplier usage by approximately one half. This is a valuable feature to incorporate into a DSP block.

28.4.5 Coefficient Storage

Coefficients are required for most DSP operations, including FIR filters and FFTs. These coefficients are normally stored in distributed memory blocks external to the DSP block. This allows for large numbers of coefficients to be used and for easy updating in the case of adaptive filters, for example. However, since most FIR filters in a hardware implementation are built using a parallel, or at least partially parallel structure, the number

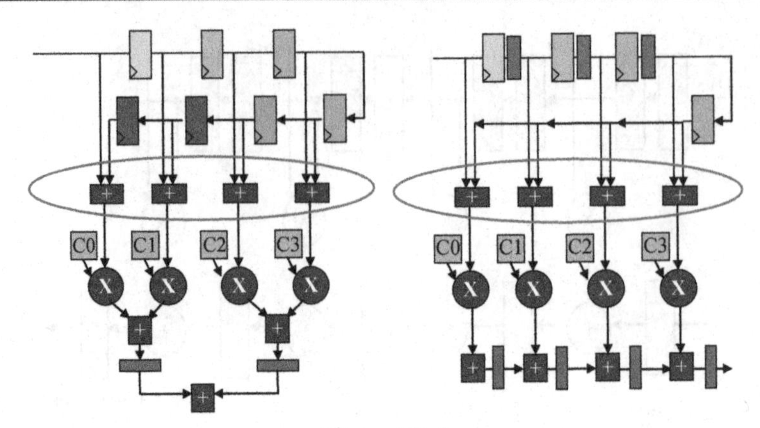

Figure 28.4
Symmetric finite impulse response filter implementation.

of coefficients used per multiplier is often fairly small. For these cases, it may be advantageous to allow internal coefficient storage, dynamically selectable on each clock cycle, within the DSP block. The advantage, besides an obvious saving of FPGA memory resources, is reduced routing congestion and logic in the region near the DSP block inputs. The logic and routing in this area of the design is usually in high demand, due to the input data routing and need to build variable length shift registers to provide proper input dataflow, especially for FIR filters incorporating preadders, interpolation, and/or decimation structures. Use of internal coefficient registers also eliminates the possibility of being unable to run at the required circuit Fmax due to routing congestion. There is a trade-off in number of coefficients, DSP block area required, and percentage of applications where the internal coefficient bank contains enough coefficients to accommodate each FIR filter. However, on the balance, this seems like a valuable feature to consider including an FPGA DSP block.

28.4.6 Barrel Shifter

Shifters are necessary in DSP processors and are an integral part of a DSP processor instruction set. Shifting is often possible in parallel with many other DSP processor operations. Surprisingly, dedicated barrel shifters are not often required in a DSP block structure. The principle justification, which is to align outputs or decimal point for 16- or 32-bit memory storage or for further processing, is driven by DSP processor architecture limitations. In programmable hardware, this can be easily achieved by simply selecting which subset of accumulator output pins to route to the next stage in an FPGA.

For the remaining cases where an adjustable shift circuit is needed, a hard multiplier circuit can be used. This is done by multiplying the operand to be shifted by 2^N, which achieves a shift by N. This can also be very useful for floating point normalizations. The low usage and the options of using either multipliers or logic as shift registers argue against hardening barrel shifter functions in the DSP block.

28.4.7 Rounding and Saturation

These functions are necessary in DSP processors, to reduce the output data width after multiply-accumulate or distributed adder circuits. This step must be implemented in hardware as there is no practical way to implement it in software. In FPGAs, only a small fraction of the DSP blocks tend to need rounding and saturation. This is partly because in a distributed adder filter, only the last stage would tend to be rounded or saturated. There are also a number of different rounding methods used by system designers in different application. Hardening this relatively silicon-costly function across thousands of DSP blocks can add significant area to the DSP block, so it may be more efficient to implement in logic in the instances it is required, with the flexibility to use any preferred rounding method. The one exception is to offer "biased" rounding, the simplest rounding method, which can be implemented almost for free in the existing accumulator/adder circuit. "Biased" is also sufficient for the majority of applications; the slight DC bias thus added is imperceptible in most cases.

28.4.8 Arithmetic Logic Unit Operations and Boolean Operations

These include ADD, SUBTRACT, AND, OR, EXOR type functions. Normally, use of these functions is not associated with multiply-accumulate operations. Therefore, it is questionable whether to use a DSP block to implement these simple arithmetic logic unit operations, since the great majority of any DSP block area is used to build multiply-accumulate circuits. Implementing in soft logic is not quite as efficient but can be easily placed exactly where needed and only in the amounts, widths, and functions necessary for a particular application.

28.4.9 Specialty Operations

Some DSP processors have special instructions for bit-reversed addressing for FFTs, modes for Viterbi FEC processing, or instructions to perform bit by bit correlation necessary in CDMA applications. FPGAs need not support these functions in a hard DSP block as they are not often used in a high percentage of DSP blocks and can be easily implemented in logic, or in the case of bit reversal, in the routing connections.

28.4.10 Tco and Fmax

Most DSP blocks in high-performance FPGAs can be clocked in excess of 600 MHz. However, delays in logic and routing limit design Fmax to 500 MHz or less in nearly all designs of large size. Further increases in DSP block Fmax may not especially be useful, unless the logic and routing can support much higher speeds, and the FPGA can handle the higher power when operating at very high Fmax.

28.5 Floating Point Implementation Using FPGAs

The great majority of DSP applications can be implemented in fixed point. However, some applications require the dynamic range and precision of floating point. Some examples with those involving matrix operations such as high-resolution radar systems, medical imaging, and MIMO channel calculations in LTE wireless systems. Division, FFTs and infinite impulse response filters can be built in fixed point, but are much easier to build using floating point. Math operators such as trigonometric, logarithmic, exponential greatly benefit from floating point dynamic range.

Single precision floating point uses a 23-bit mantissa, plus a sign bit. This requires a minimum of 24×24-bit multiplier for floating point multiplication. Floating point also requires normalization and denormalization. This requires an ability to shift a 24-bit operand within a 48-bit field. This can be implemented as barrel shifter in logic or for faster circuit operation, using a minimum of 24×24-bit multiplier and 48-bit accumulator.

In processor-based implementations, normalization and denormalization are performed at each floating point calculation. FPGA design tools and intellectual property (IP) are available, which allow for floating point datapaths to be implemented, using the fixed point DSP blocks and soft logic together. This typically requires extra mantissa bit width. The extra mantissa width eliminates the need for normalization and denormalization at each floating point calculation, which can significantly reduce routing and logic resources, allowing higher performance FPGA implementation. This technique can utilize larger multipliers, such as 27×27 precision.

Even better, some FPGA vendors now offer single precision floating point support built into the DSP blocks. Like GPUs, half precision floating point may also be supported, to allow for more efficient machine learning implementations. All the circuits to handle floating point are hardened, which eliminates the logic, routing, and power consumption penalties of implementing using soft logic. This allows FPGAs to address more applications, such as those that really require floating point. Floating point is also much easier to design with, as the designer is not required to worry about overflow, underflow, and matching the range of the signal to the bit width. The quantization noise floor is much lower than most fixed point implementations and is consistent throughout the datapath. Also, most algorithms are simulated and validated using floating point, and the product development process is greatly simplified if the implementation can also be done natively in floating point.

Future FPGAs will likely support other floating point precisions in their hardened DSP blocks. Good candidates are half precision floating point (16-bit words) for machine learning training and double precision floating point (64-bit words) for high-end compute acceleration.

28.6 Ecosystem

Similar to DSP processors, FPGA vendors also have ecosystems of tools and IP provided both by the FPGA manufacturers and third party companies. Each FPGA vendor provides comprehensive software tools to enable the various steps of synthesis, fitting, timing analysis, and many other functions. Owing to the complexity and breadth of this software, there can be considerable variation in the feature set, robustness, and ease of use between FPGA vendors. Third party companies also supply tools for the FPGA development process, especially for verification purposes.

The major FPGA vendors also supply comprehensive collections of IP modules, much of that is for DSP applications. FPGA and third party vendors also offer synthesizable "soft" microprocessor architectures, which are built from the configurable logic within the FPGA, as well as hardened microprocessors licensed from ARM. These built-in microprocessors can enable a whole new set of applications, where the microprocessor can leverage acceleration of specific processing tasks using FPGA logic. Apart from the relative merits of the FPGA silicon device itself, often the tools, IP, and reference designs can play a significant factor in choosing an FPGA platform for DSP and other applications.

28.7 Future Trends

It is generally perilous to forecast the future, but some trends are apparent and worth commenting on. In terms of their processing capabilities, DSP processors are increasing rather slowly, due to inabilities to significantly increase clock rates. Scaling, by providing more cores or more multipliers per core, increases design flow complexity and only tends to be effective on applications where the tasks can be partitioned among several cores, and dependency between tasks is limited. More cores or more parallelism within the cores also tends to make the design process more complex, partially negating the simplicity advantage DSP processors enjoy in design methodology.

FPGAs, in contrast, are increasing in processing capabilities with each succeeding new product family. While FPGA circuit clock rates are increasing slowly, as with DSP processors, the logic density of the FPGA devices is nearly doubling every 2 years. This merely increases the existing parallelism of FPGAs and does not fundamentally alter the design complexity or methodology. This is allowing ever more complex and higher computational rate algorithms to be implemented, thereby, increasing system capabilities.

Furthermore, FPGA design methodologies are also improving. Alternatives to HDL, the main flow used by FPGA designers at the time of this writing, are now available. These alternatives can provide higher productivity, by allowing the designer to work at a higher level of abstraction. One design flow is known as "Model-based" design flow. It uses the popular tool "Simulink" by the Mathworks to describe the design. The test bench can also be implemented in "Simulink," or alternatively by using another popular Mathworks tool known as "Matlab." Use of these tools can allow very rapid design changes and verification, compared to traditional HDL design flow. Recent innovations in this flow by one FPGA vender can also generate highly parallel as well as high-Fmax FPGA designs, comparable to the best HDL generated designs.

OpenCL, will be described in the next chapter, is also now an FPGA supported design flow. This flow does the best job of abstracting the hardware details of the FPGA from the designer. It does require that a "board support package" be available or developed for the target hardware board and FPGA chosen. However, once the board support package is available, either developed by the FPGA vendor, a board supplier, or the end user, an engineer with no previous experience with FPGA design can use OpenCL to implement on the FPGA.

In between is the design methodology in which multiple vendors support is known as "C to gates" or "high-level synthesis" (HLS) design. In this flow, a software description of the hardware is used, in C++ or System C. This design flow can lead to higher productivity and easier design reuse. A second advantage is a potential huge new pool of FPGA designers, as there is at least an order of magnitude more C programmers than HDL knowledgeable engineers. Support for floating point will also make FPGAs much more friendly to software engineers, who normally use only integers for functions such as counters or memory indexing. HLS can also be compatible to traditional flow, allowing the designer to develop particular functions using HLS and integrate an IP block into a traditional Verilog/VHDL design hierarchy.

Any approach to the HLS requires provisions to describe or direct not only the normal data processing steps and decision flow but also to describe what is to be implemented in parallel and serial (traditional) software flow. Products now exist to support this methodology, but at present, adoption is still in the early stages due to both tool immaturity and the performance gap between hand-optimized Verilog/VHDL and HLS implementations. Nonetheless, HLS is very likely the direction of the future.

In addition to this, many incremental improvements are occurring in the HDL design flow. Continuous reductions are occurring in the FPGA compile times. Partial recompile, which eases small updates to the design, is now available from some FPGA vendors, as is partial reconfiguration, which allow reprogramming of very small selected sections of the FPGA circuitry while the FPGA is in operation. New debugging capabilities are becoming available including what is called "hardware in the loop" or "FPGA in loop." This allows the designer to accelerate their algorithm in the simulation environment, by having the FPGA execute the computationally intensive portions in actual FPGA hardware. This can provide several orders of increase in simulation speed, which can be very helpful when simulating large data sets. Many of these changes are expected to narrow the design effort gap between DSP processors and GPUs compared to FPGAs for DSP systems and applications.

CHAPTER 29

Implementation With GPUs

Graphics processing units, or GPUs, were originally designed to perform the rendering to provide a 3D visual effect on a 2D surface (the screen or monitor). This also includes adding texturing and shading effects and involves a considerable amount of mathematical computation with a high degree of parallel processing to achieve this. Most CPUs used in computers and tablets include a GPU coprocessor to perform this function. For high-performance graphics, such as with computer gaming, a separate graphics card containing a high-performance GPU is often used. The GPU market leader, NVIDIA, came to the conclusion that the GPU could also be used for general-purpose compute acceleration, and the GP-GPU (general purpose graphics processing unit) was the result.

29.1 Characteristics of Graphics Processing Unit Architecture

A GPU is very different in architecture from a CPU, even a multicore CPU. A CPU is designed to run many tasks in parallel, with minimum latency. Each task maps to a different thread, managed by the operating system. Each task has independent instructions and is managed and scheduled explicitly. Moreover, each task has to be individually programmed. The number of simultaneous threads, or tasks run by a CPU is typically in the 10s, well under 100.

GPUs and CPUs are complementary. The CPU acts as the host, running the high-level programming and executing sequentially. If there are certain functions that are highly parallel—for example, multiplying matrices, these functions can be off-loaded to the GPU for execution. The GPU can greatly accelerate this function and is often thought of as an accelerator for the CPU. The host CPU is also responsible that the GPU has access to the data needed and may have to trigger DMA transfers of the data into GPU accessible memory.

GPUs are fundamentally designed for data parallelism. It is a SIMD-type architecture—single instruction, multiple data. The same instruction can operate on different data simultaneously. It is common for tens of thousands of threads to execute across several hundreds of compute cores. The threads are managed and scheduled by the GPU, with no S/W intervention. The programming is done for batches or groups of threads, which execute in parallel. The groups of threads executing together are referred to as a stream. GPUs process streams, not individual data variables. GPUs do not guarantee a specific latency, so are suited for problems that benefit from high degrees of parallelism

Digital Signal Processing 101. http://dx.doi.org/10.1016/B978-0-12-811453-7.00029-9

but which can tolerate indeterminate latency. This helps eliminate the need for large cache memories in the GPU, area that can be devoted to compute units instead. The fundamental compute unit is called a stream processor (in NVIDIA GPUs). The stream processors are grouped into stream multiprocessors, or SMs. Workloads are partitioned into blocks of threads among SMs. These blocks do not run until an SM is available, then it will run to completion or until stalled waiting for data. The allocations of hardware resources are done without programmer intervention. The shared memory is partitioned among blocks, and registers are partitioned among threads. Unlike CPUs, context switching happens immediately. When a thread block is stalled, waiting for data, and other thread block can immediately run.

In practice, most applications or algorithms have data dependencies that can stall the processing until some other result is available. This will cause longer latency in the computation. GPUs "hide" this latency by running many threads on the same SM. A typical number might be 128 threads per SM. The SM can switch to another group of threads when the current group is waiting for memory access, or another result is necessary to continue. Having a high number of threads is key to allowing the GPU to keep the SMs running at the great majority of the time and achieving close to the peak processing capability.

29.2 Graphics Processing Unit Programming Environment

Due to their architectural differences, GPUs are very different to program than CPUs. NVIDIA has developed CUDA (Compute Unified Device Architecture), which is a proprietary language and development environment for their GPUs, which have gained widespread use in industry. The other alternative is OpenCL, which is an open standard developed by the Khronos group, which is also gaining popularity. OpenCL is supported on NVIDIA and AMD GPUs, and more recently by Intel and Xilinx FPGAs. OpenCL can be used to access the parallel processing abilities of both GPUs and FPGAs, although the optimization techniques are quite different for GPUs and FPGAs.

CUDA and OpenCL have different terms for similar concepts.

CUDA	OpenCL	Description
Thread	Work item	Sequence of operations per data value
Warp	Wavefronts	Defined size group of work items/threads which can execute in interleaved fashion
Thread block	Work group	Grouping of warp/wavefronts on an SM/compute unit (CU)

CUDA	OpenCL	Description
Grid—2-D array of thread blocks	NDRange—3-D array of work items	Total of all threads/work items across multiple dimensions or array
Streaming multiprocessor (SM)	Compute unit (CU)	Executes instructions of work groups/blocks
Stream processor (SP)	Processing element	Functional processing units which make up SM or CU
Kernel	Kernel	S/W code to accelerate a parallel algorithm

The parallelism in an algorithm is programmed explicitly. Dependencies between different calculations or compute loops are specified using constructs in CUDA/OpenCL. The program that executes on the GPU, to accelerate a computation for the CPU, is referred to as a kernel.

A kernel is a software routine, which describes a vector processing function. A kernel is executed over a thread/work-item hierarchy, grouped into block/work groups. The blocks/work groups are organized into a multidimensional or array of grid/NDRange, The programmer writes the code to run on an SM/CU. The block/work groups are distributed to run on one of the SM/CUs, which is not under control of the programmer. Multiple block/work groups can also execute on one SM/CU concurrently.

The SM/CU creates, manages, schedules, and executes threads/work items in groups, called warps/wavefronts. The scheduler on each SM/CU evaluates all available warps/wavefronts to determine which is ready to execute an instruction. (For example, a warp/wavefront might be stalled waiting for data from memory.) In this way, the execution of the warp/wavefronts is interleaved, to hide latency restrictions in individual warp/wavefronts. However, each warp/wavefront executes its instructions in sequential order.

Each warp/wavefront executes on one SM/CU. The individual threads/work items start execution together but each is able to branch independently. If the threads/work items of a warp/wavefront diverge via a conditional branch, then the warp/wavefront will serially execute both of the branch paths. And if one thread/work-item stalls, the entire warp/wavefront stalls. There are provisions to create synchronization between warp/wavefront; for example, to stall one warp/wavefront until a certain result is available from another warp/wavefront. The next level up of hierarchy, blocks/work groups, can execute completely independently.

Each threads/work item has a unique ID, which can be identified within a kernel by the array indexing across the grid/NDRange. The indexing of each dimension is fixed within a given kernel. This ID is used to group threads/work items into warps/wavefronts and into blocks/work groups across a grid/NDrange.

One key aspect of CUDA/OpenCL is that the programmer does not need to know how many SM/CUs are available in the GPU. The GPU will automatically spread out the blocks/work groups to run across as many SM/CUs as available. Synchronization barriers will still be in force, so that algorithmically, the same processing function is supported, but will be executed with a higher degree of parallelism as the number of SM/CUs increases. The upper limit will be when full parallelism is achieved across the total of thread/work items. By having tens of thousands of thread/work items, the instances of idle SM/CUs are minimized. Usually the challenge is avoiding memory access stalls, which degrades the efficiency and computational throughput of the GPU. This means that both memory bandwidth as well as compute capability of a given GPU need to be considered when evaluating the real-world performance. The latest GPUs incorporate a type of on-chip DRAM, known as high bandwidth memory (HBM), which can increase memory bandwidth by an order of magnitude or more compared to traditional off-chip DDR memory banks.

Just like a DSP or FPGA, programming a GPU is more difficult than the linear programming style of a CPU or MCU (microcontroller unit). Getting a function to execute on a GPU is not too challenging, but the larger effort is to optimize the function to take full advantage of the GPU's compute parallelism to the extent possible for a given algorithm and memory bandwidth. Because of this, the use of preoptimized libraries is prevalent. NVIDIA and partner companies offers a very rich set of optimized libraries to run within CUDA for a wide range of applications and algorithms. AMD, Intel, and Xilinx also offer varying degrees of OpenCL libraries optimized for their devices.

29.3 Memory Hierarchy

Memory bandwidth, latency, density, and accessibility are as important as compute density and have a large impact on the effect compute bandwidth of GPUs. Like CPUs and FPGAs, GPUs take a multilayered approach to memory, and proper use of each memory type is part of the GPU optimization process.

From the programmer's point of view, there are four types of memory, with slightly different names in CUDA/OpenCL:

Global: read/write access for all threads/work items in all blocks/work groups
Shared/Local: read/write access for all threads/work items within a block/work group
Constant: read only (contains constant during execution of a kernel), visible to all threads/work items
Local/Private: read/write only for a specific thread/work item

Global memory maps to external DDR memory or to HBM in GPU devices so equipped. Memory latency to access global memory can be several 100-clock cycles. Generally, programmers should try to coalesce memory accesses, to have fewer but larger contiguous memory accesses. Random or complex address patterns will increase latency and decrease effective bandwidth.

Shared/local memory is on-chip, fast access SRAM-type memories. It is necessarily much smaller than global memory. Each SM/CU has its own dedicated local memory and is often arranged in banks. Conflicts between banks will increase access latency.

Constant memory usually maps to caches, usually one per SM/CU. This can be used for storing global constants or coefficients.

Local/private memory is on-chip register files, with specific registers allocated per thread/ work item.

29.4 Interfaces

GPUs require at least two interfaces. One is a connection to the host CPU, and the other is to external memory. The host CPU can be connected using internal buses when the GPU is copackaged on die with the CPU, as is common on smart phones and many laptops. When the GPU is a separate chip to allow for higher levels of processing, an interface known as PCI Express (PCIe) is traditionally used. PCIe uses multiple serial links running at high speed to transfer data, with a protocol that grew out of the original PCI (peripheral interconnect bus) protocol used with desktop computers. That current generation of PCIe generation 4 provides 16 lanes of serial links, each at 16 Gbps, with an aggregate bandwidth of 256 Gbps.

External memory bandwidth is also key to keep the compute engines on the GPU fed with data. High-performance GPUs traditionally had extremely high memory bandwidth, using many separate banks of the latest DDR memory to connect to the GPU chip. This has reached its limit, due to a finite the number of pins available on GPU package, and ability to locate enough memory chips adjacent to the GPU. A new solution has been developed, known as HBM, which can be integrated with the chip package, rather than externally on the printed circuit board. HBM provides on the order of 256 GBPS of memory bandwidth per modules, and multiple HBM modules can be integrated within a GPU chip. This provides an enormous increase in available memory bandwidth and is not subject to the restriction of number of pins (or balls) on the GPU chip package.

As GPUs continue to be adopted for compute acceleration in the data center, multiple GPUs can be used with one host CPU. NVIDIA has developed high-speed interconnect technology to be used between GPU chips and some CPUs, known as NVLink. This can provide a large increase in compute power across multiple GPU chips.

As GPUs are used in a greater variety of applications, embedded processors from ARM are being used to act as local host processors. Additional interfaces for cameras are also being included in some GPU-based products.

29.5 Numerical Precision

GPUs started out as integer machines, primarily 16 bits, used for graphics processing. In addition, 8- and 32-bit integer operations are also supported. As GPUs morphed into GP-GPUs, floating-point support was added. First, single precision, then double precision, and most recently half precision. Single and double precision are used for high-performance computing, accelerating algorithms running on a companion CPU. Half precision is primarily used for machine learning, specifically training for machine learning. Training is commonly done in a data center environment, where the large databases required for training are available.

Generally, the computational rates are inversely proportional to the data-word sizes, so that the memory bandwidth remains constant across the precisions. Therefore, single precision (32-bit word) computational rates are twice that of double (64-bit word), and half precision (16-bit word) computational rates are twice that of single.

29.6 Future Trends

GPUs have expanded far beyond the original application of graphics processing for gaming and computer-generated animation. Any problem with high computational needs that benefits from parallel processing is a candidate for GPU acceleration. Beyond graphics, GPUs have found usage in accelerating CPU processing of financial options pricing, genetics, radar, molecular dynamics, climate research, and other applications with high computing requirements. In the development of machine-learning algorithms and methodology, GPUs have been the platform of choice. Machine learning involved huge amounts of data and naturally required high degrees of parallel processing. This is traditionally done in servers and in data centers, used to process images, video, and databases to extract trends or information. GPUs are used in both aspects of ML: training and classification.

But GPUs are evolving beyond the controlled environment of the data center to find usage in embedded applications. For example, GPUs have found prominent use in automobiles, in driver's assistance with the eventual goal of fully autonomous driving. The GPU is an efficient and flexible platform to perform the high levels of machine-learning (ML) processing used to interpret the image data received from the cameras. These GPUs tend to be lower power and only support the lower integer precisions used for ML inference. GPUs can also be used in autonomous drones, to navigate safely and avoid collisions.

GPUs are also used in virtual reality (VR) and augmented reality (AR). AR, in particular, uses a combination of camera processing, ML, and graphics processing. GPUs can do much of this processing, whether in specialty-built AR products, or used in smart phones to run AR apps.

GPUs provide the benefits of parallel processing, without the challenges of programming FPGAs or designing purpose build hardware engines in ASICs (application specific integrated circuits). While the GPU still lacks the wide variety of hardware-interfacing capabilities of FPGAs or the ability to efficiently process data of just a few bits in width, the current trends indicate that the usage of GPU technology will continue to be adopted in an ever wider set of high compute applications, with a wider degree of interfaces.

Appendix A: Q Format Shift With Fractional Multiplication

Consider normal multiplication as below:

```
      104
    x 512
-------------------
    53248
```

In this example there is no decimal point consideration, as both numbers are integers.

Now look at the same number, but with a different decimal point arrangement.

```
      10.4
    x 5.12
-------------------
    53.248
```

How do we know where to place the decimal point, if we are doing this multiplication by hand? One way is by sanity check $5 \times 10 = 50$, so the answer of 53.248 obviously has the correct decimal point. But the way most of us learned in grade school was that the product should have the same number of digits to the right of the decimal point as the two multiplicands combined have to the right of their decimal point. The first number, 10.4, has one digit to right of decimal point, and the second number, 5.12, has two digits to right of decimal point. So the product, 53.248, should have three digits to right of decimal point, giving 53.248.

This same concept works for fractional binary multiplication. For example, one Q1.15 number multiplied by another Q1.15 number will give a 32 bit result. But where is the decimal point? By the rule above, the product should have $15 + 15 = 30$ bits to the right of the decimal point. Our result is then a Q2.30 number. If we use only the upper or most significant 16 bits, as is common in many implementations, the result is a Q2.14 number. By performing a single left shift after the multiplication, we will get a Q1.15 number using upper 16 bits, which is same format as our input data (or a Q1.31 number if we chose to keep all 32 bits). This is the reason that many digital signal processors (DSPs) have fractional multiply instructions or modes, which incorporate an extra left shift of the result. In field programmable gate array (FPGA) implementation, the user can easily choose the output format of the multiplier and take this effect into account. The example from, Chapter 1 on numerical representation is repeated here:

0x4000	value = ½ in Q.15
x 0x2000	value = ¼ in Q.15
0x0800 0000	value = 1/16 in Q31

After left shift by one

$$0 \times 1000\ 0000\ \text{value} = 1/8\ \text{in Q31—correct result!}$$

If we use only top 16-bit word from multiplier output, after the left shift we get

$$0 \times 1000\ \text{value} = 1/8\ \text{in Q15—again correct result!}$$

In DSP applications, we often perform multiply accumulate operations, where the result of many multiplies are summed to a common result (remember the FIR filter structure).

With all these additions, how can the sum can be represented without overflow occurring? How does this affect our decimal point?

The solution is to make the gain of the filter equal to one. As explained in the FIR filter chapter, the sum of all the filter coefficients is the gain of the filter. By scaling the filter coefficients (divide each coefficient by the sum of the coefficients), the gain can be easily set to unity. Then there is no possibility of overflow, and the decimal point will not be affected. The frequency response will also be unchanged, as each coefficient is being scaled equally. For example, suppose our filter coefficients are {1,3,5,3,1}. The sum of these coefficients is 13. Therefore, our scaled coefficients will be {1/13, 3/13, 5/13, 3/13, 1/13}. This allows us to represent the coefficients in fractional format, since they are all between -1 and $+1$, and it guarantees that the decimal point will not be altered in the signal as it passes through our filter. We still need the left shift at the output of the multiplier, however. Alternately, in FPGA, we could easily apply the left shift to the accumulated sum, instead of at each multiplier output.

Appendix B: Evaluation of Finite Impulse Response Design Error Minimization

We are minimizing the following expression

$$\text{Error} = \xi = \int_{-\pi}^{\pi} |\xi(\omega)|^2 d\omega$$

where,

$\xi(\omega) = D(\omega) - H(\omega)$
$D(\omega) =$ desired frequency response
$H(\omega) = \sum_{i=-\infty \text{ to } \infty} C_i e^{-j\omega i} =$ actual frequency response

We start by setting the derivative of the error ξ with respect to the filter coefficients (the one parameter that we have control over and want to optimize) equal to zero.

$$d|\xi|/dC_i = 0$$

$$d|\xi|/dC_i = \int_{-\pi}^{\pi} (d/dC_i)\{[D(\omega) - H(\omega)] \cdot [D(\omega) - H(\omega)]^*\} d\omega = 0$$

Use chain rule to differentiate each part.

$$d|\xi|/dC_i = \int_{-\pi}^{\pi} (d/dC_i)\{[D(\omega) - H(\omega)]\} \cdot [D(\omega) - H(\omega)]^* d\omega$$

$$+ \int_{-\pi}^{\pi} [D(\omega) - H(\omega)] \cdot (d/dC_i)\{[D(\omega) - H(\omega)]\}^* d\omega$$

The derivative is now a sum of two integrals. Since they are complex conjugates, the sum can only be zero when both terms are zero. So we can consider only one of the two terms, and evaluate when it is equal to zero.

$$\int_{-\pi}^{\pi} [D(\omega) - H(\omega)] \cdot (d/dC_i)\{[D(\omega) - H(\omega)]\}^* d\omega = 0$$

Now let us try to simplify the term $(d/dC_i)\ \{|\ D(\omega)-H(\omega)\ |\}^*$

$$(d/dC_i)\{|D(\omega)-H(\omega)|\}^* = (d/dC_i)\{D(\omega)^*\} - (d/dC_i)\{H(\omega)^*\}$$

Notice that $(d/dC_i)\ \{D(\omega)^*\} = 0$, since $D(\omega)^*$ does not depend on C_i. $D(\omega)$ is our desired response and not dependent on C_i.

Recall that $H(\omega) = \sum_{k=-\infty \text{ to } \infty} C_k e^{-j\omega k}$ We have replaced the index i with k since in this discussion we are already using i as the coefficient index.

Therefore

$$H(\omega)^* = \sum_{k=-\infty \text{ to } \infty} C_k * e^{+j\omega k}$$

We want to simplify

$$H(\omega)^* = (d/dC_i)\left\{\sum_{k=-\infty \text{ to } \infty} C_k * e^{+j\omega k}\right\}$$

Each coefficient C_i is independent of the others; for example, $(d\ C_1/d\ C_2) = 0$. Only when $i = k$ will there be a nonzero result. Therefore we can remove the summation and get as follows:

$$H(\omega)^* = (d/dC_i)\left\{\sum_{k=-\infty \text{ to } \infty} C_k * e^{+j\omega k}\right\} = (d/dC_i)\{C_i * e^{+j\omega i}\} = e^{+j\omega i}$$

Now let us go back and substitute the simplified result.

$$(d/dC_i)\{|D(\omega)-H(\omega)|\}^* = (d/dC_i)\{D(\omega)^*\} - (d/dC_i)\{H(\omega)^*\} = e^{+j\omega i}$$

Now we are getting close. Let us go back to the original integral and substitute again.

$$\int_{-\pi}^{\pi} [D(\omega)-H(\omega)]\cdot(d/dC_i)\{[D(\omega)-H(\omega)]\} * d\omega = \int_{-\pi}^{\pi} [D(\omega)-H(\omega)]\cdot e^{+j\omega i} d\omega$$

Since we are trying to find the solution when the integral equals zero, we can equate the two terms within.

$$\int_{-\pi}^{\pi} D(\omega)\cdot e^{+j\omega i} d\omega = \int_{-\pi}^{\pi} H(\omega)\cdot e^{+j\omega i} d\omega$$

Now let us substitute for $H(\omega)$ again. We will get as follows:

$$\int_{-\pi}^{\pi} \sum_{k=-\infty \text{ to } \infty} C_k e^{-j\omega k}\cdot e^{+j\omega i} d\omega = \int_{-\pi}^{\pi} \sum_{k=-\infty \text{ to } \infty} C_k e^{-j\omega(k-i)} d\omega$$

Our goal is to get C_k by itself on the left side, since that is what we are trying to solve for.

We can rearrange the order of summation and integration, since both are linear operations.

$$\int_{-\pi}^{\pi} \sum_{k=-\infty \text{ to } \infty} C_k e^{-j\omega(k-i)} d\omega = \sum_{k=-\infty \text{ to } \infty} C_k \int_{-\pi}^{\pi} e^{-j\omega(k-i)} d\omega$$

Let us evaluate the integral

$$\int_{-\pi}^{\pi} e^{-j\omega(k-i)} d\omega$$

There are two cases, when $k = i$ and when $k \neq i$.

$$k = i \quad \int_{-\pi}^{\pi} e^{-j\omega(k-i)} d\omega = \int_{-\pi}^{\pi} e^{-j\omega 0} d\omega = \int_{-\pi}^{\pi} 1 d\omega = 2\pi$$

$$k \neq i \quad \int_{-\pi}^{\pi} e^{-j\omega(k-i)} d\omega = -j\omega(k-i) \cdot \left[e^{-j\omega(k-i)} \Big|_{-\pi}^{\pi} \right] = -j\omega(k-i) \cdot \left[e^{-j\pi(k-i)} - e^{j\pi(k-i)} \right]$$

Let us consider two possibilities: the number $m = k-i$ is even, or it is odd:

m is even: $[\, e^{-j\pi m} - e^{j\pi m} \,] = 1 - 1 = 0$, since both of these terms will be at the point $(1+0j)$ on the unit circle.
m is odd: $[\, e^{-j\pi m} - e^{j\pi m} \,] = -1 - (-1) = 0$, since both of these terms will be at the point $(-1+0j)$ on the unit circle.

So we get a very simple result:

$$\int_{-\pi}^{\pi} H(\omega) \cdot e^{+j\omega i} d\omega = \sum_{k=-\infty \text{ to } \infty} C_k \int_{-\pi}^{\pi} e^{-j\omega(k-i)} d\omega = C_k \cdot 2\pi$$

If we substitute this into

$$\int_{-\pi}^{\pi} D(\omega) \cdot e^{+j\omega i} d\omega = \int_{-\pi}^{\pi} H(\omega) \cdot e^{+j\omega i} d\omega$$

We will finally arrive at the desired result:

$$C_k = (1/2\pi) \cdot \int_{-\pi}^{\pi} D(\omega) \cdot e^{+j\omega i} d\omega$$

The kth coefficient is found by multiplying the desired frequency response by $e^{+j\omega i}$ and integrating over the interval 2π. Each coefficient is computed independently. This method will give the lowest value of error ξ, as defined in the finite impulse response filter chapter.

Appendix C: Laplace Transform

The Laplace transform is used for continuous or analog signals. We shall use x(t) and y(t) to represent the input and output signals to an analog filter, respectively. Rather than use delays, as digital filters use, the differentiator is used instead. The input and output of the analog filter is related by the following relationship.

$$y(t) = \sum_{i=0 \text{ to } N} A_i \cdot d^i x(t)/dt^i + \sum_{i=0 \text{ to } M} B_i \cdot d^i y(t)/dt^i$$

The first term is a sum of coefficient weighted derivatives of the input, which is analogous to the FIR filter being a sum of weighted delayed inputs. The second term is a coefficient weighted sum of the derivatives of the output, which implies feedback.

These can be difficult functions to evaluate and characterize. Smart people long ago figured out a way to map things into the s-domain using the Laplace transform, where the mathematics becomes much simpler. This is an introduction to that technique.

We are going to use the exponential function again to determine the response of this filter. In this case, we will use the function e^{st}, where s is a complex variable.

Notice this property of our chosen input, which we will use to develop the s-transform.

$$de^{st}/dt = s \cdot e^{st} \text{ and } d^i e^{st}/dt^i = s^i \cdot e^{st}$$

If we apply this to our analog filter equation, we will get:

$$y(t) = \sum_{i=0 \text{ to } N} A_i \cdot s^i\, x(t) + \sum_{i=0 \text{ to } M} B_i \cdot s^i\, y(t)$$

We can therefore construct a Laplace transform-based relationship of the filter response.

We want an equation with form y(t) = func(x(t)). After doing a bit of rearranging:

$$y(t) = x(t) \cdot \left\{ \left(\sum_{i=0 \text{ to } N} A_i \cdot s^i \right) \Big/ \left(1 - \sum_{i=0 \text{ to } M} B_i \cdot s^i \right) \right\}$$

where the bracketed term is the s-transform function, denoted as $H(s)$.

$$H(s) = \left(\sum_{i=0 \text{ to } N} A_i \cdot s^i\right) \Big/ \left(1 - \sum_{i=0 \text{ to } M} B_i \cdot s^i\right)$$

The roots of the numerator give the zero locations of the Laplace transform, and the roots of the denominator give the pole locations of the s-transform. These zeros and poles will characterize the filter frequency response.

Similar to digital filters, analog filters can also be defined by their impulse response. In the digital domain, an impulse is defined as a single sample with value = 1. For the analog world, this is not as simple. The answer is to define a function called the delta dirac, known as δ(t). It is a strange function. It has infinite height and zero width, but the area under its integral is equal to 1. It is essentially a spike located at zero along the number line. This is the continuous signal equivalent to a digital impulse. We can define the impulse response of an analog filter as follows.

$$\text{Impulse response} = h(t) = y(t)\Big|_{x(t)=\delta(t)}$$

So the output is the impulse response when the input is δ(t).

Now recall that s is a complex number. We can evaluate the Laplace transform, and determine where it's zeros (values of where transform numerator goes to zero) and where it's poles (values of s where the transform denominator goes to zero), and plot these locations on the complex number plane. In this context, the complex number plane is known as the s-plane (Fig. C.1).

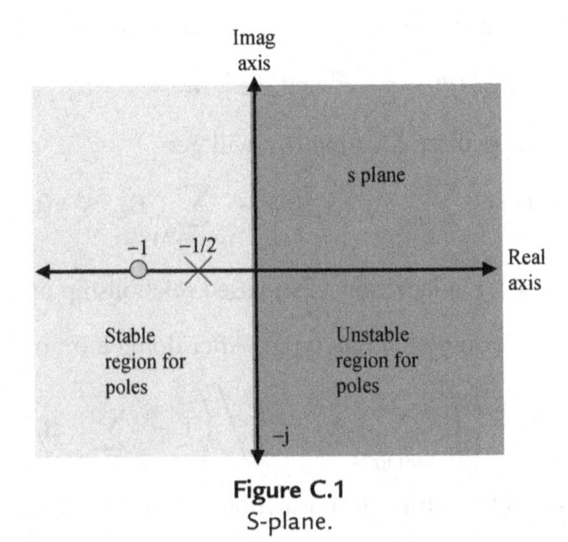

Figure C.1
S-plane.

Next, we will go through a simple example to clarify

$$y(t) = x(t) + dx(t)/dt - 2 \cdot (dy(t)/dt)$$
$$y(t) = x(t) \cdot \{(1 + s)/(1 + 2s)\}$$

We can see that there is a single zero at $s = -1$, and a single pole at $s = -1/2$. The frequency response is evaluated by setting $s = j\omega$, and evaluating for $-\infty < \omega < \infty$.

The filter's response is determined by the pole and zero locations. Use of the Laplace transform facilitates the design of analog filters.

We have mentioned that filters with feedback can be unstable. We can examine a simple case to see when this is true.

$$y(t) = A_0 \cdot x(t) + \cdot dy(t)/dt$$
$$y(t) = A_0 \cdot x(t) + B_1 \cdot s\, y(t)$$
$$H(s) = A_0/(1 - B_1 \cdot s)$$

which has a single pole at $s = 1/B_1$.

To determine stability, we compute the impulse response of the filter. Because of the simplicity of this example, the response of filter with $x(t) = \delta(t)$ can be computed directly from the differential equation. We will find that:

$$h(t) = (A_0/B_1) \cdot e^{(t/B_1)} \quad \text{for } t \geq 0 \text{ and } h(t) = 0 \ \text{ for } t < 0$$

If B_1 is negative, the exponential will decay over time—this is considered a stable response. But if B_1 is positive, the impulse response will grow to infinity over time—this is an unstable response. This leads to the following general rule with analog filters.

All poles must be on left side of s-plane.

There is no restriction on location of zeros.

Appendix D: Z-Transform

The z-transform is used for sampled signals. It is analogous to the s-transform used with continuous or analog signals. We shall use x_k and y_k to represent the input and output signals to a digital filter, respectively. Rather than using differentiators, as analog filters use, the delay function is used instead. The input and output of the IIR (recursive) digital filter is related by the following relationship.

$$y_k = \sum_{i=0 \text{ to } N-1} C_i x_{k-i} + \sum_{i=1 \text{ to } N-1} D_i y_{k-i}$$

The first term is a sum of coefficient weighted delayed versions of the input (this is the familiar FIR filter expression). The second term is a coefficient weighted sum of the delayed versions of the output, which implies feedback.

The feedback part prevents us from using our earlier method of finite impulse response (FIR) design analysis, described in Appendix B. But again, some smart people figured out a way to map things into the z-domain, where the math becomes simpler. This is an introduction to that technique.

The z-transform is very useful. It can be used to determine the frequency response of an infinite impulse response (IIR) filter. It can also be used to find the coefficients of the IIR filter from the zeros and poles locations, or to do the inverse, to find the poles and zeros of the IIR filter from the coefficients.

The z-transform uses a special exponential function z^{-i}, where z is a complex variable.

The definition of the z-transform of any given sampled time sequence x_i is

$$X(z) = \sum_{i=-\infty \text{ to } \infty} x_i z^{-i}$$

We can characterize a digital filter by its z-transform, $H(z)$. We will show that

$$Y(z) = H(z) \cdot X(z) \text{ or } H(z) = Y(z)/X(z)$$

where $Y(z)$ is the z-transform of the output sampled time sequence y_k; $X(z)$ is the z-transform of the input sampled time sequence x_k.

Let us consider a digital filter, with an impulse response of h_i. The output of this filter is given by:

$$y_k = \sum_{i=-\infty \text{ to } \infty} h_i x_{k-i}$$

where x_k is the input sequence (we developed this in the Chapter 5 of FIR filter).

Let us take the z-transform of both x_k and y_k to see if we can determine $H(z)$:

$$X(z) = \sum_{k=-\infty \text{ to } \infty} x_k z^{-k}$$

$$Y(z) = \sum_{k=-\infty \text{ to } \infty} y_k z^{-k} = \sum_{k=-\infty \text{ to } \infty} \left\{ \sum_{i=-\infty \text{ to } \infty} h_i x_{k-i} \right\} z^{-k}$$

Now we will interchange the order of summation:

$$Y(z) = \sum_{i=-\infty \text{ to } \infty} \left\{ \sum_{k=-\infty \text{ to } \infty} h_i x_{k-i} z^{-k} \right\}$$

Next remove h_i from the inner summation, as it is independent of k:

$$Y(z) = \sum_{i=-\infty \text{ to } \infty} h_i \left\{ \sum_{k=-\infty \text{ to } \infty} x_{k-i} z^{-k} \right\}$$

Next, factor z^{-k} into 2 terms:

$$Y(z) = \sum_{i=-\infty \text{ to } \infty} h_i \left\{ \sum_{k=-\infty \text{ to } \infty} x_{k-i} z^{-(k-1)} z^{-i} \right\}$$

Then remove z^{-i} from the inner summation, as it is independent of k:

$$Y(z) = \sum_{i=-\infty \text{ to } \infty} h_i z^{-i} \left\{ \sum_{k=-\infty \text{ to } \infty} x_{k-i} z^{-(k-1)} \right\}$$

The bracket term above is simply $X(z)$.

$H(z) = \sum_{i=-\infty \text{ to } \infty} h_i z^{-i}$ is the z-transform of impulse response h_i.

This demonstrates that $Y(z) = H(z) \cdot X(z)$.

Let us look at this a different way. The filter's response has a z-transform, which is equal to the z-transform of the output sequence y_k divided by the z-transform of the input sequence x_k, denoted $H(z)$.

By definition, y_k equals the impulse response when we set x_k equal to an impulse. The z-transform of the impulse function is equal to 1, so the z-transform of the impulse response of the filter is simply the z-transform of the output.

$$H(z) = Y(z)/X(z) = \sum_{i=-\infty \text{ to } \infty} h_i z^{-i}$$

where h_i is defined as the impulse response of the filter.

This sounds familiar to frequency response $H(\omega)$, and indeed the two are related. The frequency response is contained within the z-transform. The frequency response can be found from the z-transform by replacing z with $e^{j\omega}$, or when z is evaluated around the unit circle on the complex z-plane.

$$H(\omega) = \sum_{i=-\infty \text{ to } \infty} h_i e^{-j\omega i}$$

With an IIR filter, computing the impulse response sequence h_i can often be difficult. The usefulness of the z-transform is that it can be used to compute the frequency response $H(\omega)$ without first computing the impulse response h_i. The z-transform can be found easily if either the coefficients of the digital filter are available, or the poles and zeros of the digital filter are available. Then take the expression $H(z)$, replace z with $e^{j\omega}$, and then evaluate the frequency response over the interval $-\pi < \omega < \pi$.

$H(z)$ is simply a function of z. Our goal will be to express $H(z)$ as a function of two polynomials in z, in the form of

$$H(z) = A(z)/B(z)$$

With a bit of algebra, we can then rearrange the polynomials into the following form:

$$H(z) = \left(\sum_{i=0 \text{ to } N} C_i \cdot z^{-i} \right) \Big/ \left(1 - \sum_{i=1 \text{ to } M} D_i \cdot z^{-i} \right)$$

In this form, the coefficients can are available by inspection. Or, if you have the coefficients, you can easily write the z-transform of the filter.

The z-transform can also be arranged in pole—zero format. Remember, by definition, the zeros of the z-transform are the values of z where $H(z) = 0$ (the roots of the numerator) and the poles of the z-transform are the values of z where $H(z) = \infty$ (the roots of the denominator).

$$H(z) = \left[\prod_{i=0 \text{ to } M} (z - \text{zero}_i) \right] \Big/ \left[\prod_{i=0 \text{ to } N} (z - \text{pole}_i) \right]$$

To summarize, if you are given the poles and zeros, you can immediately construct the z-transform using the template above. Although it may take a bit of algebra, the z-transform can be rewritten in summation form to find the coefficients. And evaluating the z-transform in either form over the complex unit circle will give the frequency response of the filter. This is done by substituting $e^{j\omega}$ for z, and computing over interval $-\pi < \omega < \pi$,

$$H(z) = \left[\prod_{i=0 \text{ to } M} (z - \text{zero}_i)\right] \Big/ \left[\prod_{i=0 \text{ to } N} (z - \text{pole}_i)\right] =$$

$$H(z) = \left(\sum_{i=0 \text{ to } N} C_i \cdot z^{-i}\right) \Big/ \left(1 - \sum_{i=1 \text{ to } M} D_i \cdot z^{-i}\right) \text{ and}$$

$$\text{Frequency response } H(\omega) = H\left(z = e^{j\omega}\right)\omega = -\pi\Big|^{\omega=\pi}$$

As with the s-plane, we can show that pole location on the z-plane is restricted for stability reasons. All poles within the z-plane must lie within the unit circle. Zeros have no restriction on the z-plane.

Appendix E: Binary Field Arithmetic

Binary field arithmetic is used in coding. The basic rules are very simple. Addition is as follows:

$$0 + 0 = 0$$
$$0 + 1 = 1$$
$$1 + 0 = 1$$
$$1 + 1 = 0$$

This also corresponds to the exclusive OR or XOR operator in digital logic. Subtraction results are identical to addition; there is no distinction between the two operations.

Multiplication is as follows:

$$0 \cdot 0 = 0$$
$$0 \cdot 1 = 0$$
$$1 \cdot 0 = 0$$
$$1 \cdot 1 = 1$$

Division is not defined or allowed.

Linearity means that if two input codewords are input to a system, X_1 and X_2, the output result is the same as if the sum of two codewords $X_1 + X_2$ were input instead. A more mathematical way to express this idea is as follows:

$$Y1 = f(X1)$$
$$Y2 = f(X2)$$

For a linear system

$$Y2 + Y1 = f(X2 + X1)$$

Index

'*Note*: Page numbers followed by "f" indicate figures, "t" indicate tables.'

A

AC induction motor (ACIM), 163–164, 171–172, 171f, 175, 180–181, 180f
AC motor, 165, 166f, 169–170, 172, 173f, 174, 177
Airborne radar, 232, 241–243, 245, 277, 279, 284, 288–289
AlexNet graph, 350, 352–353, 352f
Aliasing, 23, 26, 27f, 36, 65–66, 70, 100, 121–123, 126, 238, 238f, 245, 246f, 249, 302
Alpha blending, 303
Analog-to-digital convertor (ADC), 21, 25–30, 37, 65, 67, 67f, 119, 121f–124f, 122–126, 175, 175f, 187, 190f–191f, 229f–230f, 255f, 260–261, 261f, 263–265, 266f, 326, 363, 371
Angle of arrival (AoA) estimation, 263, 265, 267–270, 275–276, 276f
Antenna array, 229–230, 234, 262, 277–279, 278f, 282, 285–286, 289
Assembly language, 186, 361, 367
Automotive radar, 253–276, 278–279
Autonomous driving assistance systems (ADAS), 253, 280

B

Back electromotive force (EMF), 164, 165f, 170–171
B frame compression, 339, 341–342, 345
Bi-linear transform, 78–82
Bit-reversed addressing, 115, 381
Block size, 336
Bob, 299–300, 300f
Brushless DC (BLDC) motor commutation, 168–169, 169f
Buffering, 264, 266f, 267, 339, 342–343, 367

C

Cartesian coordinates, 10–11
Cell averaging–constant false alarm rate (CA-CFAR), 270–274, 273f
Cholesky decomposition, 149, 151–157, 283
Chrominance, 297–298, 329, 330t, 331–332, 338, 340, 340t
Clark transform, 176–181, 177f, 180f
Clutter, 239, 241–242, 245–251, 253, 258, 269–274, 277–280, 277f, 282, 284, 348
Code division multiple access (CDMA), 96, 190, 191–208, 193f, 202f, 209–211, 217–218, 221–223, 225t, 381
Coding gain, 30, 130, 145, 326
Common public radio interface (CPRI), 229–230
Complementary cumulative distribution function (CCDF), 225–226
Complex conjugate, 15, 15f, 50, 152, 239, 288, 397
Complex exponential, 15–18, 16f–17f, 31–32, 34–35, 38, 45–47, 46t, 59, 101t, 103–104, 107, 109, 111, 114–115, 119, 123, 213–214, 216–217, 311, 362, 376
Complex number, 9–19, 9f, 11f, 31–32, 50, 99–100, 281–282, 291, 402
Component video, 308
Composite video blanking and sync (CVBS), 308
Compute Unified Device Architecture (CUDA), 388–390
Conjugate symmetric matrix, 151–152, 157
Constant false alarm rate (CFAR), 263, 263f, 265–267, 269–272, 274–275, 350–351
Constant memory, 391
Convolutional encoding, 135–137, 136f, 187
Convolutional neural networks (CNNs), 347–348, 350, 352–354, 357–359
Convolution layer, 348–350
Covariance matrix, 281–282, 282f, 284–285, 284f
Crest factor reduction (CFR), 220, 220f, 225–229, 230f

Cyclic prefix, 217–221, 218f–219f
Cyclic redundancy check (CRC), 146

D

DC motor, 166–169, 166f
DC scaling, 329–330
Decibels, 29–30, 259, 325–327
Decimation, 65–69, 67f, 71–72, 77, 115, 124–125, 186, 230f, 302, 362–363, 376, 379–380
Deinterlacing, 299–300, 300f–301f
Differential encoding, 319–321, 324
Digital downconversion, 119, 121–125, 121f–122f, 244
Digital predistortion (DPD), 225, 228–229, 230f
Digital-to-analog converters (DAC), 29–30, 65, 93, 119–121, 120f, 126–128, 127f, 175, 187, 190f–191f, 202, 220, 220f, 229f–230f, 243, 326, 371
Digital upconversion, 93, 118–121, 120f, 127–128, 219, 243
Digital video interfaces, 304–307
Digital visual interface (DVI), 305
Direct digital synthesizer (DDS), 260, 261f
Discrete cosine transform (DCT), 311–315, 313f–314f, 322, 329–331, 333–334, 333f, 336–337, 340, 342–343, 344f, 345
Discrete Fourier transforms (DFT), 99–116, 101t, 213–215, 293–294, 311
DisplayPort, 306–307
Doppler ambiguities, 245
Doppler frequency detection, 244, 257f, 288
Doppler frequency shift, 188, 223, 232, 241–244, 256
Doppler offset, 257, 258t
Doppler shift, 95, 189, 221, 223, 241–243, 250f, 256–258, 260–261, 279

E

Electromagnetic induction (EMI), 167
Electronic commutation, 167–171, 168f
Entropy, 315–316, 318–321, 329, 345
Entropy coding, 331–332, 333f, 340–341
Error vector measurement (EVM), 225–228
Euler equation, 16–17, 51
Exponent, 1, 7–8, 8t, 100–101, 103–105

F

Faraday's Law, 164–165
Fast Fourier transform (FFT), 99–116, 108t, 111f, 113f, 186, 214–216, 216t, 217f, 222–223, 240, 244, 255, 255f–256f, 257, 263–269, 263f, 266f, 268f, 279, 293–294, 311, 362–363, 365, 372–374, 376–377, 379–382
Field-oriented control (FOC), 163, 174, 180f
Field-oriented motor control, 163–182
Field programmable gate array (FPGA), 7, 52, 68–69, 117, 149, 152, 160–161, 253, 263, 266–267, 270, 286, 304–305, 359, 367, 371–386, 378f, 388, 390, 393, 395–396
First discrete Fourier transforms, 102–103
Fixed point representation, 2–8
Floating point representation, 7–8
Flux density, 163–164
Force vector, 164f
Forward error correction (FEC), 129, 381
Fourth discrete Fourier transforms, 105–108
Fractional representation, 2, 5, 5t–6t
Frame processing order, 339–340
Frequency division duplex (FDD), 183–185, 210–211, 220
Frequency-modulated continuous-wave (FMCW) beamforming, 261–263
Frequency-modulated continuous-wave (FMCW) Doppler detection, 256–259
Frequency-modulated continuous-wave (FMCW) implementation, 260–261
Frequency-modulated continuous-wave (FMCW) interference, 261
Frequency-modulated continuous-wave (FMCW) pulse-Doppler processing, 266–269
Frequency-modulated continuous-wave (FMCW) radar, 253–254, 264, 266–267
Frequency-modulated continuous-wave (FMCW) radar back-end processing, 269
Frequency-modulated continuous-wave (FMCW) radar front-end processing, 264–266
Frequency-modulated continuous-wave (FMCW) radar link budget, 259–260

Frequency-modulated continuous-wave (FMCW) range detection, 254–256
Frequency-modulated continuous-wave (FMCW) range-Doppler processing, 263–264
Frequency-modulated continuous-wave (FMCW) theory, 253–254
Fully connected layer, 348, 351–354
Frequency modulation, 185, 236, 260

G

Global memory, 390–391
Global system for mobile communications (GSM), 189, 207–209, 218, 225, 225t, 227
Gram–Schmidt method, 158–160, 283
Graphics processing units (GPUs), 149, 286, 359, 371–372, 382–383, 385, 387–394

H

High definition multimedia interface (HDMI), 306
Huffman coding, 316–317, 319, 321–322, 324, 332–333, 333f

I

Identity matrix, 149, 150f
I frame compression, 338–342
Ill-conditioned matrix, 149–151
Image compression, 311, 317, 329, 335, 342, 369, 373
Infinite impulse response (IIR), 45, 75–82, 76f, 366, 382, 405, 407
Interfaces, 29–30, 229–230, 305, 325–326, 363, 366, 371, 374–376, 391–393
Interference covariance matrix, 281–283, 281f
Interlacing, 299–300
Intermediate frequency (IF) subsampling, 122–128, 123f–124f
Interpolation, 65, 69–72, 69f–70f, 72f–73f, 74, 77, 93, 219–220, 230f, 255, 299, 300f, 302, 303f, 379–380
Intersymbol interference (ISI), 89, 91–92, 95, 188, 200, 217–221
Inverted covariance matrix, 284, 284f

J

Jammer, 277–278, 277f, 284, 284f
JPEG, 317, 329–334, 333f, 340–341
JPEG extensions, 333–334

L

Laplace transform, 78, 401–403
Legacy analog video interfaces, 307–308
Linear algebra, 278, 315
Linear phase, 37, 56–57, 76
Local/private memory, 391
Long term evolution (LTE), 85, 209–210, 214–219, 216t, 220f, 222, 225t, 226, 228, 365–366, 382
Lossless compression, 321–322
Low-noise amplifier (LNA), 230f, 260–261, 261f
Luminance, 297–298, 329, 330t, 331–332, 340, 340t

M

Machine learning, 8, 347–360, 367, 376, 382–383, 392
Macroblock size, 336–337, 337f, 341, 344–346
Magnetic flux, 163–165, 163f, 171–172, 175–178
Magnetism, 163–165
Mainlobe clutter, 246–247, 249
Mantissa, 7–8, 8t, 382
Markov source, 317–319
Matrix inversion, 149–162, 220–221, 283
Matrix multiplication, 149–150, 151f, 160, 351–352, 355–356
Matrix substitution, 153f
Max-pooling layer, 348, 351–352, 351f
Minimum absolute differences (MAD) method, 335–336
Minimum distance, 134–135
Minimum mean square error (MMSE) method, 335–336
Modulation, 75, 83–98, 85t, 117–118, 135, 144, 147, 169–170, 170f, 179–180, 185, 187–191, 191f, 201–202, 204–207, 210, 213–217, 219, 222, 225, 236–237, 254, 259–260, 263
Motion estimation, 336–339, 338f, 341–342, 344f–345f, 345
Motor control, 173–176, 173f, 175f
Motor torque, 169–171, 176f, 181
MPEG, 335, 338
Multiphase DC commutation, 168f
Multiple input and multiple outputs (MIMO), 149, 220–222, 376, 382

N

National Television System Committee (NTSC), 299
Noncoherent antenna, 269–270
Normalization layer, 350–351
Numerically controlled oscillator (NCO), 34–35, 120f–121f, 123–124, 127, 220f, 230f
Numerical precision, 77, 312–314, 322–323, 358–359, 392

O

Open base station architecture initiative (OBSAI), 229–230
OpenCL, 384, 388–390
Ordered sort–constant false alarm rate (OS-CFAR), 270, 272–275, 273f–274f
Orthnonormal matrices, 283
Orthogonal frequency division multiplexing (OFDM), 96, 209–211, 213–215, 217–219, 218f, 221–223, 225, 225t
Orthogonality, 198, 204, 209–214, 221, 223
Orthogonal matrix, 150–151

P

Park transform, 176–181, 179f–180f
Peak to average ratio (PAR), 223–226, 225t, 228
Permanent magnet AC (PMAC) motor, 169–170, 176–178
P frame compression, 338–341, 345
Phase-locked loop (PLL), 260, 261f, 368f, 375f
Polar coordinates, 17
Power control, 203, 205–206
Predictive coding, 319, 322, 329, 335
Progressive, 299–300, 334, 338
Proportional integral derivator (PID) controller, 173–174, 173f–174f
Pseudorandom code, 191
Pulse compression, 236–237, 248–249, 287–288
Pulse-Doppler processing, 280
Pulse-Doppler radar, 241–252, 253, 267
Pulse repetition frequency (PRF), 237–240, 243–245, 246f, 247–251, 285–286, 288, 290–295, 294f
Pulse shaping, 87–93, 202, 202f
Pulse width modulation (PWM), 169, 170f, 175, 179–180

Q

QR decomposition (QRD), 149, 157–158, 160–162, 283, 285–286, 286t
Quadrature amplitude modulation (QAM), 84–85, 84f, 85t, 89, 95–96, 96f–97f, 118, 135, 185, 189, 206–207, 213, 215–216, 215f, 220f, 226–227
Quadrature phase shift keying (QPSK), 83–86, 83f, 84t–85t, 86f, 89, 95–96, 118, 120, 120f–121f, 145f, 189, 199–201, 205, 213, 215–216, 220f, 222
Quantization, 8, 26–29, 27t–28t, 77, 114, 124, 187, 260, 322–326, 329–331, 330t, 333–334, 340, 340t–341t, 343–346, 382–383
Quantization scale factor, 340, 343–346

R

Radar memory transactions, 268f
Radar range equation, 234–235
Radio detection and ranging (RADAR), 129, 149, 231–296, 231t, 255f, 263f, 266f, 268f, 270f, 280f, 289f, 291f, 293f–294f, 350–351, 366–367, 373, 382
Raised cosine, 90–97, 90t, 91f, 94f
RAKE receiver, 200, 217, 221
Range ambiguities, 245, 288
Rayleigh fading, 188, 220
Rectified linear unit (ReLU) layer, 350
Red/Green/Blue (RGB), 298, 302, 305, 307, 311, 319, 324, 329, 333f, 348–350, 352–353
Remote radio head (RRH), 229–230
Residual quantization, 341t
Roll-off, 90–91, 90t, 91f–92f, 93

S

Second discrete Fourier transforms, 103–104
Serial data interface (SDI), 305
Shannon capacity, 146–147
Shared/local memory, 391
Side lobe clutter, 246–247
Signal-to-noise power ratio (SNR), 29–30, 123–126, 146–147, 221–222, 288–289, 323–326
SIMD-type architecture, 387–388
Sliding window data management, 271f
Soft decision decoding, 144–145
Soft handoff, 203–204, 207
Space time adaptive processing (STAP) radar, 277–286, 278f, 280f, 285f, 286t
Steering vector, 280–281, 281f, 283–285
Streaming video, 209, 304, 342
Sum of absolute differences (SAD) method, 335, 378–379
Supervised training, 347, 354
S-Video, 308
Synthetic array radar (SAR), 287–296, 289f, 291f, 293f–295f

T

Third discrete Fourier transforms, 104–105
Time division multiple access (TDMA), 96, 190f, 193, 183–190, 199–204, 206–207, 209–210, 217–218, 221–222, 225t

V

Variable rate vocoder, 202–203
Vector dot product, 150, 150f–151f, 152
Verilog, 374, 384
Very long instruction word, 363
VHDL, 374, 384
Video compression, 304, 305f, 329, 335, 338–339, 342, 344f–345f, 346
Video graphics array (VGA), 307
Video rate control, 342–343
Video readback, 342
Video storage, 342
Viterbi decoding, 129, 135–145, 187
Vocoder, 187, 189, 202–203, 207, 369

W

Walsh codes, 192–208, 211
Weave, 299–300, 300f
Webers, 163–164
WiMax, 85, 95–96, 209–210, 225t, 228
Winograd transform, 355–358, 356f

Y

YCrCb, 297–298, 302, 302t, 308, 311, 329, 333f

Z

Z transform, 41, 78–80, 405–408

Printed in the United States
By Bookmasters